ALSO BY MICHAEL J. DENTON

Evolution: A Theory in Crisis

NATURE'S DESTINY

How the Laws of Biology Reveal Purpose in the Universe

MICHAEL J. DENTON

THE FREE PRESS

New York London Toronto Singapore Sydney

*f*P

The Free Press
A Division of Simon & Schuster Inc.
1230 Avenue of the Americas
New York, NY 10020

Designed by Carla Bolte

Manufactured in the United States of America

10 9 8 7 6 5 4 3 2 1

Library of Congress Cataloging-in-Publication Data

Denton, Michael.
Nature's destiny : how the laws of biology reveal purpose in the universe / Michael J. Denton.
p. cm.
Includes bibliographical references and index.
1. Teleology. 2. Cosmology. 3. Philosophical anthropology.
I. Title.
BD541.D46 1998
124—dc21 98-3295 CIP

ISBN 0-684-84509-1

Acknowledgments

I would like first to thank my editor Bruce Nichols at the Free Press for applying his considerable skills in changing what was initially a heavy indigestible manuscript into a far more readable and accessible text. I would also like to thank David Berlinski who initially suggested the Free Press as a possible publisher. It was through David that the manuscript eventually got to Bruce Nichols's desk. I became acquainted with David through a mutual friend of ours, Professor M. P. Schutzenberger, a leading French mathematician, anti-Darwinist, and member of the French Academy. Professor Schutzenberger was known to both David and myself and to his very many academic colleagues affectionately as "Marco." It was through one of many conversations with Marco in his flat near the Bois de Boulogne in Paris, in 1989, that I first learned of Lawrence Henderson's great book *The Fitness of the Environment* and of the concept of the unique fitness of the cosmos for carbon-based life. Had Marco not brought *The Fitness* to my attention, then certainly this book would never have been written. I also owe a debt to the works of the physicists Paul Davies and John Barrow and others in the anthropic camp which stimulated and encouraged me to consider examining the fitness in the biological realm.

During the four-year gestation period when the manuscript was going through many revisions and drafts, many academic friends and colleagues provided useful criticisms and suggestions. I am particularly grateful to colleagues at the University of Otago, in particular to Dr. Mike Legge and Dr. Craig Marshall in the Biochemistry Department, and to Dr. Dorothy Oorschot in the School of Medical Sciences, who read early drafts and offered many helpful criticisms. I am also grateful to Jim Kern, a well-known nature photographer in the United States and close friend, who also read early drafts and whose support and enthusiasm for the book never wavered. While the book was in preparation, I had many interesting discussions with him regarding its content and philosophical implications while staying at his home near St. Augustine, Florida. I am also grateful to my London agent Christopher Shepheard-Walwyn for his efforts in editing an early and very unpolished version of the book and for his support for the project from the beginning.

I would like to take this opportunity to thank the following for granting permission to reprint previously published material:

Professor N. J. Berril's daughter, Lyn, for the use of several figures from her father's book *Biology in Action.* Vance Tartar's daughter, Wanda, for allowing me to use a drawing from her father's book *The Biology of Stentor.* Professor A. E. Needham's widow, Nita, for her kind permission to cite several sections from *The Uniqueness of Biological Materials.* Professor J. T. Edsall for his permission to use quotes and a figure from his book *Biophysical Chemistry.* Professor Robert Goldberg for granting me permission to use several long quotes from *Molecular Insights into Living Processes.* Professor Victor W. Rodwell for the use of several figures from *Harper's Biochemistry.* Professor Harold Morowitz for the use of a quote from *Cosmic Joy and Local Pain,* published by Ox Bow Press. Dr. Jearl Walker for permission to quote from her article in the "Amateur Scientist" section of *Scientific American.* Professor Leslie Orgel for the reproduction of a figure from *The Origins of Life on Earth.* The MIT Press for permission to reproduce Robert Fludd's *Ultriusque Cosmi Historia Oppenheim* from Bernal's *Science in History.* The Anglo-Australian Observatory for permission to reproduce the photograph of the Messier Galaxy by David Malin. Appleton & Lange for permission to reproduce figures from Milton Toporek's *Basic Chemistry of Life.* Williams & Wilkins Company for permission to copy the figure of the bronchial tree from Best and Taylor's *Physiological Basis of Medical Practice.* Garland Publishing for the use of the figure of cytochrome oxidase from *The Molecular Biology of the Cell.* Professor S. J. Singer for permission to reproduce his drawing of the cell membrane in *Science.* Professor M. L. Land for permission to use two figures of the eye of the scallop from the *Journal of Physiology.* David Scharf for allowing me to reproduce his photograph of the eye of the lobster. *Scientific American* for allowing me to redraw their illustration of the optical system of a reflecting eye. Wiley & Sons for permission to use material from I. H. Segal's *Biochemical Calculations.* Cambridge University Press for permission to quote from Paul Davies's *The Accidental Universe* and from Knut Schmidt-Nielsen's *Scaling.* W. W. Norton & Company for the use of excerpts from *Wonderful Life: The Burgess Shale and the Nature of History* by Stephen Jay Gould. Copyright © by Stephen Jay Gould, reprinted with permission of W. W. Norton & Company. W. H. Freeman & Company for the use of the quote by J. S. Lewis from *Earth* by Press and Siever, copyright © 1986 by W. H. Freeman & Company, used with permission. Random House for permission to quote from Loren Eiseley's *The Immense Journey* (1947) and from Carl Sagan's *Cosmos* (1980), copyright © 1980 by Carl Sagan Publications, Inc. International Thomson Publishing Services for the use of several sections from A. J. Gurevich's *Medieval Culture.* The Peters Fraser & Dunlop Group Ltd., for permission to quote from Julian Huxley's *Uniqueness of Man* and from Arthur Koestler's *The Ghost in the Machine.* Columbia University Press for permission to quote from *Evolution Above the Species* by Bernard Rensch. Copyright © 1959 by Columbia University Press. Reprinted by permission of the publisher.

Contents

Part 2: Evolution

Appendix

Note to the Reader

> Time out of mind it has been by the way of the "final cause," by the teleological concept of end, of purpose or of "design," in one of its many forms . . . that men have been chiefly wont to explain the phenomena of the living world: and it will be so while men have eyes to see and ears to hear withal. With Galen as with Aristotle, it was the physician's way; with John Ray as with Aristotle it was the naturalist's way; with Kant as with Aristotle it was the philosopher's way. . . . It is a common way, and a great way; for it brings with it a glimpse of a great vision, and it lies deep as the love of nature in the hearts of men.
>
> —D'Arcy Wentworth Thompson, *On Growth and Form,* 1942

The aim of this book is, first, to present the scientific evidence for believing that the cosmos is uniquely fit for life as it exists on earth and for organisms of design and biology very similar to our own species, *Homo sapiens,* and second, to argue that this "unique fitness" of the laws of nature for life is entirely consistent with the older teleological religious concept of the cosmos as a specially designed whole, with life and mankind as its primary goal and purpose.

Although this is obviously a book with many theological implications, my initial intention was not specifically to develop an argument for design; however, as I researched more deeply into the topic and as the manuscript went through successive drafts, it became increasingly clear that the laws of nature were fine-tuned for life on earth to a remarkable degree and that the emerging picture provided powerful and self-evident support for the traditional anthropocentric teleological view of the cosmos. Thus, by the time the final draft was finished, the book had become in effect an essay in natural theology in the spirit and tradition of William Paley's *Natural Theology* or the Bridgewater Treatises.

The basic thesis of the book, that the cosmos is uniquely fit for human existence, is of course not novel. For centuries before the birth of modern science, this thesis was one of the foundational axioms of medieval Christianity. More recently, it has begun to reemerge in various fields of science, most notably in physics and cosmology. Readers familiar with the views of physicists such as Freeman Dyson, Fred Hoyle, and Paul Davies will be aware that over the past few decades many physicists have pointed out that the existence of life in the cosmos is critically dependent on the laws and constants of physics having the precise values they do. The values are so critical that several well-known authors have argued that the cosmos gives every appearance of having been very finely adjusted or "prefabricated" for our existence.[1] As Paul Davies points out in his *Accidental Universe:* "If nature had opted for a slightly different set of numbers, the world would be a very different place. Probably we would not be here to see it." In his words: "The impression of design is overwhelming."[2] Because of the perceived support for the traditional teleological worldview of the major religious traditions, the views of Davies and others have received wide publicity.

There is, however, a fundamental problem with any attempt to argue for the biocentricity or anthropocentricity of nature based on evidence drawn only from physics. While such evidence may be sufficient to argue that the cosmos is arranged for "complex chemistry," solar systems, or even intelligence, it is necessarily insufficient to argue that the cosmos is in some sense uniquely fit for the specific type of *biological life as it exists on earth,* that is, for organisms constructed out of carbon compounds based in water and utilizing DNA and proteins for self-replication. And it is completely incapable of providing any support for the notion that our own species, *Homo sapiens,* has any special place in the cosmos.

Davies is careful to distance himself from any claim that humanity is central in the cosmic scheme: "Where do human beings fit into this great cosmic scheme? Can we gaze out into the cosmos, as did our remote ancestors, and declare God made it all for us? I think not."[3] And in his latest book he states explicitly that "I am not saying that we *Homo sapiens* are written into the laws of physics in a basic way."[4] And continues: "We should not expect extraterrestrial life to resemble our own in its basic chemistry. . . . There is no need, for example, to demand liquid water or even carbon. We could anticipate exotic life forms, such as creatures that float in the dense atmosphere of Jupiter or swim in the liquid nitrogen seas of Titan."[5]

Contrary to Davies and others, I believe the evidence strongly suggests that the cosmos is uniquely fit for only one type of biology—that which exists on earth—and that the phenomenon of life cannot be instantiated in any other exotic chemistry or class of material forms. Even more radically, I believe that there is a considerable amount of evidence for believing that the cosmos is uniquely fit for only one type of advanced intelligent life—beings of design and biology very similar to our own species, *Homo sapiens.* I do not agree with Davies when he claims, "The physical species *Homo sapiens* may count for nothing."[6]

To defend the postulate that the cosmos is specifically fit for biological life *as it exists on earth* necessarily involves consideration of a vast number of natural laws, phenomena, and processes which are quite outside of the areas of physics and cosmology and pertain uniquely to the biological realm, phenomena such as the thermal properties of water, the characteristics of the carbon atom, the solubility of carbon dioxide, the self-assembling properties of proteins, the nature of the cell, and so forth. Although from the evidence of physics we may be able to infer that the cosmos is uniquely fit for chemistry, stars and planets, or even intelligent beings, we cannot infer that it is specifically fit for large, air-breathing terrestrial mammals. Only through biology can our unique type of carbon-based life and especially advanced forms like ourselves lay claim to a central place in the cosmic scheme.

This book is divided into two major parts. In Part 1, evidence is presented that the laws of nature are uniquely fit for the being or existence of the type of carbon-based life that exists on earth. The chapters in this section deal with evidence drawn from many areas of the biological sciences, from molecular biology to mammalian physiology. The physical and chemical properties of the fundamental constituents of the cell, such as water, carbon dioxide, the bicarbonate buffer, oxygen, DNA, proteins, the transitional metals, the cell membrane, etc., are systematically reviewed to show that the existence of carbon- and water-based cellular life depends critically on a number of remarkable adaptations in the properties of many of life's basic constituents. What is particularly striking is that, in almost every case, each constituent appears to be the only available or unique candidate for its particular biological role and, further, gives every appearance of being ideally fit not in one or two but in all its physical and chemical characteristics. Also reviewed is evidence drawn from other areas of science that attests to the fitness of the earth's hydrosphere, the fitness of the electromagnetic radiation of the sun, and the fitness of the periodic table for the carbon-based type of life as it exists on earth. As the book also shows, the existence of some higher forms of life, such as large warm-blooded, air-breathing terrestrial vertebrates, are critically dependent on the properties of some of the basic constituents of life, such as water, carbon dioxide, and oxygen; in other words, not only are the laws of nature fit for the cell and for simple microbial life, but also for advanced complex organisms very like ourselves.

The argument developed in Part 1, that the cosmos is uniquely fit for *life's being,* leads naturally to the second argument, developed in Part 2, that the cosmos is fit also for the origin and evolutionary development of life—*life's becoming.* It is hard to escape the logic of this connection, for if the first argument is accepted, that the existence of the life forms on earth, both microscopic and macroscopic, depends on a remarkable set of mutual chemical and physical adaptations in the nature of things, the second argument, that the evolutionary development of this same set of life forms was also written into the cosmic script and directed from the beginning, is hard to refuse. Or, put another way: if the laws of nature are so finely tuned to facilitate *life's being* in the form of a unique set of carbon-based organisms, both simple and complex, on the surface of a terraqueous planet like the earth, then it seems conceivable that *their becoming* through the process of evolution might have been determined also by natural law.

At present, the evidence that the cosmos is uniquely fit for *life's being* is certainly far more convincing than the evidence that it is also fit for *life's becoming.* Nonetheless, even though direct evidence for believing that life's becoming is "built in" is lacking, there are many features of the cosmos that make sense if the becoming of life is in some way programmed into the laws of nature. Facts such as the synthesis in stars throughout the cosmos, of carbon and the more complex atoms essential for life, by intricate processes; that interstellar space contains vast quantities of organic carbon compounds[7] and some meteors such as the Murchison meteor contain considerable quantities of amino acids,[8] the building blocks of life; that planets like the earth which are probably capable of sustaining carbon-based life would appear to be very common if not almost ubiquitous throughout the cosmos[9]—all these make eminent sense if life is a natural phenomenon programmed into nature from the beginning, and fated inevitably to arise and evolve on any suitable planetary environment.

The claim that the constituents of life are uniquely designed for the roles they serve cannot be defended convincingly without detailed discussion of the relevant scientific facts. This is true of any similar type of teleological argument. If we are to argue, for example, that the components of a watch are all specially designed to function together to tell the time, the argument can only be convincing if we have some understanding of the structure and workings of the watch. We have to open up the watch, to observe the mechanisms within, particularly the reciprocal fit of the various cogs to one another and have some comprehension of the way the mechanism works overall. And we need to understand clearly, as William Paley emphasized in his famous discourse on the watch, "that if the parts had been differently shaped from what they are," the watch could never function.[10] The same is true in arguing that the constituents of the cosmos are uniquely fit for life. The argument only works if we have some knowledge of "the machinery of the cell" and some understanding of the many reciprocal adaptations in the nature of its constituents that make life possible. Consequently, the presentation of the argument in a book of this sort is quite challenging, because the nature of these mutual adaptations can only be fully appreciated by a relatively in-depth and detailed presentation of the relevant scientific facts.

However, despite the technical nature of many sections of the book, I believe that most areas covered can be easily grasped by anyone with a high-school knowledge of biology and chemistry. And even a committed reader with no scientific training should be able to grasp the essence of the argument in most of the chapters, even if this necessitates skipping some of the more highly technical sections. There are several chapters which require very little scientific background. And most chapters include at the beginning an introductory section requiring very little specialized knowledge, in which I have attempted to explain the main theme of the chapter.

I have also tried to organize the presentation of the evidence so that many of the chapters represent a fairly independent module which can be read and understood without reference to other chapters or arguments in other sections of the book. I hope this makes the book easier for a nonspecialist to handle. Finally, as mentioned above, each chapter begins with an italicized précis that may allow nontechnical readers to skip ahead.

Further, as with any such argument because the argument is essentially accumulative, deriving its power from the sheer number of the adaptations observed, it is essential that as many as possible of these are presented and discussed. The conclusion is convincing primarily because so many independent arguments, each drawn from a great number of different areas of science, all appear to point in the same direction. This inevitably involves a degree of repetition that will be a problem for some readers. However, a degree of repetition is the very essence of the whole line of attack.

Because the validity of the argument depends on so many independent lines of evidence, the conclusion is not materially threatened because the whole picture is not yet complete or because this or that phenomenon such as the origin of life or the mechanism of evolution is not understood. Just as the meaning of a jigsaw puzzle may be obvious long before all the pieces are perfectly placed, so too my argument does not necessitate that everything be explained. Nevertheless, critics of the argument will have certain clear avenues of attack. They can argue (correctly) that I have been selective in my topics. The burden of disproof will, however, rest on them to show that an area I ignored somehow opens up the possibility of either nonearthlike life in the cosmos or a superior alternative to one of the constituents of life—for example, water, carbon dioxide, etc. Or they may argue that my position merely reflects a lack of imagination and that I have not discussed possible alternatives in depth. But again, the burden of proof will be on them to offer specific alternatives. I do not see how I can be accused of omitting discussion of alternative forms of life, based in silicon or liquid ammonia or within the field of nanotechnology, when no detailed blueprints for such hypothetical life forms have ever been developed.

Although there has been little debate or interest in the question of the fitness of the cosmos for life in mainstream biology since the Darwinian revolution, and indeed the idea has been very unfashionable in many circles in the English-speaking world, interest in the question has never been completely extinguished. Throughout the twentieth century, a number of first-rate biologists have kept the tradition alive. These have included Lawrence Henderson, professor of biological chemistry at Harvard University during the first quarter of the century and author of the great classic *The Fitness of the Environment* (1913);[11] D'Arcy Wentworth Thompson, author of another great classic, *On Growth and Form* (1942);[12] George Wald, professor of biology at Harvard in the fifties and sixties, discoverer of the role of vitamin A in vision, who was one of the leading authorities on the chemistry of photoreception;[13] A. E. Needham, Oxford zoologist and author of an excellent and comprehensive review, *The Uniqueness of Biological Materials* (1965);[14] and Carl Pantin, professor of zoology at Cambridge during the sixties and author of the widely acclaimed *The Relations Between the Sciences,* published in 1968.[15]

My chapters on the properties of water, carbon, oxygen, and carbon dioxide borrow heavily from Henderson's *Fitness* and can be considered to a large degree an update of that great classic in the light of modern knowledge. Another major source cited in several chapters is Needham's *The Uniqueness of Biological Materials.*

One recent book that invites some comparison is Stuart Kauffman's *At Home in the Universe,* in which he argues that much of the course of evolution has been determined and driven by self-organizing and emergent properties of complex systems.[16] There is certainly more than a whiff of teleology about Kauffman's arguments, and his overall conclusion is consistent with my own when he claims, for example: "We will have to see that we are all natural expressions of a deeper order. Ultimately, we will discover in our creation myth that we are expected after all."[17] And further: "We may be at home in the universe in ways we have hardly begun to comprehend."[18]

Another book that also invites comparison is *Vital Dust* by the biologist and Nobel laureate Christian de Duve. De Duve has also "opted in favour of a meaningful universe"[19] and argues that the cosmos is fit for the origin and evolution of life and that the progress of evolution from simple to complex life forms was largely inevitable. However, de Duve's position falls a long way short of defending the traditional anthropocentric view of the cosmos. The unique fitness of the laws of nature for the biology of higher, air-breathing life forms such as ourselves is not discussed in any depth and nowhere does de Duve argue that the pattern of evolution was directed specifically toward the human race. Regarding man's place in the cosmos, de Duve concludes in his final chapter, "The human mind may be only a *side link* in an evolutionary saga far from completed."[20] (My emphasis.)

Because this book presents a teleological interpretation of the cosmos which has obvious theological implications, it is important to emphasize at the outset that the argument presented here is entirely consistent with the basic naturalistic assumption of modern science—that the cosmos is a *seamless unity which can be comprehended ultimately in its entirety by human reason and in which all phenomena, including life and evolution and the origin of man, are ultimately explicable in terms of natural processes.* This is an assumption which is entirely opposed to that of the so-called "special creationist school." According to special creationism, living organisms are not natural forms, whose origin and design were built into the laws of nature from the beginning, but rather contingent forms analogous in essence to human artifacts, the result of a series of supernatural acts, involving God's direct intervention in the course of nature, each of which involved the suspension of natural law. Contrary to the creationist position, the whole argument presented here is critically dependent on the presumption of the unbroken continuity of the organic world—that is, on the reality of organic evolution and on the presumption that all living organisms on earth are natural forms in the profoundest sense of the word, no less natural than salt crystals, atoms, waterfalls, or galaxies.

In large measure, therefore, the teleological argument presented here and the special creationist worldview are mutually exclusive accounts of the world. In the last analysis, evidence for one is evidence against the other. Put simply, the more convincing is the evidence for believing that the world is prefabricated to the end of life, that the design is built into the laws of nature, the less credible becomes the special creationist worldview.

Ironically, both the Darwinian and the creationist worldviews are based on the same fundamental axiom—that life is an unnecessary and fundamentally contingent phenomenon. Where the creationist sees organisms as the artifacts of God the supreme engineer, the Divine watchmaker, Darwinists see them as the artifactual products of chance and selection. That both should view life as contingent is not so surprising considering that both doctrines developed in the early nineteenth century, the heyday of the machine age, when organisms were widely seen to be analogous in some way to machines. Clearly, if life's design is indeed embedded in the laws of nature and the major paths of evolution are largely determined from the beginning, then neither creationism nor Darwinism can possibly be valid models of nature.

My argument may be unpalatable for completely different reasons to certain liberal theologians. Academic theology in the twentieth century has largely abandoned traditional natural theology. Many have held the view "that theological propositions and scientific propositions somehow occupy different epistemological realms. Hence the neo-orthodox wall between religion and science."[21] Some liberal theologians have recently explored the relationship between science and theology,[22] showing how, in Arthur Peacocke's words, "God creates in the world through what we call 'chance' operating within the created order."[23] Yet nowhere do they attempt to present a natural theology (they may even object to the term) along traditional lines. The aim of their work is to show how it is *possible to believe in God* while at the same time accepting the findings of science. It is not to argue that the *facts of science provide evidence* that the laws of nature are uniquely prefabricated for life as it exists on earth, including complex forms such as our own species.

Another final point that perhaps should be clarified here at the outset is that I am using the term "anthropocentric" throughout the text in the generic sense. The cosmic "telos" I have in mind is advanced carbon-based humanlike or humanoid life. It is not specifically our own unique species *Homo sapiens.* At present, there is insufficient evidence to argue that the laws of nature are uniquely fit for *every detail* of human biology exactly as found in our own species today. However, I believe that the current evidence points strongly in this direction and that future scientific advances will confirm the absolute centrality of mankind in the cosmic scheme.

In the last analysis, the teleological perspective presented and defended here is good for science, because it renders scientific knowledge relevant to human existence. In the doctrine of final causation, science unites man and cosmos. The pursuit of scientific knowledge becomes no longer of merely practical value but also vital and central to the spiritual and intellectual life of man.

—Michael J. Denton
Dunedin, November 1996

Microcosmos.

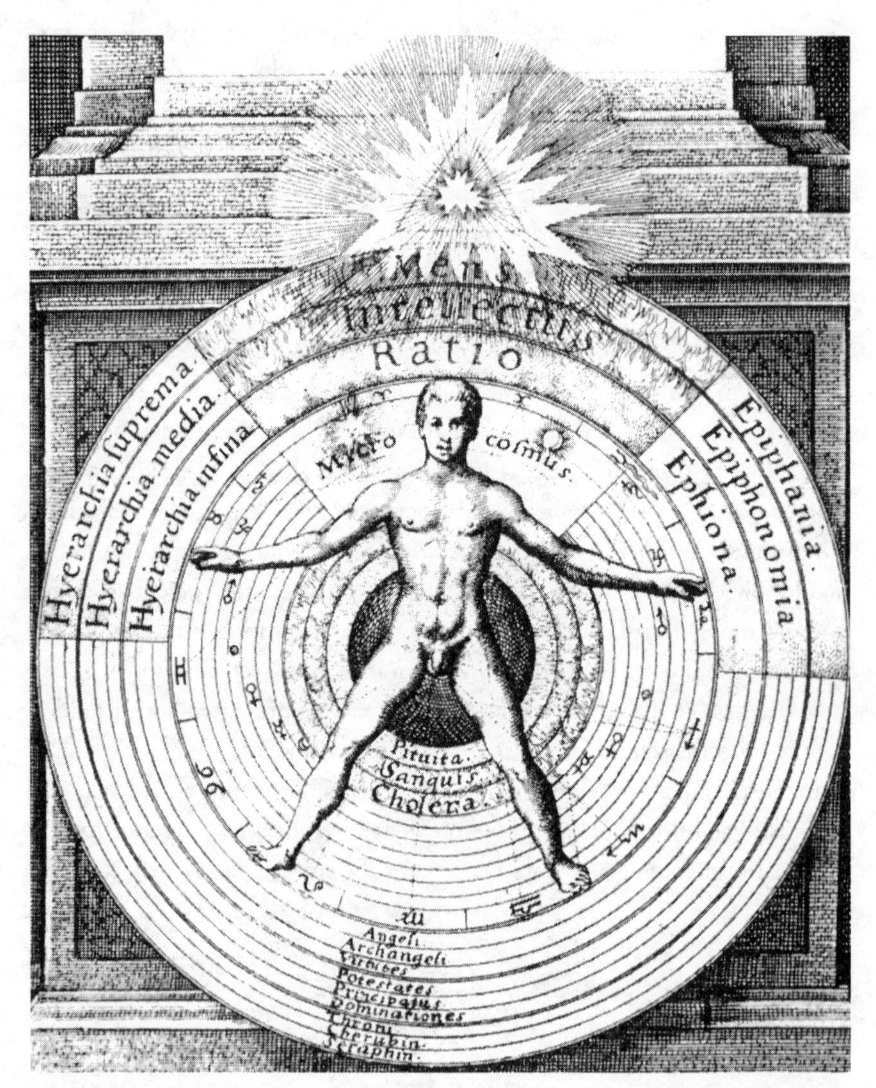

From J. D. Bernal (1969) *Science in History,* vol. 1 (London, MIT Press) p. 274. Original in Robert Fludd's *Utriusque Cosmi . . . Historia Oppenheim,* 1617–1619.

Prologue

> The ancient opinion that man was microcosmos, an abstract or model of the world, hath been fantastically strained by the alchemists, as if there were to be found in man's body certain correspondences and parallels which should have respect to all variety of things, as stars, planets, minerals, which are extant in the great world.
>
> —Francis Bacon, *The Advancement of Learning,* 1605

Living as we do in the late twentieth century, in a culture that has rejected the traditional teleological view of man as the center and purpose of the cosmos, which views our human existence as in essence a matter of profound contingency, it is fascinating to recall just how different was the medieval worldview in the late fifteenth century, shortly before the birth of modern science.

For both Christian and Islamic philosophers and theologians of the Middle Ages, the cosmos was a unique whole specially designed by God with man as its central focus and purpose. All facets of reality found their explanation in this central fact. Man was the inner microcosm. Every aspect of his being reflected the outer macrocosm, the universe in its entirety and all it contained.

For Christian scholars, the biblical revelation, and particularly the Incarnation, sanctioned the profoundly anthropocentric character of their medieval worldview. The extraordinary anthropocentricity of the culture of the Christian Middle Ages was wonderfully conveyed by Aron Gurevich in his classic work *Categories of Medieval Culture:*

> The effort to grasp the world as a single unified whole runs through all the medieval summae, the encyclopaedias and the etymologies. . . . The philosophers of the twelfth century speak of the necessity of studying nature; for in the cognition of nature in all her depths, man finds himself . . . underlying these arguments and images is a confident belief in the unity and beauty of the world, and also the conviction that the central place in the world which God has created belongs to man.
>
> The unity of man with the universe is revealed in the harmony interpenetrating them. Both man and the world are governed by the cosmic music which expresses the harmony of the whole with its parts and which permeates all from the heavenly spheres to man. *Musica humana* is in perfect concord with *musica mundana.* Everything that is measured by time is bound up with music. Music is subordinate to number. Therefore both macrocosm and man-made microcosm are ruled by numbers which define their structure and determine their motion. . . . It is in numbers that the secret of the beauty of the world lies; for the medieval mind the concepts "beauty," "orderliness," "harmony," "proportion," "comeliness," and "propriety" were very close to each other if not identical.[1]

So intensely anthropocentric was their conception of nature that, as Gurevich points out:

> Each part of the human body corresponded to a part of the universe: the head to the skies, the breath to the air, the stomach to the sea, the feet to the earth; the bones corresponded to the rocks, the veins to the branches of the trees.[2]

The presumption that the entire cosmos was man-centered, that every facet of reality and all the laws of nature reflected this central reality, was the overriding axiom upon which the whole civilization of medieval Europe was built. Not even the slightest deviation from such an all-embracing man-centered teleology was compatible with the Christian revelation. For the Bible implied that the great drama of human history was central to the purpose of God in creation. The earth was the unique and divinely chosen stage for the drama, and God himself had taken on the form of a man to bear the sins of creation.

Even after the medieval period, for many early modern thinkers such as Francis Bacon, whose scientific philosophy, with its emphasis on experiment, had an empirical tendency that was quite similar to that of modern science, mankind's teleological centrality in the natural order was presumed without question. The following section from Bacon's *De sapientia veterum* illustrates Bacon's commitment to an intensely anthropocentric framework:

> Man . . . may be regarded as the centre of the world . . . if man were taken away from the world, the rest would seem to be all astray, without aim or purpose . . . leading to nothing. . . . the whole world works together in the services of man . . . in so much that all things seem to be going about man's business and not their own.[3]

The anthropocentric perspective was not, of course, restricted to the West. It was highly developed in the Islamic world in the ninth and tenth centuries. And Judaism, Hinduism, and Buddhism also view mankind as significant in the cosmic scheme. In ancient Indian thought, for example, the general ethos was "one of an integrated man-spirit-cosmos view, a wide and comprehensive view of nature in which the *Homo sapiens,* or man, the thinker, occupied a distinct place."[4] According to the eleventh-century neo-Confucianist philosopher Shao Yung, "Man is central in the universe, and the mind is central in man. . . . Man occupies the most honoured position in the scheme of things because he combines in him the principles of all species. . . . The nature of all things is complete in the human species."[5]

The idea is practically universal, being expressed in all human cultures, as John Barrow and Frank Tipler summarize:

> the idea that humanity is important to the cosmos and indeed the idea that the material world was created for man both seem to be present in many cultural traditions; they may even be universal . . . a cursory search of the anthropological literature shows teleological notions defended in Mayan, Zuñi (New Mexican Indian) . . . Sumerian, Bantu, ancient Egyptian, Islamic-Persian, and Chinese.[6]

It is remarkable to think that only five centuries separates the current skeptical ethos in the West from this profoundly teleological view of reality.

The anthropocentric vision of medieval Christianity is one of the most extraordinary—perhaps the most extraordinary—of all the presumptions of humankind. It is the ultimate theory and in a very real sense, the ultimate conceit. No other theory or concept ever imagined by man can equal in boldness and audacity this great claim—that everything revolves around human existence—that all the starry heavens, that every species of life, that every characteristic of reality exists for mankind and for mankind alone. It is simply the most daring idea ever proposed. But most remarkably, given its audacity, it is a claim which is very far from a discredited prescientific myth. In fact, no observation has ever laid the presumption to rest. And today, four centuries after the scientific revolution, the doctrine is again reemerging. In these last decades of the twentieth century, its credibility is being enhanced by discoveries in several branches of fundamental science.

Part 1

LIFE

Chapter 1

The Harmony of the Spheres

In which evidence from physics and cosmology suggesting that the laws of physics are fine-tuned for carbon-based life, is briefly summarized. The fitness of the universe for life depends on a number of factors, including: the relative strength of the four fundamental forces (gravity, electromagnetism, and the strong and weak nuclear forces), the speed of expansion of the universe, the spacing and frequency of supernovae, the nuclear energy levels of certain atoms, etc. If these were not precisely what they are, then carbon-based life would certainly not exist. Many previous authors have covered this ground, but it bears repeating and is introductory to the theme of the book.

The spiral galaxy Messier 83, NGC 5236.

It has been shown in the preceding chapters that a great number of qualities and laws appear to have been selected in the construction of the universe; and that by the adjustment to each other of the magnitudes and laws thus selected, the constitution of the world is what we find it, and is fitted for the support of vegetables and animals in a manner in which it could not have been, if the properties of the elements had been different from what they are.

—William Whewell, Bridgewater Treatises, 1833

This now tells us how precise the Creator's aim must have been, namely to an accuracy of one part in $10^{10^{123}}$. This is an extraordinary figure. One could not possibly even write the number down in full in the ordinary denary notation: it would be 1 followed by 10^{123} successive 0's. Even if we were to write a 0 on each separate proton and on each separate neutron in the entire universe—and we could throw in all the other particles for good measure—we should fall far short of writing down the figure needed.

—Roger Penrose, *The Emperor's New Mind,* 1989

On July 4 in the year 1054 A.D., Chinese astronomers observed a spectacular event in the sky. In the constellation of Taurus, the Bull, a brilliant new star suddenly appeared. So brilliant was this new star that it was easily visible in daylight and its light at night like that of the full moon. Carl Sagan relates how on the other side of the world the ancestors of the Hopi Indians also recorded this remarkable event:

> Halfway around the world in the American Southwest, there was then a high culture, rich in astronomical tradition, that also witnessed this brilliant new star. From carbon 14 dating of the remains of a charcoal fire, we know that in the middle eleventh century . . . the antecedents of the Hopi were living under an overhanging ledge in what is today New Mexico. One of them seems to have drawn on the cliff overhang, protected from the weather, a picture of a new star. Its position relative to the crescent moon would have been just as was depicted.[1]

What the Hopi Indians and the Chinese astronomers had observed was a supernova. Supernovae are among the most dramatic of all astronomical phenomena. An entire star self-destructs in a colossal explosion, scattering all its constituent matter and energy in a gigantic wave through adjacent regions of space.

As a result of advances in astronomy and physics over the past half century, we now know that the dying of stars in these immense self-destructive explosions is intimately related to our own existence as living organisms on earth. All the elements necessary for life—carbon (C), nitrogen (N), oxygen (O) and iron (Fe), etc.—are manufactured in the nuclear furnaces in the interiors of the stars. If these elements are to accumulate in rocky planets such as earth, they must be released from the stellar interiors and dispersed widely throughout the cosmos. The crucial release and dispersal of these key building blocks of life is one of the results of a supernova explosion. It is in the dying of stars that life has its birth.

Biocentric Fine-Tuning

Over the past three decades, facts such as these drawn from astrophysics and cosmology have led many physicists to argue that the cosmos appears to be finely tuned for life. The evidence and argument has been presented many times;[2] consequently, I will not discuss it in great detail here. Nonetheless, it is the foundation upon which the argument from biology rests, so a brief review is in order.

Supernovae play another role which is critical to the existence of life. The shock waves they generate are probably important in initiating the condensation of interstellar gas and dust into planetary systems such as our own solar system. Those ancient stargazers in China and America would surely have been amazed to know that without such strange new stars, like that which so dramatically lit up the sky on that far-off July night, there would be no astronomers, no stargazers, no earth, perhaps no life of any sort.

The fateful connection between those ancient astronomers and the new star they had witnessed involves more than the mere fact that such explosions spill the atoms of life into the cosmos and set in motion the turbulence that causes the birth of planets. If that supernova had been closer to the earth, then it might have bathed the earth in a lethal radiation, obliterating life. If it had been very close, the earth might have been engulfed in a fireball and vaporized. The frequency and distribution of exploding stars are therefore also critical parameters. Supernovae are essential for life—without them none of the chemical building blocks of life will ever accumulate on the surface of a planet like earth—but they are also immensely destructive phenomena, eliminating all life on any nearby solar systems.

The distances between supernovae and indeed between all stars is critical for other reasons. The distance between stars in our galaxy is about 30 million miles. If this distance was much less, planetary orbits would be destabilized. If it was much more, then the debris thrown out by a supernova would be so diffusely distributed that planetary systems like our own would in all probability never form.[3] If the cosmos is to be a home for life, then the flickering of the supernovae must occur at a very precise rate and the average distance between them, and indeed between all stars, must be very close to the actual observed figure.

In addition, it turns out that the production of the key elements for carbon-based life not only requires the enormous energy levels within the interiors of stars but is also critically dependent on what appears to be another set of very precise conditions in the nuclear structure of certain atoms, more specifically, the nuclear energy levels of the atoms ^{8}beryllium, ^{12}carbon, and ^{16}oxygen. These energy levels affect the manufacture and abundance of carbon, oxygen, and other heavier elements essential for life. If they had been slightly different, no life-giving carbon or oxygen would have been manufactured.

That the manufacture of the key elements of life should depend on a set of such highly specific conditions is commented on by Paul Davies in his book *The Accidental Universe.*[4] Fred Hoyle considers the carbon-oxygen synthesis coincidence so remarkable that it seems like a "put-up job." Regarding the delicate positioning of the nuclear resonances, he comments:

> If you wanted to produce carbon and oxygen in roughly equal quantities by stellar nucleosynthesis, these are the two levels you would have to fix, and your fixing would have to be just about where these levels are actually found to be. . . . A commonsense interpretation of the facts suggests that a super intellect has monkeyed with physics, as well as chemistry and biology, and that there are no blind forces worth speaking about in nature.[5]

The picture that has emerged from modern physics and astronomy suggests that the formation of the chemical elements for life, and planetary systems capable of sustaining life and evolution over millions of years, are only possible if the overall structure of the universe and all the laws of nature are almost precisely as they are.

Physicists recognize four fundamental forces. These largely determine the way in which one bit of matter or radiation can interact with another. In effect, these four forces determine the main characteristics of the universe.[6] They are the gravitational force, the electromagnetic force, the strong or nuclear force, and the weak force.

An extraordinary feature of these four fundamental forces is that their strength varies enormously over many orders of magnitude. In the table below they are given in international standard units:[7]

The forces of nature.

Gravitational force	=	$5.90 \cdot 10^{-39}$
Nuclear or Strong force	=	15
Electromagnetic force	=	$3.05 \cdot 10^{-12}$
Weak force	=	$7.03 \cdot 10^{-3}$

The fact that the gravitational force is fantastically weaker than the strong nuclear force by an unimaginable thirty-eight orders of magnitude is critical to the whole cosmic scheme and particularly to the existence of stable stars and planetary systems.[8] If, for example, the gravitational force was a trillion times stronger, then the universe would be far smaller and its life history far shorter. An average star would have a mass a trillion times less than the sun and a life span of about one year—far too short a time for complex life to develop and flourish. On the other hand, if gravity had been less powerful, no stars or galaxies would ever have formed. As Hawking points out, the growth of the universe—so close to the border of collapse and external expansion that man has not been able to measure it—has been at just the proper rate to allow galaxies and stars to form.[9]

The other relationships and values are no less critical. If the strong force had been just slightly weaker, the only element that would be stable would be hydrogen. No other atoms could exist. If it had been slightly stronger in relation to electromagnetism, then an atomic nucleus consisting of only two protons would be a stable feature of the universe—which would mean there would be no hydrogen, and if any stars or galaxies evolved, they would be very different from the way they are.[10]

Clearly, if these various forces and constants did not have precisely the values they do, there would be no stars, no supernovae, no planets, no atoms, no life. As Paul Davies summarizes:

> The numerical values that nature has assigned to the fundamental constants, such as the charge on the electron, the mass of the proton, and the Newtonian gravitational constant, may be mysterious, but they are crucially relevant to the structure of the universe that we perceive. As more and more physical systems, from nuclei to galaxies, have become better understood, scientists have begun to realise that many characteristics of these systems are remarkably sensitive to the precise values of the fundamental constants. Had nature opted for a slightly different set of numbers, the world would be a very different place. Probably we would not be here to see it.
>
> More intriguing still, certain crucial structures, such as solar-type stars, depend for their characteristic features on wildly improbable numerical accidents that combine together fundamental constants from distinct branches of physics. And when one goes on to study cosmology—the overall structure and evolution of the universe—incredulity mounts. Recent discoveries about the primeval cosmos oblige us to accept that the expanding universe has been set up in its motion with a cooperation of astonishing precision.[11]

In short, the laws of physics are supremely fit for life and the cosmos gives every appearance of having been specifically and optimally tailored to that end: to ensure the generation of stable stars and planetary systems, to ensure that these will be far enough apart to avoid gravitational interactions which would destabilize planetary orbits; to ensure that a nuclear furnace is generated in the interior of stars in which hydrogen will be converted into the heavier elements essential for life; to ensure that a proportion of stars will undergo supernovae explosions to release the key elements into interstellar space; to ensure that galaxies last several times longer than the lifetime of an average star, for only then will there be time for the atoms scattered by an earlier generation of supernovae within any one galaxy to be gathered into second-generation solar systems; to ensure that the distribution and frequency of supernovae will not be so frequent that planetary surfaces would be repeatedly bathed in lethal radiation but not so infrequent that there would be no heavier atoms manufactured and gathered onto the surface of newly formed planets; to ensure in the cosmos's vastness and in the trillions of its suns and their accompanying planetary systems a stage immense enough and a time long enough to make certain that the great evolutionary drama of life's becoming will inevitably be manifest sometime, somewhere on an earthlike planet.

And so we are led toward life and our own existence via a vast and ever-lengthening chain of apparently biocentric adaptations in the design of the cosmos in which each adaptation seems adjusted with almost infinite precision toward the goal of life.

That there is indeed a deep teleological connection between the Chinese stargazers and the new star which exploded into that July night in 1054 A.D. has now been established beyond any reasonable doubt. There is simply no tolerance possible in the design of the celestial machine. For us to be here, it must be precisely as it is.

The new picture that has emerged in twentieth-century astronomy presents a dramatic challenge to the presumption which has been prevalent within scientific circles during most of the past four centuries: that life is a peripheral and purely contingent phenomenon in the cosmic scheme. These advances in astronomy and physics have established what for Newton and generations of natural theologians was only an affirmation of belief: that there is indeed a deep and necessary connection between virtually every characteristic of the cosmic stage and the drama of life. It is ironic that those very features of the cosmos that were so troubling to the astronomers of the early seventeenth century—its vast size and the apparently infinite number of stars stretched out across its immensity—which inclined Kepler to wonder, "How can all things be for man's sake?"[12] and which seemed to render the earth an irrelevant mote of dust in the cosmic scheme, have turned out to be absolutely critical and essential for our existence.

The evidence provided by modern cosmology and physics is exactly the kind of evidence that the natural theologians were looking for in the seventeenth century but failed to find in the science of their day. This can be seen in this short passage from Richard Bentley's famous "A Confutation of Atheism from the Origin and Frame of the World," published in 1692. It was prepared under the guidance of Newton and may well represent a position close to Newton's own.

> Let us now turn our thoughts and imagination to the frame of our system, if there we may trace any visible foot steps of the Divine Wisdom and Beneficence. . . . What we have always seen to be done in one constant and uniform manner; we are apt to imagine there was but that one way of doing it, and that it could not be otherwise. This is a great error and impediment in a disquisition of this nature: to remedy which, *we ought to consider every thing as not yet in Being;* and then diligently examine if it *must needs have been at all, or what other ways it might have been as possibly as the present;* and if we find a greater Good and Utility in the present constitution, than would have accrued either from the total Privation of It, or from other frames and structures that as possibly have been as It: *we may then reasonably conclude, that the present constitution proceeded neither from the necessity of material Causes nor the blind shuffles of an imaginary Chance,* but from an Intelligent and Good Being, that formed it that particular way out of choice and design. And especially if this Usefulness *be conspicuous not in one or a few only, but in a long train and series of things,* this will give us a firm and infallible assurance, that we have not passed a wrong judgement.[13] [My emphasis.]

If the existence of life had been compatible with a greater range of values for the fundamental constants, or, in other words, if the design of the celestial machine could have been different at least to some degree and yet still have sustained life, then the teleological conclusion would be far weaker. It is the necessity that it be exactly as it is—adjusted to what is in effect near infinite precision *in a long train and series of things* that makes the teleological conclusion so compelling.

As Davies comments in the last paragraph of *The Cosmic Blueprint,* "The impression of Design is overwhelming."[14] And Paul Davies is not alone.

Several well-known physicists and astronomers, among them Brandon Carter, Freeman Dyson, John Wheeler, John Barrow, Frank Tipler, and Sir Fred Hoyle, to cite only a few, have all made the point in recent publications—that our type of carbon-based life could only exist in a very special sort of universe and that if the laws of physics had been very slightly different we could not have existed. With the evidence as it now stands, it is not surprising that there now exists a significant body of opinion within the scientific community prepared to defend the idea that the universe is in some way profoundly biocentric and gives every appearance of having been specially designed for life. As a result of these discoveries, there is now a teleological intellectual current within modern physics, cosmology, and astronomy which is remarkably concordant with the older anthropocentric view and strikingly out of keeping with the antiteleological tendencies that have come to be universally associated with advances in scientific knowledge for most of the recent past.

As mentioned above, this is not the place to give a comprehensive review of the anthropic principle or to enumerate the many life-giving coincidences in the structure of the cosmos as revealed by twentieth-century astronomy and physics. The topic has been covered in a number of recent scholarly books. This brief discussion of the anthropic principle has been introduced here primarily to illustrate that the apparently triumphant antiteleological tide of skepticism which has gripped the western mind with elemental force for nearly four centuries has now decisively turned in at least one major area of science, and also because it forms a natural introduction to a book that deals with evidence for design in biology and which is in many ways an extension of the anthropic position into the biological sciences.

From Physics to Biology

Why has twentieth-century biology lagged behind physics in the rediscovery of teleology? Curiously, biology, which was so influenced by the nonbiocentric physics of the nineteenth century, has remained immune to the new biocentric-teleological physics of the late twentieth century. The prevailing view within the biological sciences is still that life and man are fundamentally contingent phenomena. This is a natural deduction from the Darwinian idea of evolution by natural selection. As Stephen Jay Gould puts it:

> *Homo sapiens* I fear is . . . in a vast universe a wildly improbable evolutionary event.[15] . . . biology's most profound insight into human nature status and potential lies in the simple phrase, the embodiment of contingency: *Homo sapiens* is an entity not a tendency.[16] . . . If you wish to ask the question of the ages, Why do humans exist? . . . We are the offspring of history, and must establish our own paths in this most diverse and interesting of conceivable universes—one indifferent to our suffering.[17]

The new anthropic vision of the physicist and the Darwinian contingent paradigm which dominates modern biology are diametrically opposed world-views. Yet, where physics led in the seventeenth century, biology eventually followed, and it is doubtful whether modern biology can for long resist the new teleological current now flowing within cosmology and the physical sciences.

This new teleological current would be challenging enough to the *contingent biology* even if the life-giving coincidences were restricted to the realm of physics and astronomy. But the coincidences do not stop at the distribution of supernovae or with the resonances of the energy levels of the carbon and oxygen atoms. They extend on into chemistry, into biochemistry and molecular biology, into the very fabric of life itself. Advances in chemistry, biochemistry, physiology, and molecular biology, commencing at the beginning of the last century, but mainly over the past fifty years, have revealed an additional set of mutual adaptations or coincidences in the chemical and physical properties of water and in many other of the key constituents of life—of precisely the kind that one might expect to find if the cosmos is indeed the biocentric whole that astronomy suggests.

Chapter 2

The Vital Fluid

In which it is argued that water gives every appearance of being uniquely fit for the type of carbon-based life that exists on earth. Every one of its chemical and physical properties seems maximally fit not only for microscopic life but also for large warm-blooded organisms such as mammals, as well as for the generation and maintenance of a stable chemical and physical environment on the surface of the earth. Some of the properties of water reviewed include its thermal properties, its surface tension, its capacity to dissolve a vast number of different substances, and its low viscosity, which allows small molecules to enter and leave cells by diffusion and which also makes possible a circulatory system. If the properties of water were not almost precisely what they are, carbon-based life would in all probability be impossible. Even the viscosity of ice is fit. If it were any greater, then all the water on earth might be trapped in vast immobile ice sheets at the poles. If the thermal properties of water were even slightly different, the maintenance of stable body temperatures in warm-blooded organisms would be problematical. No other fluid comes close to water as the ideal medium for carbon-based life. Indeed, the properties of water in themselves provide perhaps as much evidence as physics and cosmology in support of the proposition that the laws of nature are specifically arranged for carbon-based life.

The earth from space.

Courtesy NASA.

For is not the whole substance of all vegetables mere modified water? and consequently of all animals too; all of which either feed upon Vegetables or prey upon one another? is not an immense quantity of it continually exhaled by the Sun, to fill the atmosphere with Vapours and Clouds, and feed the Plants of the Earth with the balm of Dews . . . . It seems incredible at first hearing, that all the Blood in our Bodies should circulate in a trice, in a very few minutes: but I believe it would be more surprising, if we knew the short and swift periods of the great Circulation of Water, that vital Blood of the Earth which composeth and nourisheth all things.

—Richard Bentley, "A Confutation of Atheism from the Origin and Frame of the World," 1692

Although water is one of the most familiar of all substances, its remarkable nature never fails to impress. As a liquid, it accumulates on the earth's surface in bodies varying in size from the great oceans to small lakes to tiny puddles. In motion it may swirl violently down a great cataract, or flow serenely as a mature river meandering across a plain. On the surface of large bodies of water, the wind pushes up waves both great and small. Tiny droplets of the substance form the matrix of the clouds. Slightly larger drops fall through the atmosphere from the clouds to the ground as rain. As a solid, it falls as snow blanketing the earth in white, it forms the great ice sheets of the polar regions and the valley glaciers in the mountains, and it forms the frosted pattern on a windowpane in winter. In the higher latitudes water forms the entire scenery of the earth, the ice caps at the fringes of the polar continents, the icebergs floating in the restless gray and ice-cold sea, the spray carried from wave tops by the wind and frozen instantly into tiny pellets of ice in the subzero temperature and splattered like shrapnel onto the nearby ice shelves. Even the sounds associated with water are no less diverse: there is the rhythmic pounding of the surf, the deafening roar of a great waterfall, the babbling of a mountain brook, the gentle patter of summer rain, the clatter of hail against an iron roof, the grinding booms and sharp reports of an advancing glacier, and the thunder of an avalanche.

These diverse manifestations of water are remarkable indeed. But as we shall see, they are not nearly as extraordinary or amazing as the various ways in which water is so ideally and uniquely adapted to serve its biological role as the medium or matrix for life on earth.

Water has long been seen to have some special significance. That it is essential to life has been evident since the earliest of times, and many cultures have invested it with magical life-giving qualities. It is fitting that Thales, the first of the Greek philosophers, should have based his science on the assertion that water is the origin of all things, and that Bentley should describe it as "the vital blood of the Earth."[1]

Water forms the fluid matrix in which occur all the vital chemical and physical activities upon which life on earth depends. Without water, life that exists on earth would be impossible. If the vital activities of the cell are the movements of pieces on a chess board, then water would be the board. Chess is impossible without the board; life is impossible without water. Water also forms most of the bulk of most living things. Most organisms are made up of more than 50 percent water; in the case of man, water makes up more than 70 percent of the weight of the body.

The Necessity of Liquid

That life is based in a liquid medium is certainly no accident. For it is difficult to imagine how any sort of complex chemical system capable of assembling and replicating itself, of manipulating its atomic and molecular components and drawing its vital nutrients and constituents from its environment—that is, anything that displays the characteristics we attribute to life—could exist except in a liquid medium.

As A. E. Needham points out in *The Uniqueness of Biological Materials,* the other two states of matter, the solid and the gaseous, would seem to be excluded on fundamental grounds. In the case of both a crystalline solid, where the atoms are held in regular crystalline arrays and a glassy solid, where the atoms are irregularly packed, the atoms are in rigid contact with one another and there is very little scope for the occurrence of the dynamic molecular processes associated with life to occur. In gases, on the other hand, the constituent atoms are freely mobile, and consequently gases are far too volatile and labile to be considered seriously as candidates for the chemical matrix of life.[2] We are all familiar with clouds which are nebulous masses of tiny liquid droplets—or in more scientific terms, "segregations of liquid-in-gas colloids." Clouds are a rare exception to the rule that segregating subsystems are unusual in a gas. However, the very transience of cloud patterns graphically illustrates the unsuitability of gas as a medium for the support of stable, segregating subsystems.

If the laws of physics sanctioned matter to exist in our universe only in the solid or gaseous state and outlawed liquids, then life, defined above as a *complex chemical system capable of assembling and replicating itself, of manipulating its components and drawing its vital nutrients and constituents from its environment,* would almost certainly not exist. Interestingly, John von Neumann, one of the fathers of the computer, in his *Theory of Self-Reproducing Automata,* envisaged his mechanical replicators floating on an infinite lake, the surface of which was covered with all the basic constituents they required to construct themselves. In other words, the medium in which the replicators "lived" was a fluid.[3]

Because the full impact of the argument for the fitness of water is accumulative and depends on a relatively exhaustive consideration of all the individual adaptations which seem to fit water so ideally for its biological role, it is important that no adaptation is omitted, even if it is very well known, so that all the evidence is laid out as comprehensively as possible and in some detail.

Water's Unique Thermal Properties

Curiously, even as recently as the late eighteenth century, shortly after Antoine Lavoisier had first determined the chemical structure of water and shown that it was made up of two hydrogen atoms combined with one oxygen atom, its chemical and physical properties were insufficiently understood to argue that it was specially adapted for life. Just how little was known of the properties of water around 1800 is obvious from this section in Paley's *Evidences,* where he concedes that "when we come to the elements . . . we come to those things of the organisation of which, if they be organised, we are confessedly ignorant," and continues by quoting an earlier writer as observing "that we know water sufficiently to boil, . . . to freeze, . . . to evaporate . . . without knowing what water is." And as Paley notes, even after Lavoisier's discovery, "The constituent parts of water appear in some measure to have been lately discovered, yet it does not, I think, appear, that we can make any better use of water since the discovery than we did before it."[4]

The first significant consideration of water's fitness came only thirty years after the publication of Paley's *Evidences* when William Whewell, master of Trinity College, Cambridge, examined the topic in his Bridgewater Treatise entitled *Astronomy and General Physics Considered with Reference to Natural Theology,* published in 1832.[5] During those thirty years scientific knowledge had rapidly increased and Whewell was able to present the first systematic argument for the fitness of water.

Although Whewell's discussion of the properties of water is somewhat vague and nonquantitative from a modern perspective, and although he restricts his discussion to the thermal properties of water and their apparent adaptation to climatic amelioration, it nevertheless represents the first significant systematic consideration of the unique fitness of water and represents an enormous advance on Paley's *Evidences.* Beginning with its thermal properties, he points out:

> Water expands by heat and contracts by cold [but if this contraction were continued all the way to the freezing point] . . . the lower parts of water would have been first frozen and being once frozen hardly any heat applied at the surface could have melted them. . . . This is so far the case that in a vessel containing ice at the bottom and water at the top, Rumford made the upper fluid boil without thawing the congealed cake below.
>
> Now a law of water with respect to heat operating in this manner would have been very inconvenient if it had prevailed in our lakes and seas. . . . They would all have had a bed of ice, increasing with every occasion, till the whole was frozen. We would have no bodies of water, except such pools on the surfaces of these icy reservoirs as the summer sun could thaw to be again frozen to the bottom with the first frosty night. How is this inconvenience obviated?
>
> [This situation] is obviated by a modification of the law which takes place when the temperature approaches this limit. Water contracts by the increase of cold till we come near the freezing temperature; but then . . . expands till the point at which it becomes ice. Hence the water [at 4°C] will lie at the bottom with cooler water . . . above it. . . . In approaching the freezing point the coldest water will rise to the surface where congealment will take place. [But this is only part of the story.] . . . Another peculiarity in the laws which regulate the action of cold on water is, that in the very act of freezing sudden and considerable expansion takes place. . . . [Consequently, ice floats.][6]

Thus, because of these two anomalous properties, water is not bound up in vast beds of submarine ice. We now know that these two properties of water are practically unique, a fact not known in 1832, as Whewell admits: "We do not know how far these laws of expansion are connected with or depend on, more remote and general properties of this fluid or of all fluids."[7] Note that what we have here are two different characteristics of water, *both of which are mutually adapted toward the end of preserving bodies of liquid water on a planetary surface.*

One of Whewell's most interesting insights comes in a passage in which he points out that some of the thermal properties which endow water with its peculiar fitness, such as the decrease in the density of water below 4°C, and the fact that the density of ice is less than that of water, seem to be due to an apparently contrived violation of what would appear to be a natural law:

> This gradual progress of freezing and thawing, of evaporation and condensing, is produced, so far as we can discover, by a particular contrivance. Like the

> freezing of water from the top, or the floating of ice, the moderation of the rate of these changes seems to be the result of a *violation* of a law: that is, the simple rule regarding the effects of change of temperature which at first sight appears to be the law and which from its simplicity would seem to us the most obvious law for these as well as for other cases is modified at certain critical points *so as* to produce these advantageous effects.[8]

In recapitulating his argument, Whewell concludes that the various thermal properties of water, including the anomalous expansion below 4°C and its expansion on freezing, which together contribute to its remarkable fitness for the preservation of water in the liquid state, appear to be mutually independent properties. Moreover, he continues, as far as we can tell, these properties could have been different. And in a key section of his treatise, he concludes that where we see a number of natural phenomena, all of which might have been different and which also seem to be providentially arranged for the "welfare of things," this is very suggestive of design. In his own words from this classic of natural theology, he lays down the basic logic of his argument:

> All natural philosophers will, probably, agree, that there must be . . . a great number of things entirely without any mutual dependence. . . . Laws are unlike one another . . . steam . . . expands at a different rate to air . . . water expands in freezing, but mercury contracts . . . heat travels in a manner quite different through solids and fluids. We have . . . fifty substances in the world; each of which is invested with properties . . . altogether different from those of any other substance.
>
> There are, therefore, it appears, a number of things which might have been otherwise . . . substances, which might have existed any how exist exactly in such a manner . . . as they should to secure the welfare of other things . . . that the laws are tempered and fitted together in the only way in which the world could have gone on.[9]

Following Whewell, by far the most important discussion of the unique fitness of water and still the most significant to date was that of Lawrence Henderson, then professor of biological chemistry at Harvard University, in his great classic *The Fitness of the Environment,* published in 1913.[10] Henderson is remembered by every student of biochemistry and medicine as the Henderson of the Henderson-Hasselbach equation. *The Fitness of the Environment* must rate as one of the most important and influential books in the biological sciences in the first decades of the century. This is acknowledged by Joseph Needham in his *Sceptical Biologist,* published in 1929. It made, according to Needham, "unquestionably, the most important contribution to the philosophy of biology"[11] in the first quarter of the century, and this view was recently seconded by Harold Morowitz, professor of biophysics at Yale, in his *Cosmic Joy and Local Pain* of 1987.[12]

The Fitness of the Environment differs from Whewell's *Astronomy and General Physics* in two important respects. It is quantitative and comparative. Moreover, the chain of adaptations enumerated by Henderson is greater, and because something of the chemical nature of life was understood by 1900, Henderson is able to show water to be ideally fit, not only for the maintenance of global climatic stability but also to function as the matrix of living matter.

The Fitness of the Environment deals with the peculiar fitness not only of water but with other important chemical components of living things, including carbon dioxide, carbonic acid, and carbon compounds in general. The book was published in 1913. Since then and particularly over the past forty years, a vast amount of new knowledge of chemistry and molecular biology has accumulated, but this has not only entirely confirmed Henderson's position but extended it to a degree that would have seemed unimaginable in 1913.

Henderson's aim was not to present an argument for design (although his arguments could be used for that purpose) but merely to argue for an undeniable yet mysterious biocentricity in the order of things and to establish that the key components of life, including water, carbon dioxide, and bicarbonate, exhibit together a unique mutual fitness which could hardly exist in any other equivalent set of chemicals. To show, in other words, that

> in fundamental characteristics the environment [that is, the various chemicals and physico-chemical processes which constitute living things and the chemical and physical character of the hydrosphere] is the fittest possible abode for life.

He continues by admitting that

> This is not a novel hypothesis. In rudimentary form it has already a long history behind it, and it was a familiar doctrine in the early nineteenth century. It presents itself anew as a result of the science of physical chemistry.[13]

In presenting his argument for the unique fitness of water, Henderson alludes to the following thermal properties:

1. The anomalous facts (already referred to above) that water contracts as it cools until just before freezing, after which it expands until it becomes ice, and that it expands on freezing. These properties are practically unique.
2. When ice melts or water evaporates, heat is absorbed from the environment. Heat is released when the reverse happens. This is the phenomenon known as latent heat. The latent heat of freezing of water is again one of the highest of all known fluids. In the ambient temperature range only ammonia has a higher latent heat of freezing. Water's latent heat of evaporation is the highest of any known fluid in the ambient temperature range.[14]
3. That the thermal capacity or specific heat of water, which is the amount of heat required to raise the temperature of water one degree centigrade, is higher than most other liquids.
4. That the thermal conductivity of water, which is its capacity to conduct heat, is four times greater than any other common liquid.[15]
5. That the thermal conductivities of ice and snow are low.

If it were not for the properties given in point 1, most of the water on earth would be permanently frozen into vast beds of ice at the bottom of the oceans. Lakes would freeze completely from the bottom up each winter in the higher latitudes. Without those properties in point 2, the climate would be subject to far more rapid temperature changes. Small lakes and rivers would vanish and reappear constantly. Without 3, the difference between winter and summer would be more extreme and weather patterns would be less stable,[16] and the great ocean currents such as the Gulf Stream, which currently transfer vast quantities of heat from the tropics to the poles, would be far less capable of moderating the temperature differences between high and low latitudes. Without 2, again, warm-blooded animals would have a far harder time ridding their bodies of heat. Henderson was particularly struck by the adaptive significance of the cooling effect of the latent heat of evaporation in the case of warm-blooded animals. Because, as Henderson points out, "in an animal like man . . . heat is a most prominent excretory product, which has to be constantly eliminated in great amounts, and to this end only three important means are available—conduction, radiation, and evaporation."[17] But at body temperature, as Henderson continues, "very little heat can be lost by conduction or radiation and evaporative cooling is therefore the only significant means of temperature reduction." And he concludes: "To sum up, this property appears to possess a threefold importance. First, it operates powerfully to equalise and to moderate the temperature of the earth; secondly, it makes possible very effective regulation of the temperature of the living organism; and thirdly it favours the meteorological cycle. All of these effects are true maxima, for no other substance can in this respect compare with water."[18] Conversely, as the temperature falls, condensation occurs and this releases heat which tends to counteract the rate of temperature fall. Moreover, as Henderson points out, there is another aspect of the fitness of the latent heat of evaporation—the fact that as the temperature rises so does the rate of evaporation and so consequently does the cooling effect of evaporation. So the *cooling effect of evaporation increases when the usefulness of the property is most needed.* Without 4, it would be harder for cells which cannot use convection currents to distribute heat evenly throughout the cell.[19] Without 5, the protective insulation of snow and ice, essential to the survival of many forms of life in the higher latitudes, would be lost. Also, water would cool more rapidly and small lakes would be more likely to freeze completely.

And so, as Henderson argues, it turns out that not one or two, not most, but *all* the thermal properties of water are mutually adaptive not only for the maintenance of thermal stability on a planetary scale but also for the buffering of individual macroscopic life forms against sudden temperature changes. Even the low conductivity of ice is adaptive, protecting life from frost and the water below the ice from excessive cooling. Amazement mounts at the wonderful elegance and parsimony in the way the various thermal adaptations of water conspire together to achieve so many different life-sustaining ends. For example, the preservation of large bodies of liquid water on the earth's surface is ensured almost entirely by the thermal properties of water itself and of its solid form, ice. This is a particularly critical suit of adaptations because liquid water is essential to all life on earth, not only because water is the matrix in which life's chemistry occurs, but also because without bodies of liquid water no aquatic life would be possible and the evolution of complex life forms would almost certainly have been impossible. Further, the preservation of large bodies of liquid water in the oceans ensures temperature stability worldwide, which in itself ensures climatic stability on which the existence of large complex life forms depend. Moreover, complex macroscopic life forms astonishingly utilize *these same thermal properties* to buffer themselves against thermal change, which is the inevitable outcome of their metabolic processes. And so via a series of deeply interconnected and wondrously teleological thermal adaptive properties, water bestows its vital magic on earth and its living inhabitants.

The parsimony and elegance in this design is self-evident. As far as its thermal properties are concerned, water would appear to be uniquely, and in many different ways ideally, adapted for life on earth. In thermal terms, water is the unique and ideal candidate for its biological role.

Surface Tension

Of course, the thermal properties of water are by no means the only physical characteristics which make this remarkable fluid so supremely fit for its biological role. Yet another is its very high surface tension. This has many biological implications.[20] It is the high surface tension of water which draws water up through the soil within reach of the roots of plants and assists its rise from the roots to branches in tall trees. Large terrestrial plants would probably be a physiological impossibility if the surface tension of water was similar to that of most liquids. Recently, A. E. Needham commented on the utility of the high surface tension of water:

> Water has a uniquely high surface tension exceeded by few substances other than liquid selenium and this at a very much higher temperature. Water, therefore, is ideal for the formation of discrete living bodies, with stable limiting membranes. Air-water interfaces are less important, perhaps, than those between water and lipids, which likewise have high values. Other biologically useful consequences of the high tension are that materials which can lower the tension, surface active materials, tend to accumulate at the surface, and also to orientate there. Most of the biologically important carbon compounds have this property, which promotes their aggregation and concentration, as well as the formation of organised membranes.[21]

Remarkably, the very high surface tension, because it tends to draw water into the narrow cracks and fissures in the rocks, assists in the process of weathering and washing chemicals from the rocks. Also, when it freezes, the rocks are fragmented, which in turn also assists the weathering process and the formation of soils.[22] Here is another instance where a physical property of water is adapted for a role in fashioning the planetary environment for life while at the same time being adapted for a number of specific biological functions.

The Alcahest

All the various physical properties of water which endow it with such a remarkable biological fitness would of course be of no utility if its chemical properties were not similarly fit. Water could have no biological role if it was not a good solvent. The capacity to dissolve a great number of different chemical substances is presumably a criterion that must be satisfied by any fluid if it is to function as a matrix for any kind of chemical "life" remotely similar to our own.

It turns out that, as a solvent, water is indeed ideally fit, so much so that water approaches far nearer than any other liquid to the alcahest, the universal mythical solvent of the alchemists.[23] This is a property of critical importance to water's biological role. Felix Franks recently commented on the solvent action of water:

> Other remarkable properties include the almost universal solvent action of liquid water, making its rigorous purification extremely difficult. Nearly all known chemicals dissolve in water to a slight, but detectable extent.[24]

Water's power as a universal solvent is also geologically significant, as the distribution of vital minerals through the hydrosphere would be far less equitable if its solvation powers were less marked.[25]

The solvation power of water and the distribution of diverse chemical species in large amounts throughout the hydrosphere is illustrated by the vast amount of dissolved materials carried to the sea by all the rivers of the earth in one year. This quantity has been estimated to be some 5 billion tons. Henderson lists thirty-three different elements which can be found in the sea, and probably many more are present in trace amounts. To illustrate the utility of its solvation power in biological systems, he cites over fifty different compounds which are found dissolved in human urine.[26] Today one could cite many hundreds.

As one might expect from such a universal solvent, water is also a surprisingly reactive chemical. It catalyzes almost all known reactions.[27] But although quite reactive, water is far less reactive than many other liquids. Many well-known acids and alkalies are far more chemically reactive, and will dissolve substances almost insoluble in water in a matter of seconds. Yet these liquids react with the chemicals dissolved in them, exhausting themselves and consuming the solutes.[28]

Water could not fulfill its biological role if it was a highly reactive fluid, like sulfuric acid, or if it was an entirely unreactive fluid like liquid argon. It seems that, like its other properties, the reactivity of water is ideally fit for *both its biological and its geological role.*

We should note in passing that in his discussion of water Henderson omits two characteristics of water which might have been construed at the time as "defects" in its fitness for life. First, many compounds containing long hydrocarbon chains such as the lipids are virtually insoluble in it. Second, many synthetic reactions in organic chemistry can only be carried out in the absence of water. We now know, as we shall see in the following chapters, that the first of these two apparent defects, the insolubility of hydrocarbons, plays a vital role in the design of the cell system, while the other defect is circumvented by carrying out many of these synthetic reactions in special water-excluding reaction chambers in the center of proteins.

Viscosity and Diffusion

One physical property of water that was not discussed in Henderson's *Fitness* is its viscosity. The viscosity of liquids varies considerably. The viscosity of tar, glycerol, olive oil, and sulfuric acid are respectively, 10 billion times, one thousand times, about one hundred times, and twenty-five times that of water. Compared with many liquids, water has a low viscosity. Although the viscosity of water is close to the minimum known for any fluid, a few other liquids have viscosities less than water. The viscosity of ether is four times less, liquid hydrogen a hundred times less. However, as a rule, only gases have viscosities markedly less than water.[29]

The fitness of water would in all probability be less if its viscosity were much lower. The structures of living systems would be subject to far more violent movements under shearing forces if the viscosity were as low as liquid hydrogen. Shearing forces are set up in a structure when a force applied to it tends to distort its shape. A structure composed of pitch, which has a high viscosity, will tend to resist such shearing forces far more effectively than a structure composed of treacle.

If the viscosity of water was much lower, delicate structures would be easily disrupted by shearing forces and water would be incapable of supporting any permanent intricate microscopic structures. The delicate molecular architecture of the cell would probably not survive.

On the other hand, if the viscosity was much higher than it is, no fish or anything we would call a fish would be possible. One can well imagine the difficulty of attempting to sail or swim through treacle! Nor would any microorganism or cell be able to move. If the viscosity of water was higher, the controlled movement of large macromolecules and particularly structures such as mitochondria and small organelles would be impossible, as would processes like cell division. All the vital activities of the cell would be effectively frozen, and cellular life of any sort remotely resembling that with which we are familiar would be impossible. The development of higher organisms, which is critically dependent on the ability of cells to move and crawl around during embryogenesis, would certainly be impossible if the viscosity of water was even slightly greater than it is.

Viscosity also has a very important influence on the vital process of diffusion, and this has enormous bearing on the existence of our type of cellular life. It is difficult to see how else but by diffusion the necessary flow of matter into and out of any conceivable chemical self-replicating system based in a fluid medium could be maintained.

Diffusion rates in water are very rapid over short distances. Oxygen, for example, will diffuse across the average body cell in approximately one-hundredth of a second.[30] The very great rapidity of diffusion of small molecules in water over short distances explains why small microorganisms, bacteria and protozoa, and even very small multicellular organisms are able to obtain their nutrients and get rid of their waste products simply by diffusion, without the need for a circulatory system.

The rate of diffusion of a molecule in a fluid varies inversely with its viscosity. If the viscosity goes up, the rate of diffusion goes down.

If the viscosity of water had been, say, ten times greater and diffusion rates ten times less, it would be far more difficult for organisms to derive their vital nutrients by diffusion to sustain their metabolic activities. This is because the volume of a sphere is the cube of its diameter; consequently, to maintain the same level of metabolic activity, cells would have to be a thousand times smaller. In which case only the very simplest of microbial cells would be possible. If diffusion rates were a hundred times less, cells would have to be a million times smaller to maintain their metabolic activities—a volume equivalent to a sphere containing a few protein molecules.

The low viscosity of water is fit in another way because in a liquid of low viscosity the rate of diffusion of different molecules does not vary greatly from molecule to molecule.[31] Measurement of the actual diffusion rates of a variety of compounds in water shows that the diffusion rate varies inversely as the cube root of the molecular weight. This is a fascinating and important law, which is probably of critical significance. As Herbert Stern and D. L. Nanney explain in their *Biology of Cells,* "it means that the rate of diffusion is much the same for most molecules."[32] Even in the case of a molecule like a protein, of molecular weight a thousand times that of glucose, its rate of diffusion is only ten times slower. As the range of molecular weights of the great majority of key metabolites used by the cell, such as the sugars and amino acids, is no more than tenfold, the resultant variation in diffusion rates is very small.

To serve its biological role, diffusion must not be only very rapid over short distances, but its rate must be approximately the same for most of the key metabolites used by the cell. Both these criteria are satisfied by the diffusion of small metabolites in water.

The diffusion of molecules in any fluid, whatever its viscosity, including water, has an important characteristic in that it is *very rapid over short distances but very slow if there is far to go.* In fact, the diffusion time increases with the square of the diffusion distance. Thus, if the diffusion distance is increased ten times, the time taken will be increased a hundred times. The physiologist Knut Schmidt-Nielsen calculated that in the case of oxygen diffusing into the tissues, it will attain an average diffusion distance of 1 micron (one-thousandth of a millimeter) in one ten-thousandth of a second, 10 microns in one-hundredth of a second, 100 microns in one second, 1 millimeter in one hundred seconds, 10 millimeters in three hours, and *1 meter in three years.*[33]

Viscosity and the Circulatory System

Because of the increasing inefficiency of diffusion as a transport mechanism over distances greater than a fraction of a millimeter, no highly active organism more than a few millimeters thick can acquire and dispose of its metabolites by diffusion. Hence, to be viable all large organisms must have some additional means of acquiring and disposing of metabolites. In practice, this means some sort of circulatory or perfusion system.[34] In mammals billions of tiny capillaries permeate all the tissues of the body, transporting the necessary nutrients, including oxygen and glucose, to within diffusional reach of all cells where metabolic activities are occurring. Because diffusion is so ineffective over large distances, no active cell can survive in a mammal unless it is within about 50 microns from a capillary. In the active muscles of a guinea pig, there may be 3,000 open capillaries per square millimeter of muscle. This is a great number, occupying approximately 15 percent of the volume of the muscle, equivalent to 10,000 tiny parallel tubes running down a pencil lead.[35]

However, a capillary system will work only if the fluid being pumped through its constituent tubes has a very low viscosity. A low viscosity is essential because flow is inversely proportional to the viscosity. A twofold increase in viscosity causes the flow to halve. From this it is easy to see that if the viscosity of water had a value only a few times greater than it is, pumping blood through a capillary bed would require enormous pressure and almost any sort of circulatory system would be unworkable. One can readily appreciate the problem by trying to envisage pumping treacle through a narrow glass tube.

But there is a further, very striking relationship between the diameter of the tubes and the resistance to flow, one which imposes enormous design constraints on any sort of circulatory system. The resistance to flow is inversely proportional to the *fourth power* of the diameter of the tube. Which means that halving the diameter of a tube causes a sixteenfold increase in resistance to the flow of fluid through the tube.

Very little decrease in the size of capillaries could be achieved even if the viscosity of water was an order of magnitude lower. To achieve the same rate of blood flow through a capillary half the size of those which exist in the mammalian body with the same blood pressure would require either a lowering of viscosity of sixteen times, or a sixteenfold increase in pressure. In fact, no liquid at body temperature is known which has a viscosity this low. The bioengineering problems associated with redesigning a muscular pumping system like the heart to generate a perfusion pressure sixteen times as great would appear insurmountable. The smallest capillaries are about 3 to 5 microns in diameter. Given the viscosity of water, the laws which govern the flow of fluids through small tubes, and the design constraints on muscular pumping systems, the figure of 3 to 5 microns is equivalent to a physical constant—there is no way in which it could be decreased!

It is fortunate that capillaries can function down to such a small size. Because diffusion in a liquid is only effective over very small distances, the existence of higher organisms is only possible because of the existence of a myriad of tiny capillaries permeating their tissues. If the viscosity of water had been slightly greater and the smallest functional capillaries had been 10 microns in diameter instead of 3, then the capillaries would have had to occupy virtually all of the muscle tissue to provide an effective supply of oxygen and glucose. Obviously, the design of macroscopic life forms would be impossible, or enormously constrained.

The tiny diameter of the capillary also has another essential bearing on its primary function as a carrier of nutrients to the tissues. This is because the tension in the wall of a tube equals the product of the pressure within the tube and the diameter of the tube.[36] This implies that for a given pressure the tension in the wall increases in direct proportion to the radius of the tube. For this reason, as Schmidt-Nielsen points out, a large artery must have a thicker wall than a small artery. However, in the case of capillaries, "because of their very small radius a wall consisting of a single layer of cells has sufficient strength. Thus the smallness of the capillary has the important consequence that its walls can be thin enough to permit rapid diffusional exchange of material between the blood and the tissues."[37]

It seems, then, the viscosity of water must be very close to what it is if water is to be a fit medium for life. It is sufficiently high to provide some protective buffering against shearing forces for the delicate structures of the cell and sufficiently low to ensure diffusion rates fast enough to allow for material exchange between the cell and its environment. In the case of higher organisms it must be low enough to permit perfusion of the tissues via a system of capillaries down to 3 to 5 microns in diameter, which are sufficiently small to bring within diffusional distance all the tissue cells of the body without their occupying a large proportion of the volume of the tissues. If it was much higher, diffusion would be prohibitively slow, and while very simple cell systems might be possible, large, complex, metabolically active organisms would not. No conceivable set of compensatory changes—increasing the number or diameter of the capillaries, increasing the flow rate or decreasing average cell size, etc.—could be engineered to make mammalian life possible.

Non-Newtonian Fluids

There is a final and fascinating aspect of the phenomenon of viscosity related to the viscous properties of nonhomogeneous fluids which has an important bearing on the function and design of the circulatory system. Ordinary homogeneous fluids have a constant viscosity. Their flow is directly related to the pressure applied. However, as Marcus Reiner points out in his *Scientific American* article "The Flow of Matter," when a nonhomogeneous fluid, containing a suspension of particles like blood, is forced to flow through a tube, it exhibits a curious behavior: when the pressure is doubled, the rate of flow may triple. Remarkably, its viscosity becomes less as the pressure is increased. Liquids that behave in this way are called non-Newtonian.[38]

Now this apparently esoteric aspect of the phenomenon of viscosity is no triviality but rather a crucial adaptive property of blood. It means that when the blood supply to a tissue must be increased severalfold, because blood behaves as a nonhomogeneous fluid consisting of red cells suspended in a watery fluid, then as the perfusion pressure increases, *the viscosity conveniently declines.* This effect greatly facilitates the increased delivery of blood to an organ when its metabolic activity is increased. The twenty-fold increase in the perfusion of mammalian muscles as strenuous activity commences is only possible because of this characteristic of a non-Newtonian fluid.

What is particularly remarkable about this adaptive property is that the packaging of the hemoglobin (the oxygen carrying molecules in the red blood cell) in small particles, i.e., the red cells, rather than having them free in solution in the plasma, is itself adaptive, but for reasons completely unrelated to fluid flow or viscosity. These include the linking of the association and dissociation of oxygen and hemoglobin to a variety of sophisticated metabolic controls, which among other things assist in the buffering of the body against changes in its acidity and assist in the transport of carbon dioxide to the lungs. If the oxygen carrying molecules were free in solution, many of these adaptations associated with the reversible oxygenation of hemoglobin would have been in all likelihood impossible, and at the same time the advantage of the anomalous drop in viscosity when a suspension of particles is subjected to increased perfusion pressure would not accrue.

The Viscosity of Ice

Remarkably, the viscosity of ice, the solid form of water, is also adaptive for life on earth. Just as the viscosity of liquids varies greatly, the viscosity of solids also varies over many orders of magnitude. Pitch, one of the least viscous of solids, has a viscosity about 10^{12} (1 trillion) times greater than that of water, while ice, which is a crystalline solid, has a viscosity 10^{16} times that of water. The rocks which make up the crust of the earth have viscosities ranging between 10^{25} and 10^{28} times that of water. So the range of viscosities of solids is 10^{16}.[39] If the viscosity of ice had been several times lower than it is, then glacial activity would have been much less effective in grinding down the mountains and releasing vital minerals into the hydrosphere. If ice had the viscosity of pitch, then glaciers would only have been a few feet thick and would have run gently down mountainsides, making little impression on the much harder rocks that make up the earth's crust.

On the other hand, it is fortunate that the viscosity of ice is not much higher than it is. If it were anything approaching that of granite, then all the water on earth would be immobilized at the poles and on the high mountain ranges. The earth's higher latitudes would have been covered in vast sheets of granite-solid ice caps and the earth would have been sterile. There would be no liquid water on earth and no life. Today about 10 percent of the earth's water is locked up as ice in the Antarctic and Greenland ice caps. It is possible that even if the viscosity of ice had been only 100 times greater, there would have been far less liquid water on earth and the climate would have been subject to rapid fluctuations from extreme heat to extreme cold, and it is very doubtful whether life as rich as it now exists on earth would have evolved. The actual value of the viscosity of ice would appear to be yet another adaptation of "water" that ensures that large bodies of liquid water can exist on a planetary surface such as the earth's.

The Density of Water

Unlike viscosity, which varies over many orders of magnitude, the density of substances on earth varies much less. Tar, which has the same density as water, is billions of times more viscous. The density of water is 1 gram per cubic centimeter. The density of air at atmospheric pressure is about one-thousandth that of water; the density of tar and glycerol is about the same as water. The density of petrol is about 0.65 and that of many hydrocarbons ranges from between about 0.7 and 0.9 gram per cubic centimeter. Apart from the lipids and fats, many organic compounds which form the basic fabric of the cell have densities very close to that of water. Other well-known substances are more dense than water. Many common minerals have densities between three and seven times that of water. The density of two of the heaviest substances, mercury and gold, are 13.6 and 19 respectively.[40]

It is clear that as living organisms are made up largely of water, then the density of water largely determines their weight. In the case of large terrestrial organisms on a planet the size of the earth, if water were several times as dense, then the maximum size that could be attained would be only a fraction of that actually attained by existing organisms. An upright bipedal humanoid species of design similar to *Homo sapiens* would not be feasible, for the weight of the body might well prevent its being lifted off the ground and maintained in an upright position. Nor could the limbs be moved unless the proportion of muscle was greatly increased. This is because—for reasons that will be discussed in chapter 11—the power exerted by muscles per unit volume of muscle tissue cannot be much increased. If our limbs were four times their current weight, the muscles would have to be four times the volume to achieve the same level of mobility.

One set of adaptations that would theoretically be facilitated if water were less dense and organisms consequently less heavy per unit volume are those associated with flight. However, as far as aquatic life is concerned, the consequences of water having a density much less than 1 gram per cubic centimeter would be severe. In such a hypothetical world, all other things being equal, carbon-based life forms (composed of 30 percent nonaqueous materials, mainly organic carbon compounds) would tend to sink like lead balloons to the ocean floor. On the other hand, if water was just a fraction heavier than it is, all carbon-based aquatic life would be restricted to floating on the surface. It is doubtful that many life forms, particularly microorganisms, could survive the intense ultraviolet radiation that they would be subjected to if they were restricted permanently to the upper few millimeters of the sea.

Recent Discoveries

Over the last few decades additional properties of water have come to light which further confirm its remarkable fitness. Morowitz points out:

> The past few years have witnessed the developing study of a newly understood property of water [i.e., proton conductance] that appears to be almost unique to that substance, is a key element in biological-energy transfer, and was almost certainly of importance to the origin of life. The more we learn the more impressed some of us become with nature's fitness in a very precise sense. . . . Proton conductance has become a subject of central interest in biochemistry because of its role in photosynthesis and oxidative phosphorylation.[41]

As Morowitz explains, both these key processes use proton conductance and hydrated ions which are major features of water:

> Once again the fitness enters in, in the detailed way in which the molecular properties of water are matched to the molecular mechanisms of bioenergetics. A property never imagined in Henderson's time turns out to be a significant part of the fitness of the environment.[42]

Coincidence upon Coincidence

This very brief review of some of the properties of water is by no means exhaustive. There are in fact several other ways in which the properties of water are mutually fit for various biological processes, but these are more appropriately considered as they arise in consideration of the fitness of the other components of life in the following chapters. What is so very remarkable about the various physical properties of water cited above is not that each is so fit in itself, but the astonishing way in which, in many instances, several independent properties are adapted to serve cooperatively the same biological end.

Take, for example, the weathering of rocks and its end result, the distribution of the vital minerals upon which life depends via the rivers to the oceans and ultimately throughout the hydrosphere. It is the high surface tension of water which draws it into the crevices of the rock; it is its highly anomalous expansion on freezing which cracks the rock, producing additional crevices for further weathering and increasing the surface area available for the solvation action of water in leaching out the elements. On top of all this, ice possesses the appropriate viscosity and strength to form hard, grinding rivers or glaciers which reduce the rocks broken and fractured by repeated cycles of freezing and thawing to tiny particles of glacial silt. The low viscosity of water confers on it the ability to flow rapidly in rivers and mountain streams and to carry at high speed those tiny particles of rock and glacial silt which contribute further to the weathering process and the breaking down of the mountains. The chemical reactivity of water and its great solvation power also contribute to the weathering process, dissolving out the minerals and elements from the rocks and eventually distributing them throughout the hydrosphere.

The properties of water which enhance weathering.

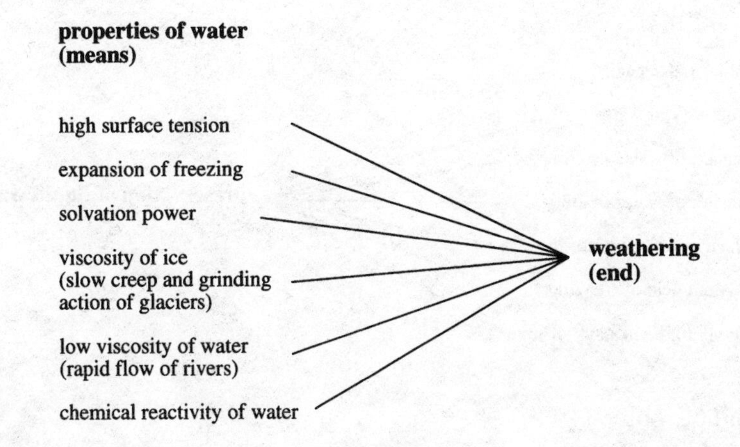

Consider also the way the various thermal properties of water cooperate to preserve large bodies of water on the surface of the earth. First, when water is cooled, its high heat capacity retards its rate of cooling. As it cools below 4°C, the coolest water rises to the surface, forming an insulating blanket on the surface which prevents further heat loss. Eventually, the surface layer freezes—a process which because of the high latent heat of fusion gives out a considerable amount of heat, thus retarding further the temperature fall. After ice has formed, because it is lighter than water, it remains on the surface, preventing, because of its poor conductivity, further cooling of the water below. Further freezing now occurs at the interface between the ice and the water so that the latent heat given out will be trapped, because of the poor conductivity of ice, *below the surface of the ice,* thereby warming the water below and thus retarding further cooling. Eventually, no matter how cold the air above the sea, the layer of ice will not increase beyond a few meters thick. Moreover, if large quantities of ice are formed in some situations because of the relatively low viscosity of ice, eventually it flows downhill or outward toward warmer temperatures or toward the sea, where it inevitably melts, generating liquid water again.

The properties of water which together tend to preserve it in the liquid state.

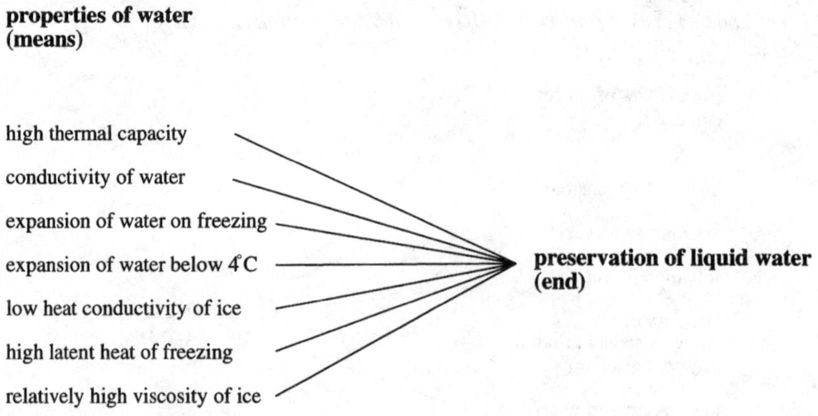

properties of water
(means)

high thermal capacity
conductivity of water
expansion of water on freezing
expansion of water below 4°C
low heat conductivity of ice
high latent heat of freezing
relatively high viscosity of ice

preservation of liquid water
(end)

Keeping Cool

As a final example, consider the way that the large heat capacity, high latent heat of evaporation, heat conductivity, and low viscosity conspire together to serve the end of temperature regulation in a large organism like a man.

All activity, including the work of machines, requires the expenditure of energy, and this necessarily involves the generation of heat. During a one-hour run over a distance of ten miles, an average man will generate a considerable amount of heat. Yet at the end of the run his body temperature will only be raised a few degrees. We are so familiar with this fact that it never occurs to us that there might be something unusual about it. But in fact it is a remarkable phenomenon.

Altogether, the work expended when a 100-kilogram man runs 10 miles in one hour will generate approximately 1,000 kilocalories of heat. If none of this heat were lost from the body during the run, it would raise the temperature of the body by 10°C. Such a temperature rise would almost certainly be fatal. If the body was constructed mainly out of, say, iron, salt, lead, or alcohol, rather than water, the temperature would be raised by 100°C, 50°C, 300°C, and 20°C respectively. The reason for the relatively modest rise in body temperature of only 10°C is that the heat capacity of water is greater than most other substances and greater than all known liquids except ammonia within the temperature range commonly encountered on the earth's surface.

But water has another unique advantage for temperature regulation. As we saw above, the latent heat of vaporization of water is the highest of any liquid in the ambient temperature range. Thus, in addition to buffering the rise in temperature, because of its very great heat capacity, the very great cooling effect when it evaporates from the skin further attenuates the rise in temperature. The evaporation of one liter of sweat from a 100-kilogram man removes about 600 kilocalories of heat from the body, lowering the body temperature by 6°C. If water was substituted for, say, alcohol or ammonia, then the cooling on evaporation would only be 2.2°C and 3.6°C respectively. Moreover, at body temperature, radiation and conduction are insufficient to rid the body of heat, so the whole burden is thrown on evaporation.

But there is much more to the story than this. Even these two nearly unique thermal properties of water would not in themselves be sufficient to maintain temperature stability unless the heat generated in the core of the body could be transported to the surface of the body. There are only two ways by which this could be done: conduction or convection.

The range of thermal conductivities among common substances is considerable. Silver and copper, for example, two of the most efficient heat conductors, have thermal conductivities more than ten thousand times greater than some of the poorest conductors, like silica gel and wood, and some thousand times greater than that of water. Liquids are poor conductors compared with metals, but of all liquids, again water is at a unique maximum, having a thermal conductivity several times as great as the vast majority of liquids at ambient temperatures.[43]

Although the thermal conductivity of water is high compared with most other liquids, it is still too low to transport heat from the center of the body to the periphery at the rate required to rid animals of the heat generated by metabolism. Only if conduction is assisted by some sort of convection mechanism is it possible to transfer heat from the core of the body to the skin. There is indeed such a convection mechanism: the circulatory system, which conveys in the average adult man nearly 6 liters of blood throughout the body, through every organ, via the capillaries, every minute carrying with it any heat generated in the body's core to the periphery. But the circulatory system, as we have seen, in turn depends on another important physical property of water, its viscosity, having almost precisely the very low value it does.

If the conductivity of water had been several times less, like that of absorbent cotton or wood, then even with the circulatory system conductivity would almost certainly have been too low to transfer heat to the surface of the body, and its elimination from the body, especially in situations of strenuous exercise, would pose insurmountable problems. The body would seize up like an overheated car engine. On the other hand, if the thermal conductivity of water was many times more, like that of copper, then the temperature of living things would equilibrate very rapidly with their environment, so that temperature regulation would be far more difficult to achieve. Changes in the environmental temperature would be rapidly conducted (as is the case with a piece of metal) throughout the body of the organism, which consequently would suffer continual swings of temperature. Small warm-blooded animals would probably be impossible, and even a large organism would experience difficulties in drinking a large quantity of cold water. To be fit for macroscopic life the thermal conductivity of water must be close to what it is.

We see, then, that the very modest rise in body temperature after strenuous exercise is no ordinary phenomenon. It turns out to be dependent on the unique fitness of water as a buffer against changes in temperature. This fitness is dependent on four quite different physical properties of water that all exhibit a coincidental mutual fitness and which together perfectly fit water for this biological role.

No other liquid is known which can even remotely approach the fitness of water for temperature regulation of a large terrestrial carbon-based form of life at the ambient temperature range of 0°C to 50°C. And, moreover, although some liquids such as ammonia and liquid sodium exhibit some of the thermal properties of water, none possess quite the same set of mutually adaptive properties. At certain temperatures liquid sodium, for example, exhibits a higher latent heat of evaporation than water but its thermal conductivity is very many times more than water, too high to permit any theoretical organism based in that medium to maintain a steady temperature in the face of environmental challenges.

Heat loss in our marathon runner involves the following adaptations:

The properties of water involved in temperature regulation.

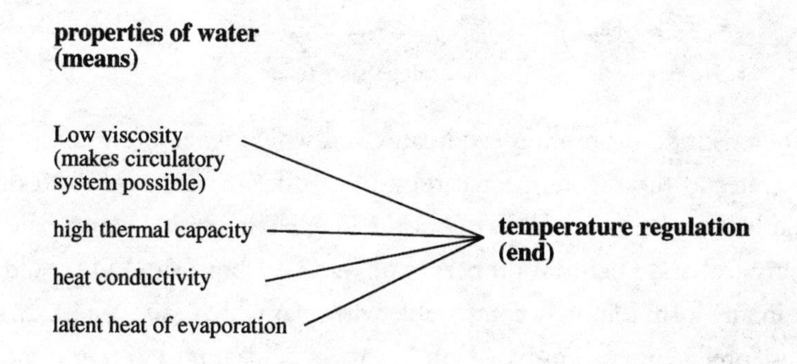

Conclusion

The evidence reviewed in this chapter indicates that water is uniquely and ideally adapted to serve as the fluid medium for life on earth in not just one, or many, but in *every single one* of its known physical and chemical characteristics.

The unique and ideal fitness of water can be illustrated in graphic form by plotting all known fluids against their utility for carbon-based life as shown on page 46.

As Henderson concluded,

> If doubts remain, let a search be made for any other substance which, however slightly, can claim to rival water as the milieu of simple organisms, as the milieu intérieur of all living things or in any of the countless physiological functions which it performs.[44]

The unique fitness of water.

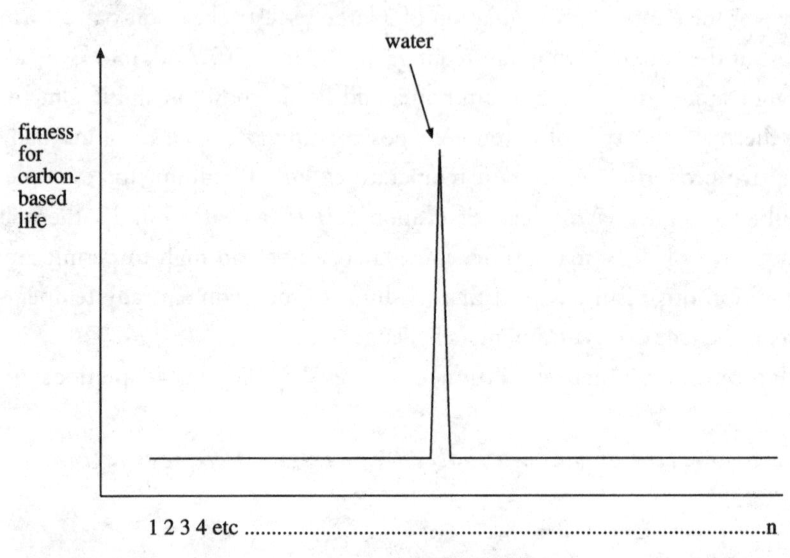

There is indeed no other candidate fluid which is remotely competitive with water as the medium for carbon-based life. If water did not exist, it would have to be invented. Without the long chain of vital coincidences in the physical and chemical properties of water, carbon-based life could not exist in any form remotely comparable with that which exists on earth. And we, as intelligent carbon-based life forms, would almost certainly not be here to wonder at the life-giving properties of this vital fluid. And if there is life like our own anywhere in the cosmos on some other earth, there will also be water and in all probability there will be seas and rivers and clouds and rain. There will be storms and waterfalls and icebergs, and surf will break on the beaches of that distant world.

In the many mutually adaptive properties of this most remarkable of all fluids, we are brought dramatically face-to-face with an extraordinary body of evidence of precisely the sort we would expect on the hypothesis that the laws of nature are uniquely fit for our own type of carbon-based life as it exists on earth.

Chapter 3

The Fitness of the Light

In which it is shown that the electromagnetic radiation reaching the surface of the earth is uniquely fit for carbon-based life. The sun's radiation is mainly in the visual range—from the near ultraviolet to the near infrared. Not only is most of the electromagnetic radiation outside this tiny range harmful to life, but the energy levels within the visual spectrum are precisely fit for photochemistry. Remarkably, the atmospheric gases, including water vapor and liquid water, absorb virtually all the harmful radiation outside the visual range and transmit only this tiny band of biologically useful radiation. These coincidences provide convincing evidence of nature's fitness for carbon-based life. On top of its utility for photochemistry the wavelength and energy levels of visual light are fit for biological vision with a camera-type eye of the sort utilized in higher vertebrates, including man. Like water, the light of the sun appears to be of optimal biological utility.

The electromagnetic spectrum.

The range of electromagnetic wavelengths is 10^{25}. As shown in the figure, the shortest gamma rays have wavelengths of 10^{-16} microns and the longest radio waves have wavelengths of about a kilometer or 10^{9} microns (one micron is one-millionth of a meter or one thousandth of a millimeter).

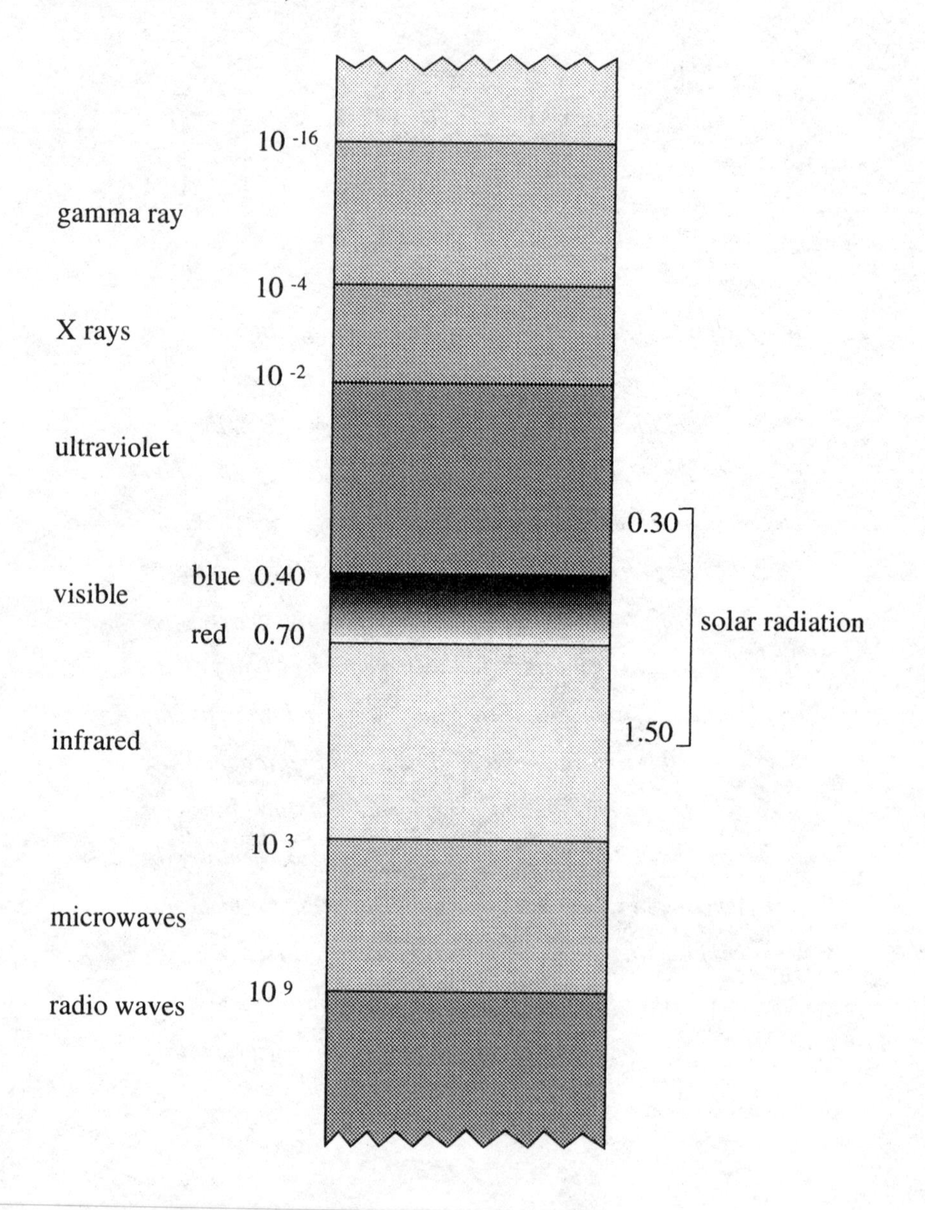

Think of the Sun's heat on your upturned face on a cloudless summer's day; think how dangerous it is to gaze at the Sun directly. From 150 million kilometers away, we recognize its power. What would we feel on its seething self-luminous surface, or immersed in its heart of nuclear fire? The Sun warms us and feeds us and permits us to see. It fecundated the Earth. It is powerful beyond human experience. Birds greet the sunrise with an audible ecstasy. Even some one-celled organisms know to swim to the light. Our ancestors worshipped the Sun and they were far from foolish.

—Carl Sagan, *Cosmos,* 1980

It may form an interesting intellectual exercise to imagine ways in which life might arise and having arisen might maintain itself, on a dark planet; but I doubt very much that this has ever happened, or that it can happen.

—George Wald, *Scientific American* magazine, 1959

As Sagan so rightly comments, ancient man was right to worship the sun as the giver and sustainer of life.[1] The sun provides heat and light, both of which are essential to life.

It is the heat provided by solar radiation in the infrared region of the spectrum which warms the earth, keeping the mean temperature of the earth above the freezing point of water and within the temperature range where the chemical reactions upon which life depends can occur. It is the sun's heat which energizes the great water cycle, drawing water by evaporation from the oceans into the atmosphere which then, via the precipitation of rain and snow, forms rivers and glaciers which carry the evaporated water back again to the ocean. And it is the energy provided by solar radiation within the visual region of the spectrum which drives the process of photosynthesis through which light energy is utilized to synthesize the fuels of life, the sugars and fats, which power the activities of virtually all complex forms of life on earth.

The sun's radiation is essential in two ways: it provides the heat energy which keeps the earth's temperature within the appropriate range for life and it provides the light energy necessary for photosynthesis.

The Electromagnetic Spectrum

Both heat and light are forms of radiant energy known as electromagnetic radiation. All the various different types of electromagnetic radiation, including heat and light, flow through space in the form of energy waves analogous to ripples on the surface of a pond. And just as the waves on the surface of a pond may have different wavelengths—small ripples may be only a few centimeters from crest to crest, while large waves might be more than a meter—so similarly the wavelength of the various types of electromagnetic radiation also varies, but over a vastly greater range.

The longest electromagnetic waves are radio waves, which measure several kilometers across (10^9 microns), while the smallest are the gamma rays, which measure less than a trillionth of a centimeter (10^{-16} microns) across. The wavelength of microwaves is around 1 centimeter (10^4 microns). The wavelength of visible light ranges from about one ten-thousandth of a centimeter (0.70 microns) at the long, or red, end of the spectrum to about one twenty-five-thousandth of a centimeter (0.40 microns) at the short, or blue, end of the spectrum. The wavelength of infrared, or heat, radiation ranges from one ten-thousandth of a centimeter (1 micron) to one-tenth of a centimeter (10^3 microns).

Note that the wavelength of the longest type of electromagnetic radiation is unimaginably longer than the shortest by a factor of 10^{25}, or 10,000,000,000,000,000,000,000,000. Some idea of the immensity this figure represents can be grasped by the fact that the number of seconds since the formation of the earth 4 billion years ago, is *only* about 10^{17}. To count 10^{25} seconds we would have to keep counting every day and night through a period of time equal to *100 million times the age of the earth!* If we were to build a pile of 10^{25} playing cards, we would end up with a stack stretching halfway across the observable universe.

Remarkably, although the wavelength of electromagnetic radiation in the cosmos varies over such a colossal range, 70 percent of the electromagnetic radiation emitted from the surface of the sun is concentrated in an exceedingly narrow radiation band extending from the near ultraviolet (0.3 microns) through the visible light range into the near infrared (1.50 microns). This minute band represents the unimaginably small fraction of approximately one part in 10^{25} of the entire electromagnetic spectrum—equivalent to one playing card in a stack of cards stretching halfway across the cosmos, or one second in 100 quadrillion (100,000,000,000,000,000) years.

There is nothing exceptional about this amazing compaction of the sun's radiant energy into such a small radiation band. The spectrum of radiation emitted by a star is determined by its surface temperature. The temperature of the sun's surface is close to 6,000°C. Because the sun is an "ordinary, even mediocre, star,"[2] being about in the middle of the range of stellar temperatures and sizes, many stars have surface temperatures close to this value and emit nearly all their radiation in this same very small band.

Solar Radiation

When electromagnetic radiation interacts with matter, energy is imparted. If the radiation is highly energetic in the X-ray or gamma-ray regions, this can tear atoms and molecules apart. On the other hand, radiation in the radio region imparts so little energy that it passes through matter with hardly any detectable effect. Only radiation in this tiny band—in the visual and infrared region—interacts gently enough with matter to be of utility to life. Consequently, the fact that the sun's radiation (and that of many main sequence stars) is compressed into this tiny region of the spectrum is of enormous biological significance.

Atoms and molecules only react together when they possess energies equal to or greater than a particular threshold value, which is known as their energy of activation. For the great majority of chemical reactions which occur in living things, the activation energy of the reactions (the amount of energy needed to cause a chemical reaction between two molecules) generally lies between 15 and 65 kilocalories per mole. The common way of expressing these energy levels is in terms of quantity of heat—i.e., calories—per quantity of matter (moles). These energy levels are provided by electromagnetic radiation between 0.80 microns and 0.32 microns, more or less exactly that provided by visible light.[3] Radiation at wavelengths slightly longer than 0.70 microns is too weak to raise molecules into energy states which can activate chemical reactions. But on the other hand, radiation in the ultraviolet (shorter than 0.30 microns) is too energetic and causes disruption of life's delicate molecular structures.

The correspondence between the "energy needs" of biological chemistry and the "energy levels" of solar radiation was discussed by George Wald in a well-known *Scientific American* article entitled "Life and Light." As Wald comments, "the radiation that is useful in promoting orderly chemical reactions comprises the great bulk of that of our sun. The commonly stated limit of human vision—400–700 millimicrons—already includes 41% of the sun's radiant energy."[4] The diagram below illustrates the intensity of solar radiation between 0.20 microns and 1.50 microns.[5]

Infrared radiation is also essential to life but for a different reason. When radiation in the infrared region of the spectrum interacts with matter, energy is imparted, which causes the random movement and vibration of atoms and molecules to increase. This we register as heat. As already mentioned, it is the heat imparted to the earth by radiation in the infrared region of the spectrum that keeps the earth's hydrosphere warm, keeps water a liquid, and drives the climatic systems and the water cycle.

Moreover, heat energy is important in another way. At least some heat is necessary for chemical reactions, because to interact chemically with one another, atoms and molecules must come into contact and this can only occur if they are in motion and may collide. Note, however, that heat energy is only of utility to the orderly chemical processes of life in a narrow temperature range—approximately that in which water is a liquid, a point discussed again in chapter 5. The heat imparted to the earth's hydrosphere by infrared

The solar spectrum, showing the intensity of the sun's radiation between 0.1 and 1.50 microns.

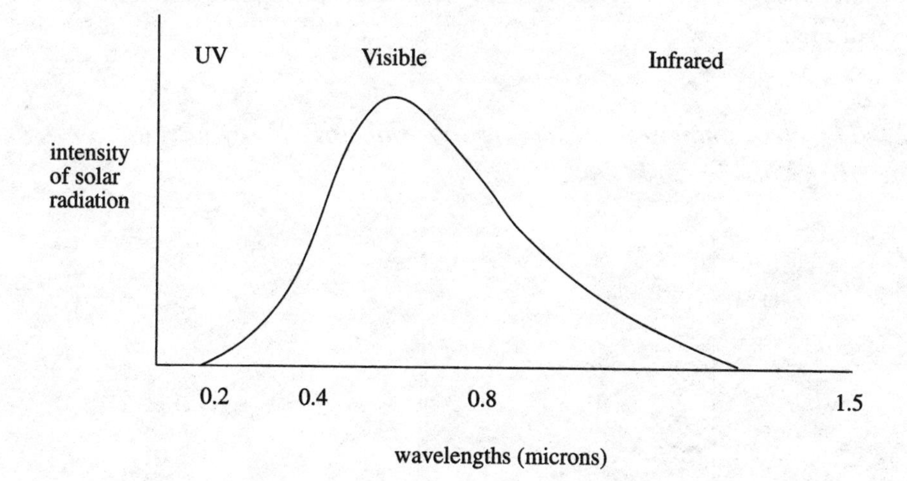

radiation from the sun would, if unchecked, eventually raise the temperature of the earth by many hundreds of degrees. At such temperatures the velocity of atomic and molecular collisions becomes such that the orderly chemistry of life is impossible. The earth escapes this fate because—via a variety of feedback systems (see next chapter) and other not fully understood controls—over time heat loss exactly balances heat gain.

That the radiation from the sun (and from many main sequence stars) should be concentrated into a minuscule band of the electromagnetic spectrum which provides precisely the radiation required to maintain life on earth is a very remarkable coincidence described as "staggering" by Ian Campbell in *Energy and the Atmosphere.*[6] Note that the compaction of solar radiation into the visible and near infrared is determined by a completely different set of physical laws to those that dictate which wavelengths are suitable for photobiology.

Our amazement grows further when we note that not only is the radiant energy in this tiny region the *only radiation of utility to life* but that radiant energy in most other regions of the spectrum is *either lethal or profoundly damaging.* Electromagnetic radiation from gamma rays through X rays to ultraviolet rays is all harmful to life. Similarly, radiation in the far infrared and microwave regions is also damaging to life. Just about the only region of the electromagnetic spectrum which is harmless to life apart from the visible and the near infrared is the region of very long wavelength radiation—the radio waves. So the sun not only puts out all its radiant energy in the tiny band of utility to life but virtually *none, in those regions of the spectrum which are harmful to life.* This coincidence is expressed in graphic form in the two diagrams below.

The electromagnetic spectrum and radiant energy output of the sun.

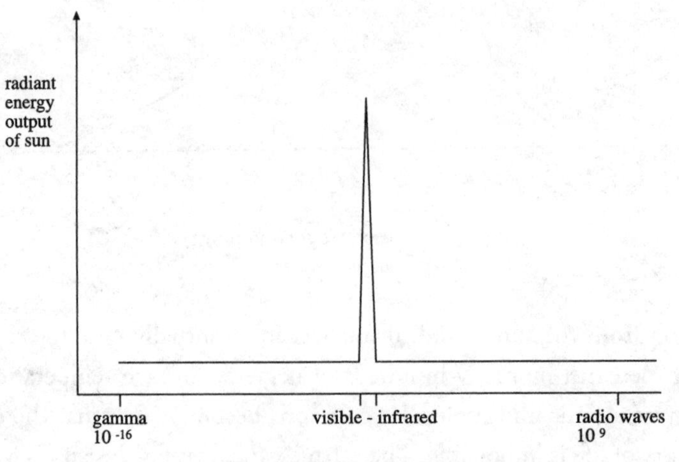

The region of the electromagnetic spectrum of utility for photochemistry.

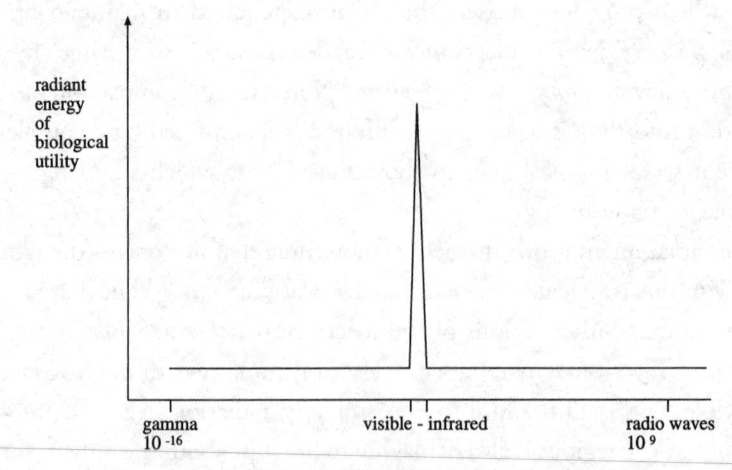

The Absorption of the Atmosphere

Now this is a remarkable enough coincidence in itself. But there are further coincidences to consider. To be of any utility to life, the radiation of the sun has to reach the surface of the earth. To do so it must pass through the atmosphere. Necessarily, any atmosphere surrounding a terraqueous planet containing carbon-based life is bound to contain some carbon dioxide gas, water vapor, at least some nitrogen, and for advanced highly active life forms considerable concentrations of oxygen. It is difficult to see how the actual concentrations of these gases could be very different from what they are in any atmosphere supporting a carbon-based biosphere (see discussion in chapter 6). At the temperature range that exists at the earth's surface, there is bound to be water vapor in considerable amounts in the atmosphere.

The fact that the atmospheric gases oxygen, nitrogen, carbon dioxide, and water vapor transmit 80 percent of the sun's radiation in the visible and near infrared and allow it to reach the earth's surface is another coincidence of enormous significance. The great majority of all atoms and molecular substances are completely opaque to visible light and radiation in the near-infrared region of the spectrum. Window glass, an example of a transparent solid which transmits light in the visible region, is exceptional. If the atmosphere had contained gases or other substances which absorbed strongly visible light, then no life-giving light would have reached the surface of the earth. In the case of nearly all solid substances, layers only a fraction of a millimeter thick are sufficient to prevent the penetration of light. Even the atmospheric gases themselves absorb electromagnetic radiation very strongly in those regions of the spectrum immediately on either side of the visible and near infrared. The diagram below indicates the spectral regions absorbed by the atmosphere.[7] Note that the only region of the spectrum allowed to pass through the atmosphere over the entire range of electromagnetic radiation from radio to gamma rays is the exceedingly narrow band including the visible and near infrared. Virtually no gamma, X, ultraviolet, far-infrared, and microwave radiation reaches the surface of the earth.

Despite these three remarkable coincidences, life would still not be possible without a fourth coincidence—the fact that liquid water is highly transparent to visible light.

The regions of the electromagnetic spectrum absorbed by the atmosphere.

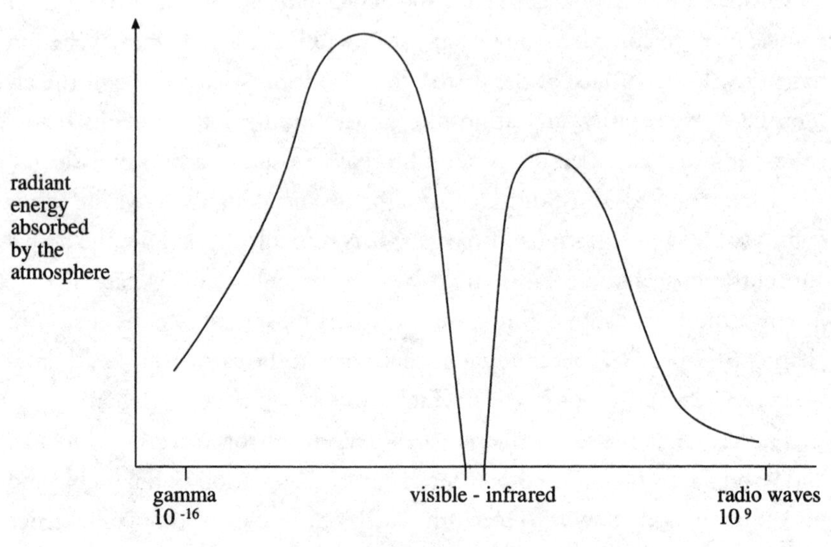

The Absorption of Water

The significance of the transparency of water to light cannot be exaggerated. All biological chemistry occurs in liquid water. If the energy of sunlight is to sustain life in the ocean then it must be capable of penetrating some distance below the surface of the sea. Even on land if light energy is to reach the chemical machinery of the cell it must invariably penetrate a thin layer of water.

Nearly all electromagnetic wavelengths are strongly absorbed by water, except radio waves and light within the visible spectrum.[8] Even far ultraviolet and infrared radiation, the two bands immediately adjacent to the visible band, are absorbed readily by water and only penetrate a fraction of a millimeter below the surface. The absorption of visible light by water varies markedly across the visible spectrum. No red light can be observed below 18 meters. Yellow light only penetrates to 100 meters. By 240 meters most of the green and blue light has been absorbed. The absorbency spectrum of liquid water is shown in the diagram below.[9]

The very remarkable fact that the only region of the spectrum allowed through the atmosphere and allowed to penetrate liquid water is the tiny range of the spectrum useful for life is commented on in the latest edition

The regions of the electromagnetic spectrum absorbed by water.

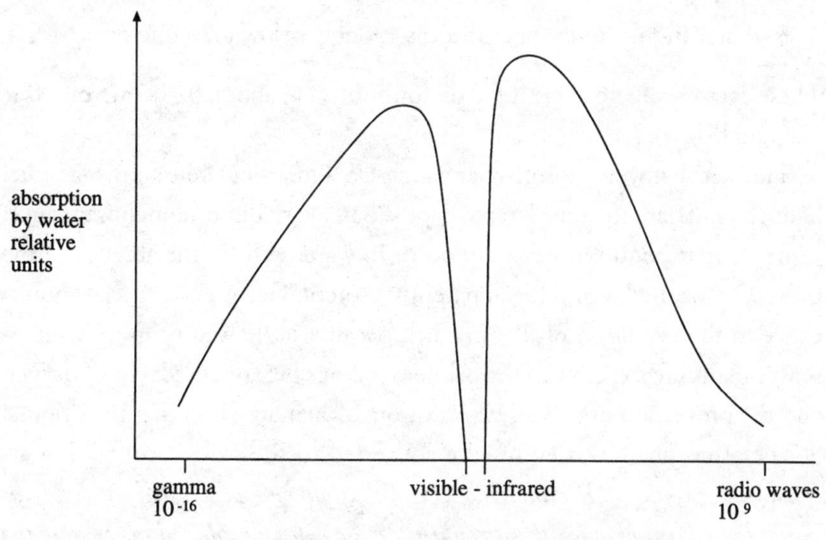

(15th) of the *Encyclopaedia Britannica:* "Considering the importance of visible sunlight for all aspects of terrestrial life, one cannot help being *awed* by the dramatically narrow window in the atmospheric absorption . . . and in the absorption spectrum of water."[10] (My emphasis.)

The fact that light at the blue end of the spectrum penetrates further into water than light at the red end may explain why chlorophyll so strongly absorbs light in the blue region. The question as to why in a biocentric world chlorophyll should absorb light in a region of the solar spectrum where the incident radiation is not maximum is raised in chapter 9. We have here in the differential penetration of light a possible answer to the conundrum. Photosynthesis in water, especially several meters below the surface, is only possible if chlorophyll absorbs light in the blue end of the spectrum.

The fact that water absorbs light in the far ultraviolet is of obvious biological significance, as it acts as another device to shield life from the damaging influence of ultraviolet radiation. Note that there are three independent mechanisms attenuating the UV flux reaching biological systems:

1. The radiant output of the sun falls dramatically from 0.40 microns to 0.30 microns so that very little ultraviolet radiation leaves the sun in the first place.

2. Ozone in the upper atmosphere absorbs UV light strongly below 0.30 microns.
3. Water (liquid and vapor) absorbs strongly below 0.20 microns.

These factors together create a discontinuity at about 0.30 microns (see diagram below).

The fact that water absorbs damaging UV radiation while allowing visual light to penetrate to considerable depths has a very important consequence; it means that photosynthesis can occur in water even in the absence of the protective ozone layer. This apparently esoteric fact was of crucial significance to the evolution of life on earth because it allowed primitive microscopic plant life to thrive in the primeval oceans before the advent of oxygen and the protection of ozone. It was through their activities that the original anoxic atmosphere was eventually converted to the oxygen-containing atmosphere of today. *If damaging UV radiation had penetrated as deeply as visual light, then photosynthesis could never have been exploited to generate the current oxygen-containing atmosphere.* And as we shall see in chapter 6, without oxygen in the atmosphere, actively metabolizing complex multicellular life forms would not have been possible, and lacking the protection of ozone, terrestrial plant and animal life would have been enormously constrained if not impossible. In a very real sense, the existence of all complex life on earth is dependent on the relative absorbance of UV and visual light in water being very close to what it is.

The very small amount of ultraviolet radiation that does reach the earth's

Mechanisms attenuating the ultraviolet flux reaching the earth's surface.

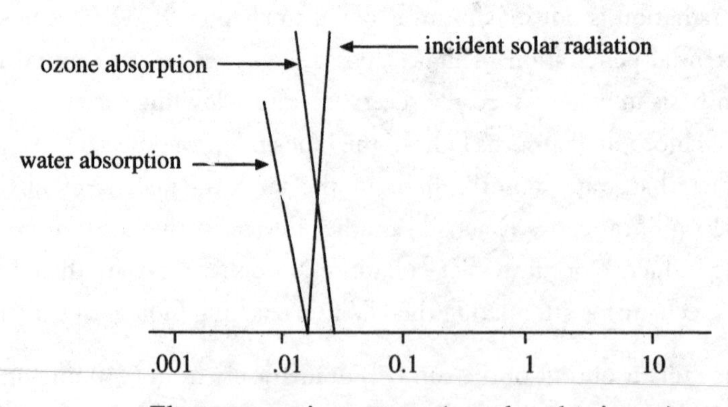

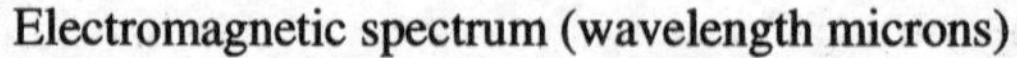

surface has clearly not hindered the evolution and development of life on earth. The spectacular success and persistence of life over the past 4 billion years indicates that life can thrive when subjected to at least some ultraviolet radiation and that the ultraviolet reaching the earth's surface must have had little, if any, deleterious effect on life in general. In fact, very small amounts of ultraviolet may have played a significant role in evolution by raising slightly the average mutation rate. Without mutations, there can be no evolutionary change, and it is possible that the raised levels of mutation caused by the ultraviolet flux could have played a critical role in the evolutionary history of life.

In man and other vertebrates ultraviolet radiation (between 0.29 and 0.32 microns) is essential for the synthesis of vitamin D. This occurs in the skin when it is irradiated by ultraviolet light. As vitamin D is vital for the maintenance and control of calcium levels in the body and for the formation of bone in all vertebrates, and as the only means of its synthesis is via ultraviolet light in the skin, then as far as humans are concerned, ultraviolet light is essential to life. The fact that a vital vitamin is produced in this way raises the question as to whether low levels of ultraviolet radiation may have other important biological influences.[11] Another important role of near UV light may be the photochemical release of nitrogen from aquatic dissolved organic matter. This plays an important role in the recycling of nitrogen in marine and freshwater ecosystems.[12]

As well as strongly absorbing ultraviolet light, liquid water also strongly absorbs radiant energy in the infrared region. Very little radiation in the infrared region penetrates more than a few millimeters into water, which means that all the heat reaching the surface of the sea is absorbed and retained in the surface layers.[13] As was mentioned in chapter 2, water, like all other fluids, is a poor conductor of heat. This helps retain the heat in the surface layers of the ocean. Consequently, in most parts of the sea there is a relatively thin surface layer about 20 meters thick of warm, less dense water. Below 100 meters the temperature of the sea in all parts of the globe falls rapidly till it reaches about 4°C at 1,000 meters below the surface. The retention of heat in the thin surface layer facilitates the transference of heat to the air and winds and by surface currents to colder regions of the ocean, thereby assisting in the maintenance of global temperature equilibration. In the higher latitudes this retention of the heat at the surface of the ocean tends to slow down the rate of freezing.

So both the transparency of water to visible light and its opacity to infrared radiation contribute to the fitness of the earth's hydrosphere for life. The radiation needed for photobiology, radiation in the visible region of the spectrum, penetrates to depths of 100 meters, while the infrared needed to warm the hydrosphere and drive the climate and the water cycle is retained in the surface layers where its utility is greatest. And in addition, all highly destructive ultraviolet radiation below 0.20 microns is almost entirely absorbed in the first few millimeters.

Another fascinating aspect of the fitness of the electromagnetic spectrum for life is the fact that both types of useful radiation, the visible and the infrared, are adjacent in the spectrum. What we have in effect are two adjacent playing cards back-to-back in a deck which extends across the cosmos. Just as the transparency of water to visible light and the fitness of the solar radiation for photochemistry are of necessity, so the close proximity of these two vital types of radiant energy gives every appearance of also being of necessity. If these two vital types of radiation were far apart in the spectrum, the possibility of prearranging nature so that they could both reach the surface of a watery planet in appropriate quantities from one unique source, such as the sun, would in all probability have been impossible.

We should indeed be *awed* and *staggered* by this series of coincidences: that the electromagnetic radiation of the sun should be restricted to a tiny region of the total electromagnetic spectrum, equivalent to one specific playing card in a deck of 10^{25} cards stretching across the universe; that the very same infinitely minute region should be precisely that required for life; that the atmospheric gases should be opaque to all regions of the spectrum except this same tiny region; that water should likewise be opaque to all regions of the spectrum save this same infinitesimally tiny region, etc. It is as if a cardplayer had drawn precisely the same card on four occasions from a deck of 10^{25}.

Even with all these coincidences, unless the sun's radiation reaching the earth's surface had remained virtually constant throughout the past 4 billion years, life could never have survived and evolved as it has. The sun is fit as an energy source for carbon-based life forms not only in providing radiant energy with precisely the levels necessary for life, but also because it has provided that vital and necessary energy at an almost perfectly constant intensity for unimaginable eons of time. Even the slightest change in the output of radiant energy from the sun at any stage during the history of life would have had disastrous consequences.

Alternatives to Light

Life does not depend on light. There is a vast diversity of microbial species which can survive in total darkness and which derive energy from the oxidation of substances such as hydrogen, hydrogen sulfide, and ferrous iron that are generated by geochemical processes in the earth's crust. It is even possible that a considerable proportion of the earth's biomass may consist of bacteria in the crustal rocks. Near the deep-sea hydrothermal vents a quite complex community of multicellular species is sustained by nutrients synthesized by bacteria which derive their energy from the oxidation of sulfides released from the vents. Geothermal springs are another site where a complex biota of microorganisms derive energy from the oxidation of hydrogen sulfide.[14] However, it is difficult to imagine how these other potential sources of energy could sustain complex aquatic or terrestrial ecosystems on the surface of a planet like the earth. And it is difficult to envisage the evolution of advanced and complex life forms in such restricted environments as microscopic fissures in the crustal rocks, thermal springs, or deep-sea hydrothermal vents.

In the last analysis, there is no alternative to stellar radiation to sustain a rich and diverse biosphere over billions of years on a planetary surface like the earth's. If the radiant energy of stars was not fit for photochemistry and to provide the heat to energize a hydrosphere like that which exists on earth, then the cosmos would in all probability be only capable of sustaining carbon-based biospheres far less complex than our own.

Fitness for Vision

The light of the sun is uniquely fit in yet another way for life on earth—the energy levels and wavelength of electromagnetic radiation in the visual spectrum are both uniquely fit for high-resolution vision.

One reason that visual light is fit for biological vision is that if an eye is to "see" it must be able to detect the type of radiation forming the image. Light radiation is the only type of electromagnetic radiation that has the appropriate energy level for detection by biological systems. UV, X ray, and gamma rays are too energetic and are highly destructive, while infrared and radio waves are too weak to be detected because they impart so little in energy interacting with matter. Moreover, most electromagnetic radiation outside the visual region is strongly absorbed by water and other biological materials, so it is difficult to imagine what biological substances could be utilized for the construction of a lens sufficiently transparent to transmit and focus the radiation. Our ability to discriminate between different wavelengths in the visual spectrum—i.e., to see colors—is also dependent on the perfect correspondence between the energy levels of electromagnetic radiation in the visual region and those required for photochemical detection by biological systems.

Not only are the *energy levels* and absorbance characteristics of light waves fit for detection by biological systems, but the *actual length of the waves* in the visual region of the spectrum is perfectly fit for the high-resolution camera-type eye of the precise design and dimension as that found in all higher vertebrate species, including man.

Instrumental Constraints

The wavelength of light imposes constraints on the design and dimension of the eye mainly because of the phenomenon of diffraction, which is an inevitable consequence of the wave nature of light itself. Like the ocean swell entering a small harbor, when light waves pass through a small aperture, they suffer dispersion or diffraction. As a consequence, when light from a point source in the visual field (A or B, in the diagram below) passes through the pupil and is focused on the retina, instead of being focused to a point, it spreads out into a tiny disc surrounded by concentric rings (a, b). The disc is known as the Airy disc.(See opposite.)

And again just like an ocean swell entering a harbor, where the degree of diffraction is determined by the size of the swell, in the case of light, the larger the wavelength, the larger the diffractional effect, the larger the Airy disc, and the poorer the resolution. At any wavelength the phenomenon of diffraction and its consequence, the Airy disc, imposes a limit on the resolving power of any sort of optical instrument, camera, or eye.

In addition to wavelength, the other factor which influences diffraction is aperture. And again in terms of the ocean swell–harbor analogy, just as the greater the opening of the harbor the less the swell is diffracted, so the greater the aperture the less the diffraction in the case of an optical instrument.

In short, the size of the Airy disc is determined by two primary factors—the *wavelength of the radiation* itself and the *size of the aperture* of the instrument.

The Airy disc.

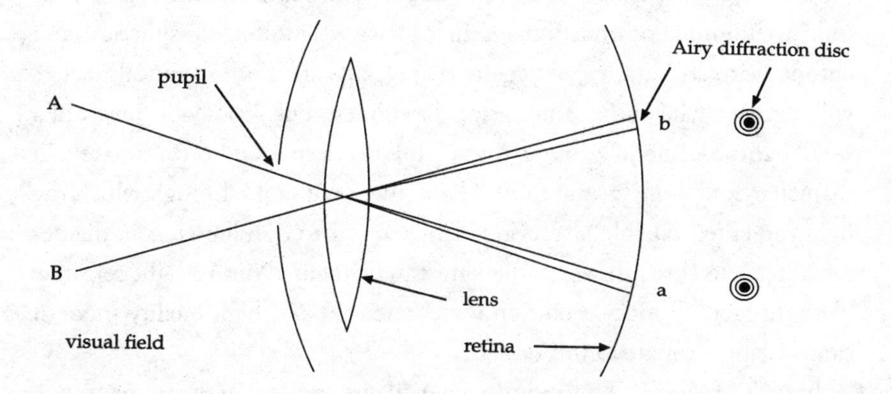

To achieve optimal resolving power the effect of diffraction must be minimized. One way this may be achieved is by increasing the size of the aperture. The resolving power of the Mount Palomar telescope is a thousand times greater than that of the human eye, mainly because its aperture (the mirror) is approximately a thousand times wider than the pupil of the human eye.

Unfortunately, in the case of the eye, which focuses an image by refraction through a lens, because chromatic and spherical aberration also increase as the aperture is increased, increasing the size of the aperture can only bring limited improvement in resolution. Therefore, minimizing the effect of diffraction involves the satisfaction of several conflicting criteria: if the aperture of an eye or a camera is increased, then the diffraction may be decreased and resolution improved but spherical aberration is also increased, thereby minimizing any improvement that might have been gained; if the pupil or aperture is decreased, then although spherical and chromatic aberration is decreased, diffraction effects increase and the resolution is decreased while at the same time the illumination of the retina (or photographic plate in the case of a camera) is decreased, and so on.[15]

Although large telescopes have very much greater resolving power (aperture) than the vertebrate eye, their field of view is far smaller and only distant objects can be brought into focus. In the case of an eye, any advantage that might be gained in resolution would be more than offset in the case of an organism by the decrease in the field of view and the inability to bring objects close at hand into focus.

Consideration of the many conflicting criteria which must be satisfied in attempting to optimize the resolution of a camera-type eye utilizing light of a wavelength of about 0.5 microns, including minimizing diffraction, maximizing illumination, maximizing field of view, minimizing spherical and chromatic aberration, etc., suggests that all high-resolution optical devices will necessarily be of the same design and dimensions. Each will consist of a small lightproof hollow rounded structure between 1 and 6 centimeters in diameter, containing at the front an aperture or "pupil" through which the light can enter, capable of varying from about 1 to 8 millimeters in diameter, and a lens through which the light can be focused onto a light-sensitive plate. In fact, all high-resolution vertebrate eyes and high-quality modern cameras approximate to this design.

In such an eye, the minimum-sized Airy disc formed on the retina from a point source of light will be about *2 microns* in diameter, which is the size of the Airy disc in many vertebrates, including man. This cannot be reduced and represents the optimum resolving power of the vertebrate eye. As one of the foremost authorities in this area, Horace Barlow comments:[16] "It is easy to verify that the highest resolving power achieved by man is quite close to the limiting resolution set by the diameter of the pupil and the wavelength of light."

Note that although many factors influence the resolving power of the vertebrate eye, it is the *actual wavelength of light* itself, or more specifically the degree of diffraction that inevitably occurs when radiation of that specific wavelength passes through an aperture, which more than any other factor determines the actual dimensions of the high-resolution vertebrate camera eye. For example, if the wavelength of light had been ten times less (0.05 microns), then an eye with the resolving power of the human eye would need to be only a few millimeters in diameter. Conversely, if the wavelength of light had been ten times greater (5 microns), then an eye with the same resolving power would need to be 25 centimeters in diameter—larger than the human head.

Note that the optimized vertebrate camera eye is quite a large object on the scale of all biological structures on earth, which range from about ten-thousandths of a micron, the size of small subcellular structures, to 100 meters (1 billion microns) in the case of a California redwood. Clearly, if the minimum possible size of the camera-type eye constructed out of biological materials had been only ten or twenty times more, then no large organism would have possessed high-acuity vision. To begin with, its shape must be held in an approximately globular conformation. This is partly achieved in the case of many vertebrates by placing the eye in a round bony socket (which also preserves the eye's delicate structure from damage). Another adaptation which helps maintain the spherical shape of the eye is the maintenance of a relatively high hydrostatic pressure in the interior of the eye.[17] But even with these adaptations, the globe of the human eye still suffers some degree of distortion.[18] In the case of some whales, which have the largest eyes of any mammal, more than half the mass within the globe is made up of an immensely thickened tough fibrous coat. So thick is this coating that it may make up three-quarters of the length of the anteroposterior axis of the eyeball. This is considered by some to be an adaptation to maintain the approximately spherical shape of the eye against distortion when the whale dives to great depths.[19] The size of the lens, which is made up of living tissue and which must obtain its nutrients by diffusion, must also place an upper limit on the size of the vertebrate camera-type eye. From these and many other considerations it seems likely that the size of the vertebrate camera eye approximates closely to the maximum possible size of a camera-type eye constructed out of biological materials.

But although the high-resolution camera eye is quite a large structure, it is still small enough for function and anatomical placement in a very wide variety of large terrestrial organisms, including the human species.

Thus, various factors, including the wavelength of light, diffraction, the size of aperture, and chromatic and spherical aberration, together impose what we might term, after Horace Barlow, an *instrumental* limit on the resolution of the camera-type eye. However, this *instrumental* limit is not the only limit to the resolving power of the eye.

Micro-optical Constraints

Clearly, no eye can resolve images smaller than the diameter of its individual photoreceptor units, and because of the inevitable constraints on cell design it is difficult to envisage photoreceptors much smaller than a few microns in diameter. (Most cells in higher organisms are between 10 and 50 microns in diameter.) This suggests that the cellular limit cannot be far removed from the *instrumental* limit of 2 microns imposed by the various factors alluded to above. Interestingly, the smallest photoreceptors in the vertebrate retina are in fact about 2 microns across.[20]

Moreover, recent work in the field of fiber optics suggests that even if photoreceptors could be theoretically reduced in size below 2 microns in diameter, no improvement in resolving power would be gained because there are other constraints arising from the wavelength of light which impose a minimum size on photoreceptor cells.[21] As one authority comments: "Optical cross talk between receptors would occur if they were smaller and packed more closely together than about 2 microns. . . . On this view, limiting retinal resolution, expressed in linear units, is the same in all eyes and results from the properties of light passing through a set of waveguides. As expected on this view, cones smaller than 2 microns across have never been found."[22]

This is another remarkable coincidence. The two optical limits to the resolving power of the eye—the one set by the waveguide optics of the photoreceptor, which we might refer to as the *micro-optical* limit; the other set by the laws of classical optics relating to the resolving power of a camera-type eye, the *instrumental* limit—*have the same value: 2 microns.* Moreover, both these limits imposed on the resolving power of the eye, the micro-optical and the instrumental limit, one imposed by the laws of fiber optics, and the other by the laws of classical optics, correspond very closely with a third—the *cellular* limit—imposed by purely biological constraints which limit the minimum size of any sort of feasible cellular receptor system. Our high-resolution vision is only possible because these three limits precisely coincide.

There is almost certainly yet another factor which imposes a limit on the minimum possible size of the photoreceptor cell—the need of the cell to respond to exceedingly small quantities of light energy. The amount of light falling on the retina from a distant star is 1 trillion times (10^{12}) less than that from a brightly lit snowfield.[23] The extreme sensitivity of the eye to very small quantities of light makes vision possible at night and in other situations of low light intensity.

The eye is able to detect extremely small quantities of light because each photoreceptor is able to respond to a single photon of light—the smallest possible quantity of light energy.[24] A photon of light is detected in the photoreceptor cell when it interacts with a single light-sensitive molecule, called rhodopsin. The photoreceptor cell is able to interact with a single photon only because it is packed with many millions of rhodopsin molecules, which maximizes the probability that an individual photon will be absorbed.[25] The need for many millions of rhodopsin molecules in an individual photoreceptor cell, if the cell is to detect a single photon, is another factor imposing a lower limit on the size of the photodetector cell. Only relatively large photoreceptor cells could detect individual photons and make night vision possible. The energy change resulting from the interaction of a single rhodopsin molecule with a single photon of light is extremely tiny, and the cell is only able to detect it after it has been massively amplified via a complex chain of enzymic reactions, which ultimately change the electrochemical state of the cell.[26] Because of the complexity of the molecular system needed to amplify the tiny initial signal, it is hard to believe that any photoreceptor cell capable of responding to a single photon of light, and of therefore "seeing" in the dark, could be much smaller than the photoreceptor cells in the vertebrate retina.

The energy levels and the wavelength of light are of course only two aspects of the natural order that must possess precisely the properties and values they do for high-resolution vision to be possible. There are other features as well: there is the transparency of water to light; there is the low refractive index of water; there is the diffusion rates of small organic compounds in water to nourish the lens, which is a living tissue; there is the necessity for a large nervous system to analyze the visual data; and so on.

Seeing Outside the Visual Spectrum

Finally, let us consider some of the hypothetical problems that would be met in attempting to construct a biological high-resolution camera-type eye to "see" in wavelengths outside the visual region of the spectrum.

For example, to obtain, with infrared or radio waves, the same degree of resolution as that which can be achieved with the human eye and many other optimized vertebrate camera eyes would require eyes of vastly increased size, even if a biological detector device could be constructed. To obtain the same resolving power as that of the human eye with, say, radio waves with a wavelength of 100 centimeters would require a reflector disc 10 kilometers in diameter. Even far-shorter-frequency radio waves, or microwaves of 1 millimeter wavelength, would require a lens or disc with a diameter of 10 meters to equal the resolving power of the human eye. So in addition to the profound difficulty that would be encountered in designing biological detector devices capable of measuring the low energy levels of infrared and radio waves, there would also be the tremendous engineering challenge of having to construct immense telescopes of planetary dimensions. Evidently, the infrared and radio regions of the spectrum are totally unfit for biological vision.

Curiously, not only is light far fitter for biological seeing and particularly for high-resolution vision than infrared or radio waves, it would also seem to be fitter for nonbiological eyes such as the radio and infrared telescopes used by astronomers. Even though a vast diversity of materials can be recruited for constructional purposes, no telescope has been built with a resolving power remotely equal to the human eye using wavelengths longer than those in the visual spectrum. A cursory comparison of the many photographs taken of galaxies and other astronomical objects through light, infrared, and radio telescopes shows clearly the far greater resolving power of optical devices in the visual region of the spectrum.[27]

UV, X ray, and gamma rays are also unfit for biological vision. One reason already alluded to is that the energy of the radiation at these shorter wavelengths is so high that it would be bound to destroy any biological materials used in the construction of a hypothetical eye designed specifically to see in these regions of the spectrum. And again (as in the infrared and radio regions of the spectrum), the design of imaging devices to function in the UV, X-ray, and gamma-ray regions poses severe engineering problems even outside of biology! In the UV region, for example, a lens must be made of special materials that are transparent to UV light. Substances such as quartz and metal fluorides are commonly used. Unfortunately, because nearly all substances absorb in the far UV, focusing an image requires reflection optics in a vacuum chamber.[28] The use of X rays is also problematical. Because X rays are only reflected if they hit a metal surface at a very shallow angle, the design of X-ray telescopes is necessarily complex, involving a series of highly polished nickel-coated surfaces of first a paraboloid and then a hyperboloid surface.[29] Similarly, gamma rays cannot be focused through a lens because all substances absorb them. Even reflection cannot be used because gamma rays are far smaller than the atoms making up any mirror. To detect the radiation in this region of the spectrum, a variety of electronic systems are used, including scintillation counters, proportional counters, and microchannel plates.[30] Despite the progress made by astronomers over the past thirty years to construct telescopes which utilize the various short-wavelength regions of the spectrum, to date their resolving power is still far less than conventional light telescopes.

Assuming, for the sake of argument, that we could get around the detection problem and find some biological material capable of focusing UV and X rays, the theoretical resolving power in the far UV, where wavelengths are a hundred times less than in the visible band, would be a hundred times better. For X rays ten thousand times less, the resolving power would be theoretically ten thousand times better. However, no improvement in visual acuity could be gained by having such eyes, because the smallest possible biological photodetector cell would be many orders of magnitude larger than the tiny UV or X-ray images that could theoretically be focused on such a hypothetical "retina."

Compared with the visual spectrum, the other regions of the electromagnetic spectrum are not only totally unfit for biological vision, they would also appear to be far less fit for nonbiological vision. Even today, despite the development of radio and X-ray telescopes, much of our knowledge of astronomy has come from observations made through light telescopes. The following diagram summarizes some of the conclusions discussed above.

And so it would appear that for several different reasons the visual region of the electromagnetic spectrum *is the one region supremely fit for biological vision* and particularly for the *high-resolution vertebrate camera eye of a design and dimension very close to that of the human eye.*

The fitness of the visual spectrum for vision.

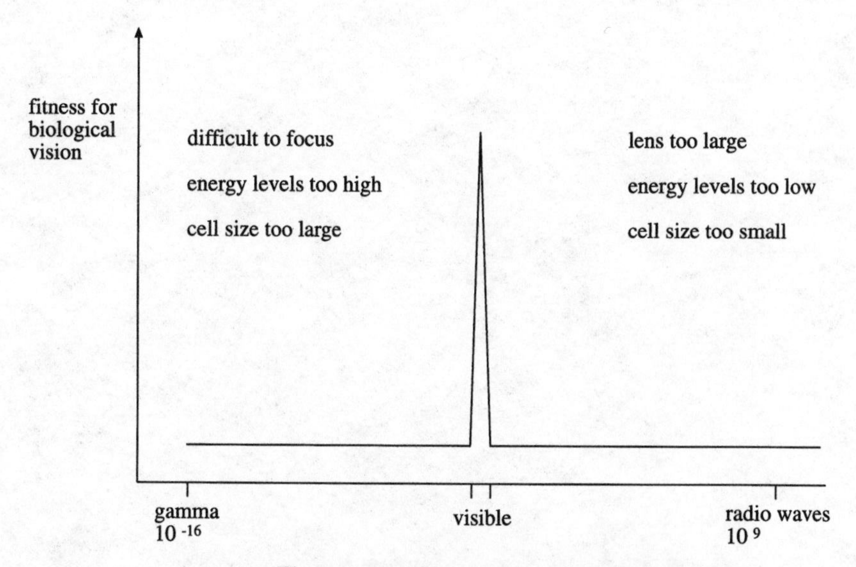

Conclusion

While high-quality vision may not be essential to all life on earth, human existence would be inconceivable without it. While other species may be as reliant on seeing as humans are for survival, our uniquely human desire for knowledge could only have been satisfied, as Aristotle rightly points out in the opening paragraph of his *Metaphysics,* by the gift of sight. Virtually all our knowledge of the world, and particularly scientific knowledge, acquired over the past four centuries has been largely dependent on our possession of eyes of very high resolving power, or visual acuity, and capable therefore of bringing us a very detailed and information-rich image of our surroundings.

We saw in the previous chapter that water, in many fascinating and highly intriguing ways, is uniquely and ideally fit for the type of carbon-based life that exists on earth, not just for simple unicellular microbial life but also for large terrestrial organisms. The evidence presented in this chapter shows that the light of the stars is also, no less than water, supremely fit for life, again in a multitude of different ways. Moreover, again, as in the case of water, this fitness is not merely for simple microbial life, but for large complex organisms such as ourselves. It is fit to provide the warmth upon which all life on the earth's surface depends. It is fit for photosynthesis, which generates the reduced carbon fuels, whose oxidation provides energy for all complex life on earth, and it is fit for vision, the key adaptation through which our own species gained knowledge of the world.

Chapter 4

The Fitness of the Elements and the Earth

In which the biological significance of various elements of the periodic table is examined. The fitness of the cosmos for carbon-based life is highlighted by the fact that the cosmic abundance of the elements corresponds to their abundance in living organisms and that the space between the stars is filled with immense quantities of organic compounds. Further representatives of every class of atoms in the periodic table are necessary for life. Even uranium atom 92 is essential for life, providing the heat and energy required for tectonic activity and the turnover of the earth's crustal rocks, which in conjunction with the water cycle ensures the chemical constancy of the earth's surficial layers. The properties of some of the minerals which play such a vital role in the maintenance of this chemical constancy are examined. The fact that recent astronomical studies suggest that solar systems not too dissimilar to our own, containing rocky planets somewhat like the earth, may be relatively common can be taken as further evidence of nature's fitness for carbon-based life. It is concluded that habitats like the earth which are so fit for a rich complex carbon-based biosphere are not freakish events but rather the inevitable end of natural law.

The periodic table of the elements.

The first atom in the periodic table is hydrogen (H), atom 1, the next is helium (He), atom 2, and the third is lithium (Li), atom 3, and so on. Sodium (Na) is atom 11, potassium (K) is atom 19, and cesium (Cs) is atom 55. The last atom is uranium (U), which is atom 92. The atoms from scandium (Sc) to zinc (Zn) are known as the transition metals. The atoms from lanthanum (La) to ytterbium (Yb) are known as the rare earth metals and are chemically very similar. The atoms can be classified into groups numbered 1 to 18 (see the diagram). All the members of each family are quite similar chemically. For example, all the members of group 1—lithium (Li), sodium (Na), potassium (K), rubidium (Rb), cesium (Cs), and francium (Fr)—are all metals and are highly reactive. On the other hand, all the members of group 18—helium (He), neon (Ne), argon (Ar), krypton (Kr), xenon (Xe), and radon (Rn)—are all inert gases. Most of the elements are metals. Metals tend to give up electrons in chemical combination while nonmetals tend to accept them. The nonmetals occupy the upper right-hand side of the table. Familiar nonmetals are the gas nitrogen (N), which makes up 80 percent of the atmosphere, silicon (Si), one of the major components of the granite rocks, carbon (C), sulphur (S), and chlorine (Cl), which is one of the atoms in common salt (Na Cl). Iron (Fe) is one of the commonly used metals and also forms the molten core at the center of the earth.

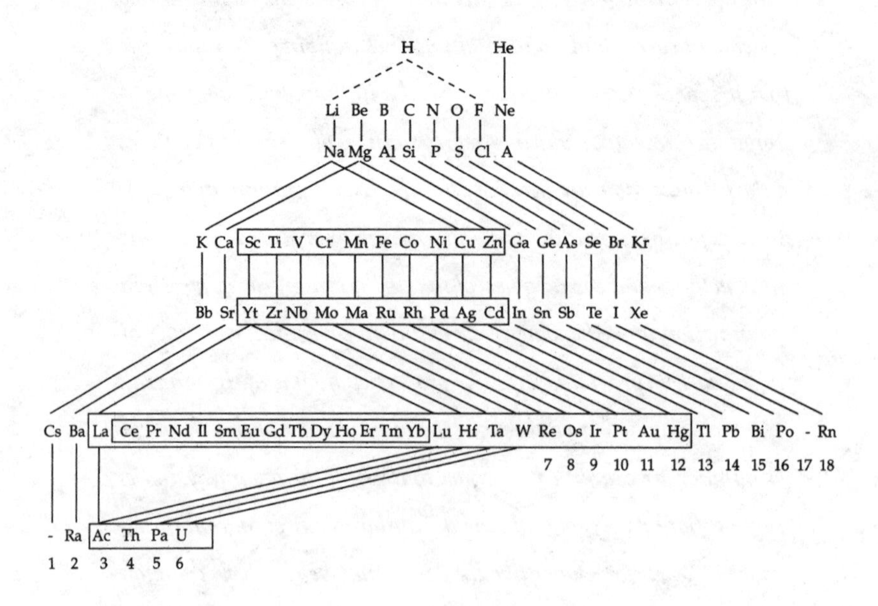

This, as most other of the Atheists' Arguments, proceeds from a deep Ignorance of Natural Philosophy; for if there were but half the sea that now is, there would be also but half the Quantity of Vapours, and consequently we could have but half as many Rivers as now there are to supply all the dry land we have at present, and half as much more; for the quantity of Vapours which are raised, bears a proportion to the Surface whence they are raised, as well as to the heat which raised them. The Wise Creator therefore did so prudently order it, that the seas should be large enough to supply Vapours sufficient for all the land.

—John Ray, *The Wisdom of God Manifested in the Words of Creation,* 1701

The earth, "with its atmosphere and oceans, its complex biosphere, its crust of relatively oxidised, silica rich, sedimentary, igneous, and metamorphic rocks overlying [a magnesium silicate mantle and core] of metallic iron, with its ice caps, deserts, forests, tundra, jungles, grasslands, fresh-water lakes, coal beds, oil deposits, volcanoes, fumaroles, factories, automobiles, plants, animals, magnetic field, ionosphere, mid-ocean ridges, convicting mantle . . . is a system of stunning complexity."

—J. S. Lewis, in F. Press and R. Siever, *Earth,* 1986

The entire cosmos and all the vast diversity of objects which it contains—the stars, the earth, elephants, clouds, supernovae, volcanoes, waterfalls, mountains, and automobiles—is ultimately made up of 92 atoms (see the periodic table of the elements on page 72). These 92 atoms are the basic elemental units or building blocks which, by combining with each other according to the laws of chemistry, form all the vast diversity of substances and materials with which we are so familiar—granite, wood, plastic, agate, salt, proteins, hair, gasoline, penicillin, and so on.

The Structure and Classes of Atoms

Chemists usually refer to atoms by an abbreviated symbol consisting most often of one or two letters. Thus, sodium is Na, oxygen is O, carbon is C, chlorine is Cl. Each atom consists of a nucleus containing one or more subatomic particles known as protons, which carry a positive charge, and neutrons, which carry no charge, and a series of 7 concentric electron orbits, or shells, each at a different radius from the center of the atom (electrons carry a negative charge). Each orbit or shell may contain only a certain maximum number of electrons. The first shell may contain a maximum of 2 electrons, the second 8, the third 18, the fourth 32, the fifth 50, and so on.

The simplest atom is that of hydrogen, which contains only one proton in its nucleus and one electron in the first electron shell. Carbon contains 6 electrons in its electron shells, and 6 protons and 6 neutrons in its nucleus. The inner or first shell is complete, containing the maximum of 2 electrons. The second shell is incomplete, containing only 4 electrons out of the maximal permitted number of 8. Uranium (U), which is the largest of atoms that occurs naturally, contains a total of 92 electrons and 92 protons and more than 100 neutrons in its nucleus.

Hydrogen (H) and helium (He), which make up the main substance of the stars, are by far the most abundant atoms in the universe. The least abundant elements are 10^{12}, or a trillion, times less abundant than hydrogen. Of the 24 most abundant elements, all are either essential for life or are utilized by some organism for some vital process. The only exceptions to this are the 3 inert gases argon (Ar), neon (Ne), and helium (He). The 5 most abundant elements are hydrogen (H), helium (He), oxygen (O), carbon (C), and nitrogen (N), and these, with the exception of helium, form the bulk of all the macromolecules utilized in living organisms.

As discussed in chapter 1, all the elements, including those key atoms of life such as carbon and oxygen, are synthesized in the interior of the stars. The process starts with hydrogen, and through a process of fusion whereby atoms combine with each other in various ways, gradually all the atoms of the periodic table are built up. The fact that only a select number of atoms are manufactured results from a set of generative rules, which restrict stable combinations of protons, neutrons, and electrons to a few unique permissible patterns. These patterns represent the 92 naturally occurring atoms of the periodic table. Interestingly, the atoms are only stable up to atom 83, bismuth (Bi). Beyond this element the atoms are unstable and are continually breaking down into smaller elements. Thus, uranium (U), element 92, decays via the short-lived radioactive elements thorium (Th), protoactinium (Pa), radium (Ra), radon (Rn), polonium (Po), astatine (At) into bismuth (Bi) and finally lead (Pb). During the process a variety of different types of highly energetic particles are given off. If the 7 short-lived radioactive atoms are discounted, this leaves only 85 stable elements. The biological significance of the radioactive atoms will be discussed below.

Because of their structure and particularly because of the characteristics of their surrounding electronic shells, atoms can be grouped into classes which exhibit unique chemical and physical properties. For example, there are metals and nonmetals. The nonmetals occupy the upper right-hand side of the periodic table and the nonmetals the rest of the table. Within the metals, group 1, including sodium (Na), potassium (K), and cesium (Cs) are known as the alkali metals and all have many properties in common. Then there are the alkaline earth metals in group 2, which include calcium (Ca) and magnesium (Mg) and which also have many properties in common. They are all highly reactive, with low melting points. Likewise, the so-called halogens in group 17, including fluorine (F), chlorine (Cl), bromine (Br), and iodine (I), are all reactive nonmetals and also closely resemble each other in many of their chemical properties. It was the fact that the elements could be grouped into distinct families which led the Russian chemist Dmitri Mendeleev to his discovery of the periodic table of the elements in 1869.

As we shall see in the following chapters, many different atoms are used in living things, and in many cases life is critically dependent on these atoms having precisely the properties they possess. Of the 92 naturally occurring atoms, 25 are presently considered essential for life. Of these 25, 11 are present in all living things and in approximately the same proportions. These are hydrogen (H), carbon (C), oxygen (O), nitrogen (N), sodium (Na), magnesium (Mg), phosphorus (P), sulfur (S), chlorine (Cl), potassium (K), and calcium (Ca). Together these atoms make up 99.9 percent of the human body. Another 14 atoms are present in very small amounts in most living organisms, but often in varying amounts, and are known as trace elements. These are vanadium (V), chromium (Cr), manganese (Mn), iron (Fe), cobalt (Co), nickel (Ni), copper (Cu), zinc (Zn), molybdenum (Mo), boron (B), silicon (Si), selenium (Se), fluorine (F), and iodine (I). Another 3 atoms, arsenic (As), tin (Sn), and tungsten (W), are known to be essential in many organisms, but in many cases their biological role is obscure.[1]

Yet another 3 atoms, bromine (Br), strontium (Sr), and barium (Ba), are found in many species, although their role and whether or not they are essential is not yet clear. There are even organisms which accumulate unusual quantities of aluminum and lithium, atoms which are generally considered toxic to most life forms.

Thus, life processes utilize atoms from nearly all the groups in the table. As J. J. R. Fraústo da Silva and R. J. P. Williams comment:

> The biological elements seem to have been selected from practically all groups and subgroups of the periodic table (the only exceptions are groups III A and IV B, besides that of the inert gases) and this means that practically all kinds of chemical properties are associated with life processes within the limits imposed by environmental constraints.[2]

Cosmic and Biological Abundance of the Elements

Most of the atoms actually utilized in living organisms occur in the first half of the periodic table from hydrogen (H), atom 1, to molybdenum (Mo), atom 42. After molybdenum only selenium (Se), iodine (I), and tungsten (W) play any role in living things, and even these atoms are not essential in most organisms. The fact that the atoms in the first half of the table are also the most abundant fits well with the notion that the atom-building system is designed specifically to generate the elements of life. Note that the atoms from carbon (C) to iron (Fe), which are the most important atoms utilized by living things, are all relatively abundant.

The chart below shows the approximate relative cosmic abundance of all the naturally occurring elements from atom 1, hydrogen (H), to atom 92, uranium (U).[3]

The cosmic abundance of the elements.

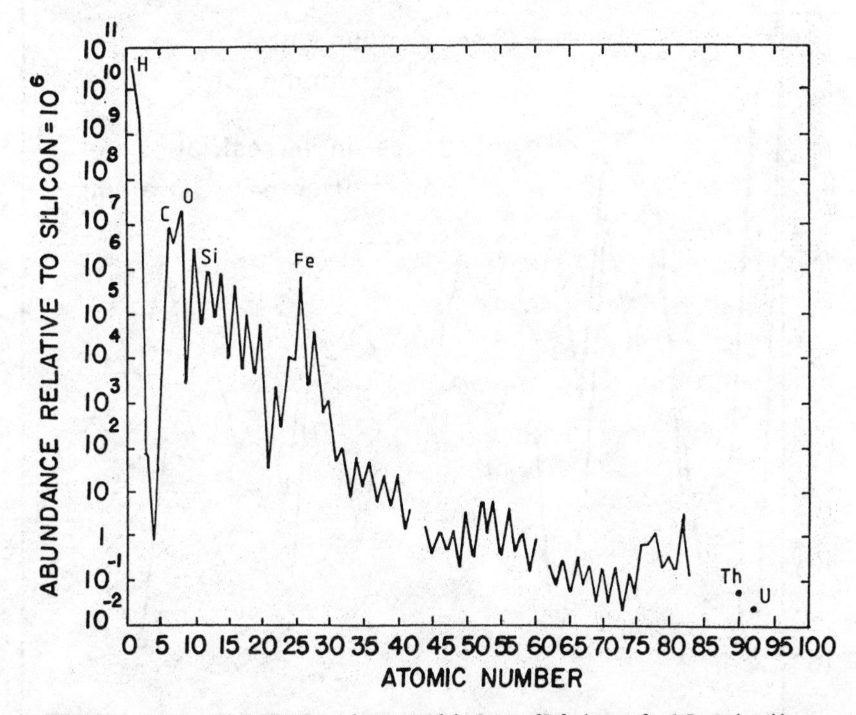

From C. Ponnamperuma, ed. (1983) *Cosmochemistry and the Origin of Life,* chap. 1, fig. 6. Reproduced by permission of Kluwer Academic Publishers, Holland.

Some of the common and important atoms are indicated in the chart using their conventional chemical symbols. The gaps in the chart are due to atoms 43, technetium (Tc); 63, promethium (Pm); 84, polonium (Po); 85, astatine (At); 86, radon (Rn); 87, francium (Fr); 88, radium (Ra); and 91, protoactinium (Pa), which are short-lived radioactive elements and only occur in nature in vanishingly small amounts. The two dots at the right end of the chart are the two long-lived radioactive elements: atom 90, thorium (Th), and atom 92, uranium (U).

As the chart below indicates, there is a very striking correlation between the abundance of the elements and their utility for life.[4]

Relative abundance of the first thirty-one elements in the cosmos and in living organisms on the earth's surface.

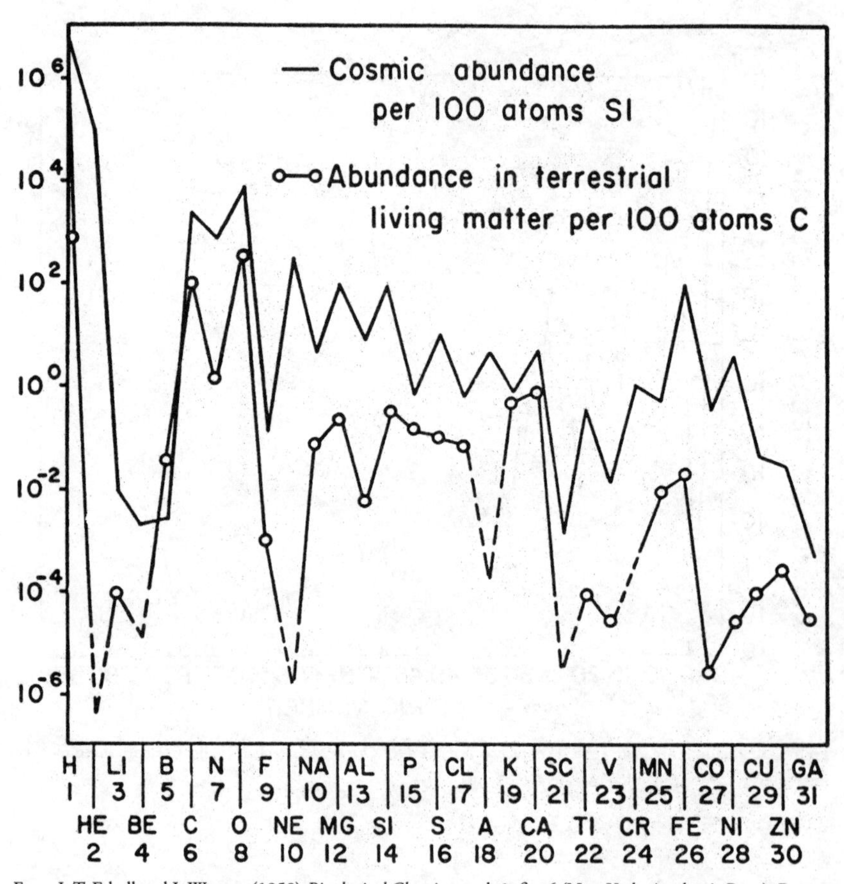

From J. T. Edsall and J. Wyman (1958) *Biophysical Chemistry,* vol. 1, fig. 6 (New York: Academic Press). Reproduced by permission of the publisher.

And in addition to this striking correlation, recent studies have revealed that interstellar space is filled with vast quantities of water, methane, ammonia, carbon monoxide, and many other organic compounds. The quantities are immense—up to 200 million solar masses in our own galaxy. The cosmos is literally overflowing with the basic constituents of carbon-based life.

The Biological Significance of Radioactivity

Nearly all the atoms in the second half of the table which are nonessential to life are also very rare. As can be seen in the chart above, the abundance of the atoms from atom 44, ruthenium (Ru), to atom 92, uranium (U), are all very close to 10^{10} times less common than hydrogen and about a million times less common than most of the essential atoms, such as carbon (C), oxygen (O), nitrogen (N), sulfur (S), and iron (Fe). However, although very rare, the fascinating question arises; why, if the cosmos is specifically fit for life, are there many atoms apparently not utilized in carbon-based life? What possible relevance do the atoms of the second half of the table have to the phenomenon of life?

This question cannot be dodged in any argument for the biocentricity of nature. If the atom-building rules have indeed been arranged to construct a set of atoms with a peculiar utility for the carbon-based life forms that exist on earth, then surely all the atoms, right up to uranium, must be in some way "adapted" for life. Although the number of atoms essential for life is continually rising as research proceeds, it is very doubtful if any biological role will ever be found for the majority of the atoms in the second half of the table. A striking feature of the periodic table is that a great many of the nonessential atoms in the second half of the table have very similar chemical properties. For example, the so-called rare earth metals, atoms 58 to 70 in the periodic table—cerium (Ce) to ytterbium (Yb)—can hardly be distinguished from one another; their chemical and physical properties are virtually identical.[5] This in itself suggests that the utility for life of most of this series of thirteen elements could only be minimal. It is doubtful that biochemical systems could be designed to distinguish between them on chemical grounds.

There are some immediate, at least partial, answers to these questions. We can readily explain the existence of the six inert gases—helium (He), neon (Ne), argon (Ar), krypton (Kr), xenon (Xe) and radon (Rn)—as the inevitable result of the rules which govern the assembly of atoms from the three basic subatomic particles: the proton, the electron, and the neutron. These rules include, for example, a number of stringent restrictions on the configuration of the electron shells. The existence of these rules necessarily means that some atoms will have complete electron shells and will be unable to undergo chemical interactions with other atoms.

The existence of these noble gases which play no role in biology is therefore of necessity. Given the atom-building rules as they are, then inevitably some atoms will be inert and will not enter into chemical combination with other elements. Similarly, the existence of a vast number of possibly lifeless planetary systems unsuitable for carbon-based life is an inevitable consequence of the general laws of nature which generate planetary systems. Even in our own solar system there are a total of sixty-two planets and moons, although more than half the moons are very small objects, being in many cases only 10 to 100 kilometers in diameter. Although Mars and other planets may have allowed primitive life to flourish briefly in the past and some of the moons of the giant outer planets may even contain primitive life at present, only the earth is fit for a biosphere containing a rich and complex variety of carbon-based life forms.

Given the laws of chemistry, which are in fact basically a set of rules that govern the way atoms may combine, an almost unlimited number of compounds and minerals will be formed of necessity. The fact, for example, that carbon adopts a crystalline form—a diamond, if compressed under pressures at which no living thing could survive—has no direct relevance to life but is an inevitable result of the properties of carbon when subjected to the natural conditions which exist within the earth's interior. The properties of steam at very high temperatures and pressures are irrelevant to life but inevitable given the basic chemical and physical properties of water. Again, the fact that carbon dioxide forms a solid at temperatures slightly lower than any found on earth is of no direct relevance to life. All such phenomena are the result of the operation of the general laws of nature and the inherent nature of matter in conditions of temperature and pressure far outside that compatible with life. If there are to be any laws of chemistry and physics, then inevitably there are bound to be a vast number of chemical and physical phenomena of no direct utility to life. Even in the field of artificial life, where experimental worlds sometimes called "toy universes" are generated in computers from sets of rules, many phenomena are inevitably generated which are not of any direct relevance to the major focus of interest, which is the creation and behavior of the artificial life forms.[6]

It is impossible to speculate as to what sorts of alternative atom-building systems might have been used to avoid the existence of chemically unreactive atoms. But on very general principles it is hard to see how any such generative system derived from the application of a simple set of building rules or laws of nature could avoid at least some degree of redundancy and unwanted atoms. This sort of redundancy is inherent in all combinatorial mechanisms for generating complexity. We see it in the vast diversity of sentences permitted by the rules of grammar and in the vast diversity of chemical compounds generated by the rules of chemistry. We see it in the vast number of carbon compounds which chemistry permits and in the vast number of gene sequences that can be derived by combining the four bases in the DNA.

While it is possible to rationalize the existence of the six inert gases as "of necessity," given the atom-building rules actually utilized by nature, the question still remains, why does the atom-building process continue right up to and through the long list of near-identical, rare earth elements, and on to the radioactive elements from atom 84 to atom 92, the last naturally occurring atom, uranium (U)? One answer to this riddle may lie in the critical role that uranium (U) and the other radioactive elements have played in the geophysical evolution of the planet and in those processes which have created on its surface a unique physical and chemical aqueous environment known as the "hydrosphere," which is supremely fit to support carbon-based life over vast periods of time.

The role played by the heavier unstable elements in fashioning the surface of the earth into a stable home has only become apparent as a result of advances in the earth sciences over the past three decades. We now know that uranium and the other radioactive elements have played a critical role in the evolution of the earth because of the heat provided by their radioactive decay. As Frank Press and Raymond Siever point out in their well-known textbook *Earth:* "The heavy elements uranium and thorium . . . are not very plentiful on earth. Their occurrence is measured in a few parts per million. . . . Yet these elements had a profound effect on the evolution of the earth because of their *radioactivity.*"[7] As Press and Siever calculate, the amount of heat produced in one year by radioactive decay in the granite which forms the major component of the earth's crust "is equal to 1,000 times the energy released each year by earthquakes and about 250,000 times the energy of a 1-megaton nuclear explosion."[8] Interestingly, the heating of the earth was assisted by the poor thermal conductivity of rock. If rock had not been such a good insulator, the earth may never have warmed.

It was this radioactive heat which was responsible for warming the interior of the earth shortly after the formation of the planet.[9] As the warming continued, eventually the temperature reached the melting point of iron, when drops of iron began to form and began falling to the center of the earth, releasing huge amounts of gravitational energy that was eventually converted to heat, which raised the temperature further to about 2,000°C and caused a large fraction of the earth to melt.[10] This crucial event marked the beginning of the process referred to as differentiation, which converted the earth from a largely homogeneous body, with roughly the same kind of material at all depths, to a zoned or layered structure with a dense iron core, a crust composed of lighter material with lower melting points, and between them, the mantle.[11] Without differentiation, no life; without radioactivity, no differentiation.

An important aspect of differentiation was the process of chemical zonation, which led to the current distribution of the elements and their compounds from the core to the crust. This process did not lead to a vertical arrangement of the elements based entirely on their relative weights. The reason is that the elements combined into a variety of compounds, and it was the physical and chemical properties of these compounds—properties such as melting points, chemical affinities, and densities—that governed the distribution of elements, rather than the properties of the elements themselves. For example, it is because of the properties of the silicates rather than their constituent atoms—silicon (Si), oxygen (O), calcium (Ca), magnesium (Mg) and aluminum (Al)—that the crustal rocks are largely composed of these five atoms. The silicates melt at relatively low temperatures and when molten are relatively light. Necessarily, they rose during zonation and accumulated in the crust. While the silicates rose, other elements fell. Gold (Au) and platinum (Pt), which have little chemical affinity for silicon (Si) or oxygen (O) and are heavy, are very rare in the crustal rocks.[12]

Although many of the heavy elements such as gold (Au), platinum (Pt), silver (Ag), and mercury (Hg) have little affinity for oxygen and are rare in the crust, uranium (U) and thorium (Th), on the other hand, readily form oxides and silicates and are far more common than gold in the earth's crust.[13] The fact that the majority of the uranium and thorium rose to the surface layers during differentiation is fortunate and may be of great significance. Rock is not a good conductor, and it is possible that unless a major fraction of these radioactive elements had floated to the surface, the heat generated by radioactivity may not have been so easily lost by conduction from the crustal layers. Being trapped in the earth's interior may have caused a very large increase in temperature over time, and with no conductive escape this might have led to a far more violent level of volcanism and turbulence in the earth's center. This may well have repeatedly destabilized the crustal layers and the hydrosphere in violent and explosive episodes of volcanism, rendering the earth's surface far less fit as a habitat for life.

The actual cosmic abundance of the radioactive elements has also been critical to their geophysical role in heating the earth. If too abundant, the earth-sized planets would be molten for eons of time; if too rare, no heating would ever have occurred.

Plate Tectonics

The heat provided by radioactive decay was not only responsible for triggering the initial differentiation and chemical zoning of the planet, the outgassing of water, and the formation of the hydrosphere, it has also been responsible for the massive convection currents deep in the Earth's mantle which have been continuously moving the great crustal plates over the surface of the Earth for the past 4 billion years.[14] The development of the theory of plate tectonics was perhaps the major revolution in the earth sciences in the twentieth century. As Hazel Rymer points out in *Nature,* before the 1960s "the Earth was seen as a sphere with a thick, inert, rocky mantle encasing a central molten core. The complex pattern of land and sea masses was believed to be essentially static. By the end of the [plate tectonic] revolution, only the core remained. The mantle had become a solid yet flowing region, convecting heat from within the Earth through a thin, strong and brittle shell that was broken into a few large plates moving laterally on and with the mantle."[15]

A detailed description of plate tectonics is beyond the scope of this chapter. But briefly, when two tectonic plates collide, one plate, "the overriding plate," is crumpled and uplifted into great mountain chains, while the other, "the underlying plate," is forced down into the Earth's interior. This remarkable process results in the continual recycling of the Earth's crustal material, including the many elements essential for life.

In itself the tectonic cycle would be insufficient to maintain an environment fit for life. It is only through the integration of the tectonic cycle with another great geophysical cycle, the hydrologic cycle, that the physical and chemical constancy of the environment is ensured. The water cycle is so familiar that it needs little further comment here except to say that it is almost entirely due to the weathering by water that the elements in the uplifted crustal rocks are returned again to the sea. The extraordinary mutual fitness of these two cycles for the maintenance of the constancy of the environment is self-evident (see below). Like two gigantic cogwheels engineered to fit perfectly together, these two great cycles have turned together in perfect unison for billions of years, ensuring the continual turnover and essential cycling of the vital elements of life.

Thus, the chemical and physical stability of the Earth's surficial environment is the result of the two great interacting geological recycling systems: the "cold system," or the water cycle, operating on the surface of the planet and energized by the sun, via which the mountains are continually worn and washed into the sea and deposited ultimately as sediments on the ocean floor; and the hot system, the "tectonic cycle," energized by the heat produced by radioactive decay, via which the materials deposited in these oceanic sediments are thrust up and returned to the surface again through volcanism or through the uplift generated by colliding plates.[16]

There is nothing contingent about the existence of these two remarkable and mutually fit interacting cycles. The existence of both systems is probably inevitable in any planet of the same size, elemental composition, dis-

The tectonic and water cycles.

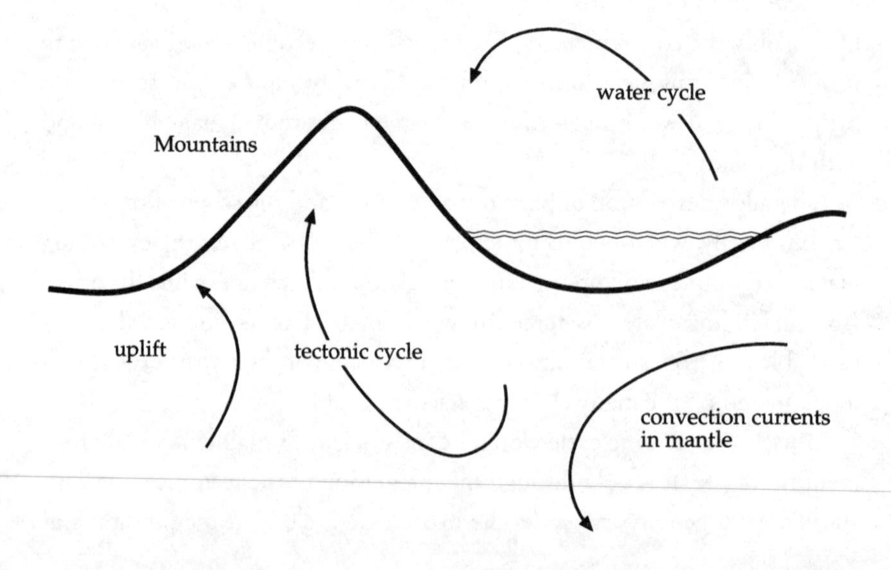

tance from its sun, and history as the Earth. The water cycle is inherent in the properties of water, while the vast quantities of water which make up the oceans, rivers, etc., on earth and which make the cycle possible are themselves the inevitable result of the outgassing of the volatiles from the Earth's interior as it heated up shortly after its initial formation. Likewise, the existence of the great tectonic plates made up mainly of lighter silicate material, and the great convection currents deep in the Earth's mantle which drive the crustal plates across the Earth and empower the whole tectonic cycle of crustal renewal, are no-less-inevitable characteristics of a planet like the Earth. It is now known that similar geophysical processes have played a role in shaping the surfaces of the moon, Mars, and Venus. And in the case of Mars, there is clear evidence, in the existence of sinuous, long dried-out riverbeds and great canyons like the Valley of the Mariners, that the ancient Martian surface suffered erosion by water in ways strikingly similar to that which is still occurring on earth.

The water cycle has long been a source of wonderment and a popular topic of natural theology. Yi-Fu Tuan in his *Hydrologic Cycle and the Wisdom of God* points out that the concept of the hydrological cycle was from 1700 to 1850 the "handmaiden of natural theology as much as it was a child of natural philosophy."[17] And, as he continues:

> Of the three classes of evidence—astronomical, terrestrial and biological—terrestrial evidence proved in some ways to be the most difficult to draw upon in support of the notion of a wise and providential God. Until the concept of the hydrologic cycle was introduced and elaborated, it was difficult to argue convincingly for a rationality in the pattern of land and sea, in the existence of mountains, in the occurrence of floods etc.
>
> The hydrologic cycle served as an ordering principle, and when combined with the geologic cycle, it assumed a grandeur of inclusiveness that makes some of our modern efforts to describe the earth look like a medley of disjointed facts and ideas.[18]

The Magnetic Shield

The earth has a magnetic field which shields it from ionizing radiation from the sun and cosmic radiation originating from the depths of space. The way in which the field is generated has not yet been fully worked out, yet it seems likely that the movement of molten iron in the center of the planet, driven by convection currents, plays a critical role.[19] About 1 billion amperes of current is needed to produce the earth's magnetic field. This is nearly as much electric current as the total amount generated by man throughout history.

We now know that the magnetic field deflects away from the earth, the solar wind, cosmic radiation, and the intense ionizing radiation which periodically bursts from the surface of the sun, preventing all but about 0.1 percent from reaching the earth. The biological significance of this protective shield is controversial. Even without the shield, the atmosphere would absorb most of this radiation before it reached the earth's surface. Moreover, during periods when the earth's magnetic field reversed, the earth was temporarily left without any protection, perhaps for a duration of several years, and such reversals have occurred repeatedly.

Another possible "biological role" of the magnetic field may be to protect the ozone layer, which prevents most of the damaging ultraviolet radiation from reaching the earth's surface. Ionizing radiation, which reaches the upper atmosphere, is known to be a major cause of nitrous oxide (NO) production, which in turn causes the destruction of ozone. Without the magnetic field, it is doubtful if more than a fraction of the current ozone levels could be maintained.[20]

Silicates and Clay

Another feature of the earth which is ultimately the result of the heat generated by radioactivity is the fact that the surface rocks of the earth are largely *silicates.* This apparently esoteric fact is of great biological significance because the end product of silicate weathering by water and carbon dioxide is clay, which forms a major component of soil and plays a vital biological role by absorbing and retaining water and the key elements for plant life.

This water- and ion-absorbing characteristic of clay resides in its unique layered structure—like the pages of a book—in which each page consists basically of a layer of silicon and oxygen atoms. These atoms carry charges which attract other charged atoms (ions) and water so that the whole structure acts as a great reservoir holding ions in the soil and preventing their being leached out by water as it percolates through the soil. The total internal surface area of clays vastly exceeds the area of their external surfaces.[21] As two leading soil scientists point out: "*Next to photosynthesis and respiration, probably no process in nature is as vital* to plant life as the exchange of ions between soil particles and growing plant roots. These cation and anion exchanges take place mostly on the surfaces of the finer or colloidal fractions of both the inorganic matter—clays and humus."[22] "*Cation exchange joins photosynthesis as a fundamental life-supporting process.* Without this property of soils terrestrial ecosystems would not be able to retain sufficient nutrients to support natural or introduced vegetation."[23] (My emphasis.)

There seems little doubt that were it not for the almost universal occurrence of clay minerals in soil, there would be no large terrestrial plants on earth and consequently no large terrestrial mammals.[24] In a very real sense our existence depends on the fact that the most common crustal rocks weather to what would appear to be an ideal material for the growth of plant life, absorbing both water and the essential nutrients needed for growth.

It is surely a "coincidence" of great significance that the very rocks which by virtue of their viscosity and density will inevitably form the crustal rocks on a planet like the earth are weathered by the two substances water and carbon dioxide, the key ingredients of any carbon-based biosphere, into a substance that forms an ideal substratum for the growth of plants.

It is clear from this brief excursion into the fields of geophysics and geochemistry that uranium is in a very real sense, *no less than carbon, essential for life.* Moreover, to get to uranium from hydrogen, obeying the atom-building rules which nature has decreed means that inevitably a considerable number of "intermediate atoms" such as the "rare earths" will necessarily be created, albeit in very small amounts. So we can at least tentatively conclude that the whole periodic table is in essence biocentric, from alpha to omega, from hydrogen to uranium, the last naturally occurring element. Moreover, although many of the elements in the second half of the table seem, at present, of no direct relevance to life, we should recall that the origin of life is still mysterious, and it is certainly conceivable that the properties of many of the elements considered nonessential today may eventually prove to have some exotic but perhaps critical biological role in the process.

The Properties of Minerals

It is obvious that a great many of the physical and chemical properties of the minerals that make up the earth's crust and mantle must be very close to the observed values or the whole crustal recycling system would be untenable; indeed, the earth would be, in all probability, unrecognizable and quite incapable of supporting carbon-based life.

Consider viscosity—not of water, but of rock. The fact that rock subjected to the high temperatures and pressures found in the earth's mantle has a relatively low viscosity is clearly critical to the whole design of the tectonic system. All substances, even the most solid, are viscous to some degree. The viscosities of various familiar substances range over twenty-eight orders of magnitude, from water to granite. The viscosity of the crustal rocks is so great that such rocks flow only very slowly. But they do flow. They erode more quickly than they flow, but they flow.[25]

If the crustal rocks were less viscous, say, like pitch, the mountains would have melted into vast flat plains and nothing we would call a mountain chain would exist on the surface of the earth. If the viscosity of the rocks of the mantle had been substantially less, the convective turbulence would have been immense and the surface of the earth subject to daily movements and volcanism. If the viscosity of the mantle had been much greater than it is, on the other hand, the convection currents would have ceased and the tectonic system would have ground to a halt.

And it is not only the viscosity of the minerals which must be very close to the observed values. If the elements essential for life are to be effectively recycled through the crustal rocks and hydrosphere, the solubility of the various compounds in which the key elements occur in the crustal rocks must also be close to the observed values. For example, the solubility of silicate minerals (which contain the element silica) is thousands of times lower than the solubility of the carbonate minerals (which contain carbon) such as calcite and dolomite, which make up limestone. If the carbonates had been less soluble, then all the carbon on earth would have been locked up in the limestone sediments and there would have been insufficient carbon in the hydrosphere to support life. On the other hand, if the silicates had been as soluble as the carbonates, then the hydrosphere would have been overwhelmed with vast quantities of potassium, aluminum, silica dioxide, calcium, chloride, and other elements converting the sea into a supersaturated viscous sludge.

Every one of the cycles essential to life on earth—the carbon cycle, the oxygen cycle, the nitrogen cycle, the phosphorus cycle, the sulfur cycle, the calcium cycle, the sodium cycle, and so on—involves a host of different chemical compounds and processes which carry the essential elements from the rocks to the sea, where they are deposited in the oceanic sediments, incorporated into the crustal rocks, and then via tectonic uplift and volcanism carried again to the surface, where the weathering cycle can recommence.[26]

Given the vast diversity of chemical compounds and the vast range of chemical and physical properties they exhibit, it is remarkable that so many of the elements can be so efficiently cycled. It is very easy to imagine changing the properties of only one key compound in any one of the critical cycles to see that carbon-based life would be impossible. We saw in chapter 2 that if the latent heat of evaporation and condensation of water had been much less than they are, then small pools of water would completely evaporate after even a small rise in temperature and small lakes would only exist as ephemeral phenomena. Similarly, if for example, limestone had been as insoluble as quartz, then almost certainly carbon would not have been available in the waters of a planet like the earth.

Moreover, although each element is recycled via a unique set of processes and reactions, these cycles are also interdependent. The iron cycle is intimately influenced by the cycles of other metals, and of phosphorus, sulfur, dioxygen, and carbon. Even the intensity of light reaching the surface of the earth influences the flux of iron throughout the hydrosphere.[27] And in the case of the carbon cycle, the level of carbon in the sea is dependent on the level of calcium, which itself is influenced by the level of phosphate ions and so on. Given the complexity of the geochemistry of the earth's crust and hydrosphere, the constancy of so many variables—the mean temperature of the sea, the carbon dioxide concentration in the air, the salinity of the sea, the annual rate of deposition in the sea of about twenty-five or so different elements for nearly 4 billion years—is surely as extraordinary as any topic discussed in this book.

The maintenance of the approximately constant levels of each of the twenty-five or so elements essential to life in the hydrosphere over the past 4 billion years via a set of interlocking cycles—the water, carbon, iron, magnesium, tectonic cycle, and so on—conjures up the image of a vast terrestrial clock with the size and configuration of all its component cogs superbly tailored to fit perfectly together to ensure that the whole turns harmoniously and fine tuned to ensure that the individual cycles turn at the appropriate rate to maintain the required level of each of the elements, essential to life, in the hydrosphere.

Gaia

The constancy of the chemical and physical characteristics of the hydrosphere, maintained as it is by a complex and exquisitely integrated set of interlocking geochemical cycles, has led a number of authors to regard the earth as a homeostatic system analogous to a living organism. As Siever comments in a *Scientific American* article entitled "The Dynamic Earth": "In spite of all the changes that are observed at many different scales of space and time, the Earth as a whole stays remarkably constant. . . . it has become apparent . . . that the core, the mantle, the crust, the oceans and the atmosphere can be . . . viewed as a complex, interacting system in which there is a cyclic flow of materials from one reservoir to another. . . . The Earth as a vast recycling system has its counterpart in the physiological model of dynamic equilibrium known as homeostasis."[28]

Consider two of the earth's feedback systems. One well-known example regulates the temperature of the earth. When the temperature rises, more clouds are formed. These clouds reflect back more of the sun's radiation into space, which has the effect of lowering the temperature—an example of negative feedback control.

Another so-called negative feedback system is the control of the levels of CO_2 in the atmosphere via the weathering of silicate rocks. As the CO_2 level goes up, the temperature increases due to the greenhouse effect. However, the rate of uptake of CO_2 by the weathering of silicate rocks also increases, removing CO_2 from the atmosphere, which tends to cause the temperature to fall. This is widely considered to be one of the principle long-term feedback controls on atmospheric CO_2.[29]

A particularly fascinating aspect of the role of silicate in CO_2 is the necessity for tectonic uplift if the silicate is to be exposed in any quantity to the atmosphere. Without uplift and the formation of mountain chains, the silicates would remain buried and inaccessible to weathering except for a thin superficial layer.[30]

The way in which these various factors—tectonic uplift, the crustal silicates, etc.—work together to ensure temperature stability and constant carbon dioxide levels over millions of years is very striking. The fact that (1) the silicates are the major crustal rocks, that (2) their weathering by the two major components of a carbon-based biosphere, water and carbon dioxide, produces a substance—clay—which is an ideal substratum for the growth of plants, and that (3) at the same time, the very same weathering process controls, via a negative feedback loop, both global temperature and carbon dioxide levels (which must be stringently controlled if plant life or any form of life is to thrive on earth) is further striking evidence of the fitness of the cosmos for carbon-based life.

Some of the feedback systems involve the integration of physical and biological phenomena. For example, when the amount of carbon dioxide in the atmosphere rises, this has a warming effect. However, the warming influence stimulates plant growth which extracts some of the increased carbon dioxide from the air. And in the sea the microscopic organisms that make up the plankton increase in concentration and carry increased amounts of carbonate in their minute shells to the bottom of the ocean, thus lowering the overall carbon dioxide levels in the hydrosphere.

Many other instances in which the constancy of some aspect of the earth's environment is maintained by the integration of biology and chemistry are described by James Lovelock in his book *Gaia.* In this work he develops the concept of the earth as a living organism "with the capacity to keep our planet healthy by controlling the chemical and physical environment."[31] If the Gaia hypothesis is correct, then Gaia would be, as Lovelock points out, "the largest living creature on Earth"[32] and we and all other living things would be parts and partners of a vast being who in her entirety has the power to maintain our planet as a fit and comfortable habitat for life.

Although the position taken here differs from Lovelock's Gaia hypothesis, there are some obvious parallels and the two viewpoints are not, of course, mutually exclusive. From the teleological position advocated here, when biology interacts with chemistry to maintain the constancy of the environment, this is the result of a *preexisting* mutual fitness of carbon-based life and the earth's hydrosphere. Thus, the constancy of the environment does not arise because the earth is itself a "living, self-regulating entity" but rather because the laws of nature are fit to that end.

However, Gaia is a hypothesis which cannot be so easily dismissed. Are the cells of our body aware of the organism they serve? Does an ant know it is part of a greater whole? Do most humans consider themselves components of a global ecological system? Recent work has shown that trees communicate with each other via chemical messages, or pheromones, which are carried from one tree to another in the air. When the trees on the edge of a forest suffer some injury, which might be from insects or microorganisms, they send messages across the forest warning the other trees of the impending attack. Forewarned of the danger, the trees preempt the attack by secreting chemicals that are harmful to the invading insects or microorganisms. In the case of the African Acacia, the pheromone is ethylene gas.[33] It is possible to think of all the individual members of a particular bacterial species as being members of a superorganism spread out all over the earth.[34] This is not so far-fetched as it seems, because all the members of a bacterial species are in continuous genetic communication by the exchange of genetic material via the plasmid system. If one individual bacteria acquires resistance to an antibiotic, it is this genetic communication system which spreads resistance very rapidly throughout the world to all the other bacterial members of the same species.

The Earth's Fitness

This excursion into the earth's sciences has shown that the criteria that must be satisfied by a planet if it is to possess a stable hydrosphere fit for carbon-based life are quite stringent. Press and Siever comment:[35] "Life as we know it is possible over a very narrow temperature interval . . . this interval is perhaps 1 or 2% of the range between the temperature of absolute zero and the surface temperature of the sun." And they note that this range of temperatures is only found on a planet at approximately the distance that the earth is from the sun. Continuing, they comment on the size of the Earth:[36]

> Earth's size is just about right—not too small that its gravity was too weak to hold the atmosphere and not so large that its atmosphere would hold too much atmosphere including harmful gases. . . . the Earth's interior is a delicately balanced heat engine fuelled by radioactivity. . . . were it running too slowly . . . the continents might not have evolved to their present form. . . . Iron may never have melted and sunk to the liquid core, and the magnetic field would never have developed. . . . If there had been more radioactive fuel, and therefore a faster running engine, volcanic dust would have blotted out the Sun, the atmosphere would have been oppressively dense, and the surface would have been racked by daily earthquakes and volcanic explosions.[37]

At present, astronomers know of only one planet, Earth, which satisfies all the necessary criteria and is fit for a rich diverse biosphere of carbon-based life. Mars and some of the moons of the outer planets, such as Europa, may harbor life at present or may have harbored it in the past. Mars may have been once far warmer and wetter than it is today. But at present in our own solar system, none of the other eight planets, including Mars, nor any of their fifty-four moons which surround them, are remotely as fit for complex carbon-based life as is Earth. None have provided anything resembling the wondrously stable watery and terrestrial environment for the thousands of millions of years during which life has existed and evolved on earth.

However, from everything that we now know of planetary evolution, there are no grounds for concluding that the peculiar fitness of the earth is a matter of contingency. On the contrary, the general character of the earth, its division into core, mantle, and crust, its hydrosphere, the plate tectonic system and the recycling of the crust, its chemical composition and particularly the characteristic distribution of the elements in the various layers—all these features are, as far as we can tell, the inevitable result of the interaction of a number of natural processes and phenomena including gravity, the unique physical and chemical properties of the atoms themselves, and the actual cosmic abundance of the elements in the starting material that coalesced together as the planet formed.

The proportions, for example, of the various atoms making up the various layers—core, mantle, etc.—were largely determined by their original abundance in the dust ball from which the earth formed and by their physical and chemical properties. The fact that iron, silicon, and oxygen are among the most common of the elements in the cosmos and form many nonvolatile compounds explains why together they form 80 percent of the mass of the earth. The heaviness of iron and the relative lightness of silicates (composed mainly of oxygen and silicon) explains why the earth's core is nearly 100 percent iron and why only about 6 percent of the crust is composed of iron and why the two atoms silicon and oxygen make up nearly 70 percent of the crust. The relative rarity of hydrogen, carbon, and nitrogen on earth, despite their cosmic abundance, is also explained by the fact that many of their compounds are either light or form volatile compounds which were presumably lost to space during the early history of the planet.

The size and mass of the earth has also been a critical determinant in its evolution. The fact that large planets such as Jupiter and Saturn are largely hydrogen and helium is probably the result of their larger gravitational fields, which were sufficiently strong to hold the lightest elements. Their composition reflects the cosmic abundance of the elements in the dust and gas cloud from which they initially formed. It seems likely that only planets with a relatively small mass and a weak gravitational field can lose their hydrogen and helium and other lighter volatile atoms and compounds. On the other hand, planets with a mass considerably less than that of Earth are probably too small to maintain a hydrosphere such as we have on Earth. From this it seems probable that the range of mass and hence strength of gravitational field, compatible with the formation of a solid rocky planet with a hydrosphere like Earth's, is quite small and very close to that of Earth itself.

The impression gained from these considerations is that there is nothing unusual about Earth and that, given the cosmic abundance of the elements, the laws of nature will generate a planet with chemical and physical characteristics very similar to those of Earth, with a hydrosphere supremely fit for life. The fact that the other rocky planets, Mars, Mercury, and Venus, and the Moon appear to have undergone analogous changes serves to support the conclusion. Recent studies of the voluminous data brought back by the various space missions to Mars since the 1970s, reviewed by Jeffrey Kargel and Robert Strom in *Scientific American,* suggest that in the past Mars may have been a world remarkably similar to Earth: "with flowing rivers, thawing seas, melting glaciers and perhaps abundant life."[38] The evidence suggests that Mars has experienced a complex climatic history punctuated with many relatively warm episodes. The evidence for glaciation on Mars consists of geological features which closely resemble those on Earth: "bouldery ridges of sediment left by melting glaciers at their margins and meandering lines of sand and gravel deposited beneath glaciers by streams running under the ice . . . and apron-shaped lobes of rocky debris seen on the flank of some Martian mountains [which are probably] 'rock glaciers' like the ones that form within the Alaska Range."[39] And even the recent pictures beamed back to earth by NASA, from the Mars Rover, are reminiscent of a typical desert scene on Earth today.

We still have insufficient knowledge of planetary evolution to be able to provide a really convincing explanation of why the evolution of the atmospheres on Mars, Venus, and Earth turned out so differently. Why, for instance, did Mars cool down and lose its seas? Why is Venus so hot? But reasonably plausible explanations can be provided. In answer to the question "How did the three planets—especially Earth—get to their present-day states?"[40] Ann Henderson-Sellers suggests: "The most important parameter, by a long way, is the mean global surface temperature at the time when an atmosphere began to form. This determines where the water goes to, and that in turn determines the evolution of the planetary system from then on. . . . Crucially the temperature of Venus then was high enough for water to be kept in its vapour state trapping infrared radiation, eventually producing a runaway greenhouse effect. . . . The intermediate position of our planet resulted in temperatures that ensured condensation of water vapour released into the atmosphere, forming large oceans, permitting carbon dioxide solution and leading to the formation of sedimentary rocks. . . . With carbon dioxide removed from the atmosphere the path of future temperature evolution was determined."[41]

Although the criteria which must be satisfied to form a planetary surface fit for life may indeed be stringent, nature gives every appearance of having been specifically arranged to that end. From what we now know of the history of the solar system and of the various geophysical and geochemical processes that molded the planets since they first formed out of a mass of cosmic dust some 4 billion years ago, the characteristics of Earth and in particular the maintenance of the hydrosphere and its relatively constant chemical composition—i.e., an environment fit for life—is not a matter of chance. Indeed, given a rocky planet with the mass of Earth, formed out of the matter of the cosmos about the same distance from its sun, it seems inescapable that it will turn out to be very similar to Earth. Given gravity, the cosmic abundance and properties of the atoms, the properties of the minerals formed by the combining of atoms, the phenomenon of radioactivity, the viscosity of silicate rocks, etc., then an earthlike planet with a stable hydrosphere, with oceans and rivers and rain, with mountains and volcanoes, with clay soils, with calcite rock, with a silicate crust, with plate tectonics, may be an almost inevitable end of geophysical evolution. The fact that two adjacent planets in our own solar system, Mars and Earth, are so strikingly similar, provides strong evidence in support of the notion that life-supporting planets are the inevitable end of natural law.

Other Solar Systems

If the cosmos is indeed uniquely fit for life as it exists on earth, then the existence of planetary systems capable of harboring life should be relatively common. Over the past few years techniques capable of detecting large planets the size of Jupiter and Saturn have for the first time provided convincing evidence that other planetary systems do in fact exist and may also be quite common.[42] If the views of Israeli astrophysicist Noam Soker are correct, and "stars like the sun, with several gas giant planets are the rule rather than the exception," then perhaps a significant proportion of planetary systems will resemble very closely our own—having large gaseous planets in the outer regions and small rocky planets in the inner regions.[43] As Sagan comments: "For a range of plausible initial conditions, planetary systems—about ten planets, terrestrials close to the star, Jovians on the exterior—recognizably like ours are generated."[44] Just what proportion of planetary systems will closely resemble our own is, however, controversial.[45] But even if only one in a hundred or one in a thousand solar systems resembles our own, this would still mean that there might be unimaginable numbers of planets very similar to the earth throughout the cosmos, all capable of sustaining carbon-based life very similar to that which exists on Earth. Indeed, as the authors of a recent *Nature* article comment: "Our inference . . . suggests that planetary systems are abundant in the Galaxy. We speculate that if life arises readily on terrestrial planets, then life, too, may be abundant. The recent announcement that rocks from Mars may contain evidence of life would, if confirmed, support this speculation. Our nearest neighbours may be very near indeed."[46] There may even be life in the oceans of Europa, perhaps drawing energy from geothermal sources like the hydrothermal fauna on the ocean floor of our own planet.

And there is another final and intriguing twist to the story. The fact that a significant proportion of all planetary systems may contain large Jupiter-sized gaseous planets in the same approximate position they occupy in our solar system has further teleological significance: first, because recent theoretical modeling of the dynamics of solar systems suggest that a large gaseous planet occupying the same position as Jupiter does in our own solar system confers dynamical stability to the whole planetary system, ensuring that the orbits of the other smaller planets are stable over billions of years and, second, because as planetary scientist George Wetherill points out, "without a large planet positioned precisely where Jupiter is, the earth would have been struck a thousand times more frequently in the past by comets and meteors and other interplanetary debris."[47] Wetherill continues that if it were not for Jupiter "we wouldn't be around to study the origin of the solar system."[48]

The evidence increasingly suggests that not only are there planetary systems surrounding nearly every sun, but many may resemble our own—containing rocky planets of just the right size and at just the right distance from their sun to sustain a carbon-based biosphere like that of Earth over billions of years. Moreover, many may also often contain Jupiter-sized planets of just the right size and in just the right position to confer long-term dynamical stability to the system and to protect the rocky life-bearing planets from the destructive effect of meteor bombardment. The emerging picture is entirely in keeping with the teleological preassumption that nature is ordered to generate terraqueous planets closely resembling Earth—uniquely fit for the origin, and evolution, of carbon-based life.

As we did with water and carbon, we can represent again the unique fitness of the earth for our kind of carbon-based life in the form of a graph plotting all known planetary environments against their utility or fitness for carbon-based life. What we get is a unique optimum indicated by the uniqueness and sharpness of the peak. This is perfectly consistent with the hypothesis: that there is one environment determined by the laws of nature (the hydrosphere of a planet of the same size and distance from its sun as Earth) that is uniquely and ideally fit for carbon-based life. If there had been

The unique fitness of the earth's hydrosphere for carbon-based life forms.

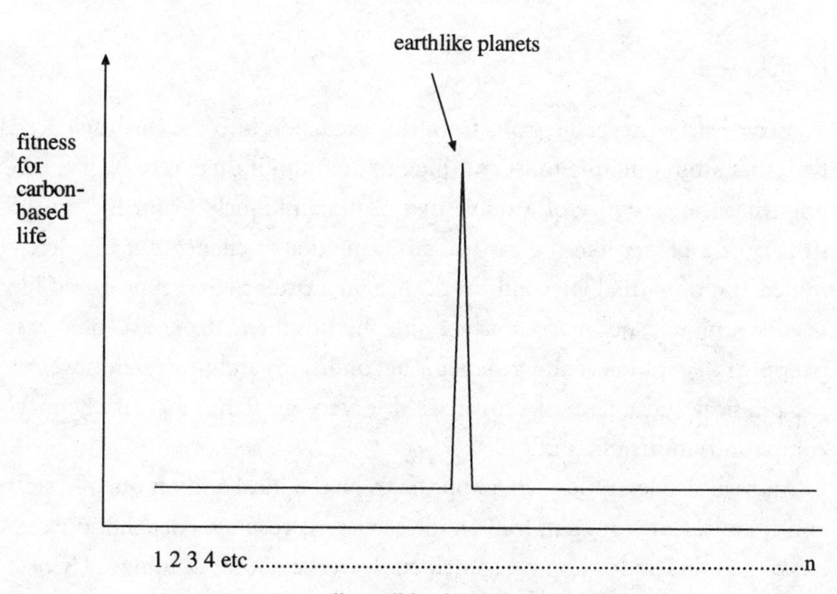

several other types of environment having some fitness for carbon-based life, so that the plot resembled the pattern seen in the graph below, the design hypothesis would have been effectively disproved.

The comparative fitness of the earth's hydrosphere in a nonbiocentric cosmos.

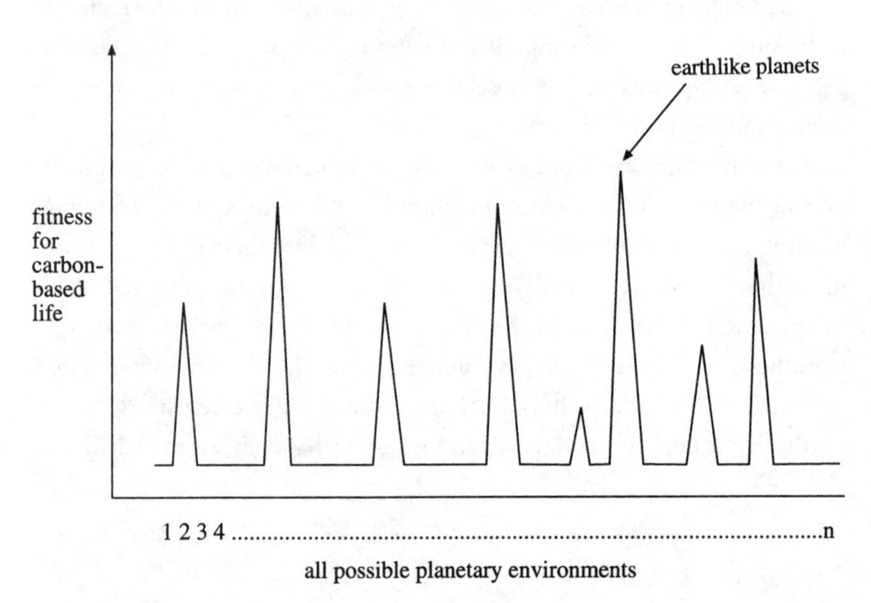

Conclusion

We have learned several lessons from this excursion into the earth sciences: first, that atom building must continue to uranium if there is to be life; second, that the existence of a stable hydrosphere uniquely fit for life on the surface of a planet like the earth is not a matter of chance but the determined end of natural law; and third, that the existence of carbon-based life in this unique and marvelously stable hydrosphere depends on a vast panoply of geophysical and geochemical conditions and processes as well as the physical and chemical properties of a very great number of chemical compounds, minerals, and gases.

And we also learn that what appears to be the ideal and unique physical and chemical environment for life, the earth's hydrosphere, depends on a series of genuine and profound coincidences in the nature of things. There is the coincidence that main sequence stars like the sun provide a uniquely constant and ideal source of radiant energy to energize the water cycle on which life itself depends while at the same time emitting visible light of just the required energy levels for photobiology. Then there is, first, the coincidence that planets the size of the earth have just the proper mass to heat up sufficiently to cause, by outgassing, the formation of a hydrosphere shortly after their formation; second, that this mass provides sufficient gravitational force to retain the atmosphere and hydrosphere after the initial formation; and, third, a planet of a mass equal to the earth's has the required geophysical properties to drive the crustal tectonic cycle, which itself is so perfectly fit to function in unison with the water cycle.

It is hard to escape the feeling that planets fit for our type of life will not only have seas and booming surfs and gentle rain, they will also have volcanoes and great mountain chains on which glaciers will form and from which rivers will emerge and carry the vital nutrients of weathering into the seas and throughout the hydrosphere. There will be continental drift and plate tectonics. It is a familiar picture, and not in the least contingent, but rather the inevitable and determined outcome of natural law.

Chapter 5

The Fitness of Carbon

In which evidence is presented for believing that the chemical properties of the carbon atom are uniquely fit to form the complex molecules required for life. Silicon, which is carbon's sister atom in the periodic table, falls far short of carbon in the diversity and complexity of its compounds. The fitness of carbon compounds for life is maximal in the same temperature range that water is a fluid. Both the strong covalent and the weak bonds are of maximal utility in this same temperature range. Such coincidences are precisely what one might expect to see in a cosmos specially adapted for carbon-based life.

The compounds of carbon.

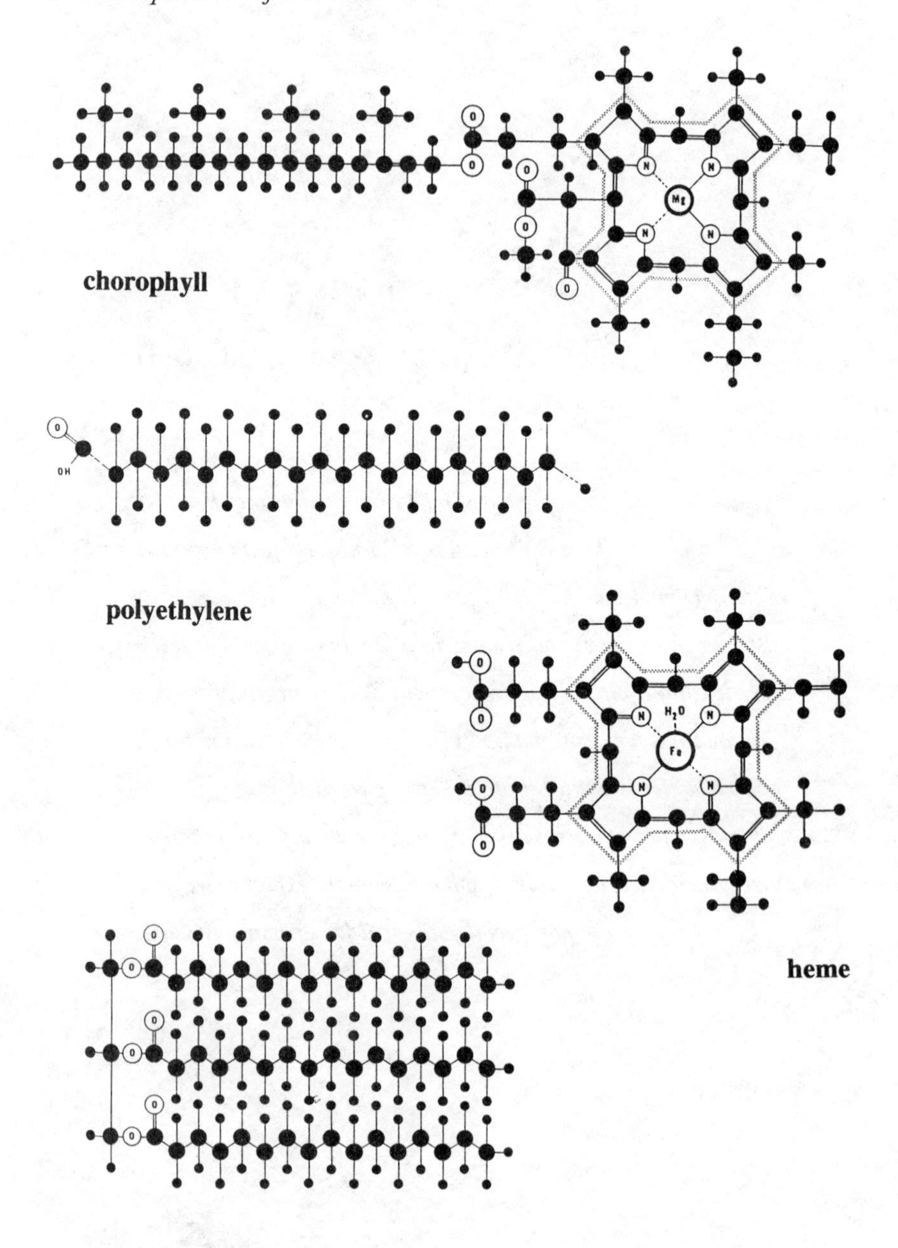

From N. J. Berrill (1966) *Biology in Action* (New York: Dodd, Mead), fig. 3.8 and fig. 3.16. Reproduced by permission of N. J. Berrill.

Nature has been kinder to us than we had any right to expect. As we look out into the universe and identify the many accidents of physics and astronomy that have worked together for our benefit, it almost seems as if the universe in some sense must have known that we were coming.

—Freeman Dyson, *Scientific American* magazine, 1971

I should consider that I know nothing about physics if I were able to explain only how things *might* be, and were unable to demonstrate that they *could not be otherwise.*

—René Descartes, letter to Mersenne, 1640

We saw in chapter 2 that water is wonderfully fit for its biological role, giving every appearance of being ideally and uniquely "prefabricated" in all its chemical and physical properties to serve as the matrix of carbon-based life as it exists on earth. And we have also seen, in the previous two chapters, that both the light of the sun and the earth's crustal rocks and hydrosphere are also remarkably fit for life in many different ways. Together they create a unique, chemically and thermally stable environment which has been bathed in a limitless and constant source of radiant energy for 4 billion years.

However, the hypothesis that the cosmos is uniquely and ideally fit for life cannot be secured by showing that one or two, or even several, of the conditions necessary for life appear to be ideally adapted to the ends they serve. If the hypothesis is true, then we should expect to find that all the basic conditions for life and all of the components of living organisms are ideally and uniquely adapted to the particular biological ends they serve, "as if the Creator has given us a kit of prefabricated parts ready made for the work in hand," in the words of Robert E. D. Clark.[1] We shall see in this chapter, and the following chapters, that this does indeed appear to be the case.

Organic Compounds

A house is built up from wood, brick, stone, and metal components. A car engine is built up mainly from metal components. Many household goods are constructed out of plastic components. Computers are composed of various components made up of plastics, metals, and silicon chips. In the case of living organisms, the basic chemical building blocks utilized in their construction are organic compounds—molecules composed of the atom carbon (C), in combination with a handful of other atoms which include hydrogen (H), oxygen (O), and nitrogen (N).

The world of life is very much the product of the compounds of carbon. All the machinery of the cell, and all the vital structures of living organisms from the molecular to the morphological level, are constructed from the compounds of carbon. Structures as diverse as the cell membrane, the horns of an elk, the trunk of a redwood, the lens of the eye, the venom of a spider, the petals of a flower, the DNA helix, and the blood pigment hemoglobin, are all composed almost entirely of compounds made up of combinations of the carbon atom with hydrogen, oxygen, and nitrogen.

The possibility that living things might be some sort of carbon-based chemical machine had already been raised in the late eighteenth century when Antoine Lavoisier and Pierre Laplace established that water and carbon dioxide are the products of animal and human respiration and that the oxidation of carbon and hydrogen was the source of animal heat and an essential process of life.[2] However, the critical role of carbon and its compounds in the design of life was only fully appreciated in the second half of the nineteenth century. A major distraction in the early nineteenth century in the way of a true appreciation of the marvels of carbon chemistry was the vitalistic doctrine prevalent at the time. According to vitalism, living substances were in some essential way different from nonliving, and their synthesis in living things was the result of some mysterious vital force. Chemistry was therefore divided into two divisions: inorganic, which dealt with the substances and compounds of the inorganic world and which were amenable to scientific analysis, and organic, dealing with the substances of life. The nature and formation of organic substances was imagined to be beyond scientific analysis.

Vitalism remained unchallenged throughout the period between 1780 to 1828, when chemistry was being established as a science and chemical knowledge rapidly increased. By 1820, forty different elements were known, and the great Swedish chemist Berzelius had himself described the preparation, purification, and exact quantitative analysis of two thousand different compounds. It was also well established by then that the element carbon was a key constituent of life.[3] It was only after the synthesis of urea by Friedrich Wöhler in 1828, and shortly afterward of many other undoubted constituents of animals and plants, that the way was open to bring the study of biochemicals fully into the scope of science.

Shortly after Wöhler and the collapse of the vitalistic doctrine, the English chemist William Prout suggested for the first time in his 1834 Bridgewater Treatise entitled *Chemistry, Meteorology, and the Function of Digestion* that the carbon atom may be uniquely fit for life because of its potential to form vast numbers of diverse compounds.[4] After 1840 the development of organic chemistry as a unique branch of chemistry gathered momentum. And as Henderson points out, by the turn of the nineteenth century it was clear that, in the number and diversity of its compounds, carbon is without peer among the other 92 elements. At least 100,000 different compounds were known.[5]

The Carbon Atom

The reason for the unique diversity and number of carbon compounds lies in certain unique characteristics of the carbon atom, atom 8 in the periodic table.[6] As the British chemist Nevil Sidgwick explains in his classic textbook *Chemical Elements and Their Compounds:*

> Carbon is unique among the elements in the number and variety of the compounds which it can form. Over a quarter of a million have already been isolated and described, but this gives a very imperfect idea of its powers, since it is the basis of all forms of living matter. Moreover it is the only element which could occupy such a position. We know enough now to be sure that the idea of a world in which silicon should take the place of carbon as the basis of life is impossible. . . .[7]

Two of the reasons for the great diversity of carbon compounds compared with silicon given by Sidgwick are, first, that "silicon compounds have not the stability of those of carbon, and in particular it is not possible to form stable compounds with long chains of silicon atoms," and, second, because "the affinity of carbon for the most diverse elements and especially for itself, for hydrogen, nitrogen, oxygen and the halogens, does not differ very greatly: so that even the most diverse derivatives need not vary very much in energy content, that is thermodynamic stability."[8] Yet another characteristic of carbon is its capacity, by sharing two or more of its electrons with another atom, to form what are known as multiple bonds. Silicon does not share this capacity, and its chemistry is consequently much less rich and diverse.[9]

Covalent Bonds

When carbon combines with other atoms to form organic compounds, the bonds between the atoms are known as covalent bonds or "ordinary chemical bonds." In all the molecular structures shown in figure 5.1, the atoms are all linked together by covalent bonds. Covalent bonds are formed when atoms share electrons in their outer electron shell in an attempt to complete the shell. We saw in the previous chapter that each shell can contain up to a certain maximum number of electrons. The innermost shell can contain up to 2 electrons, the next shell can contain up to 8 and the next, up to 18.

In the diagram below, which shows the atomic structure of water (H_2O), note that by sharing electrons the outer shells of both the oxygen and the two hydrogen atoms are complete. Oxygen has only 6 electrons in its outer shell and requires 2 additional electrons to fill this shell with its maximum permitted number.

The structure of water.

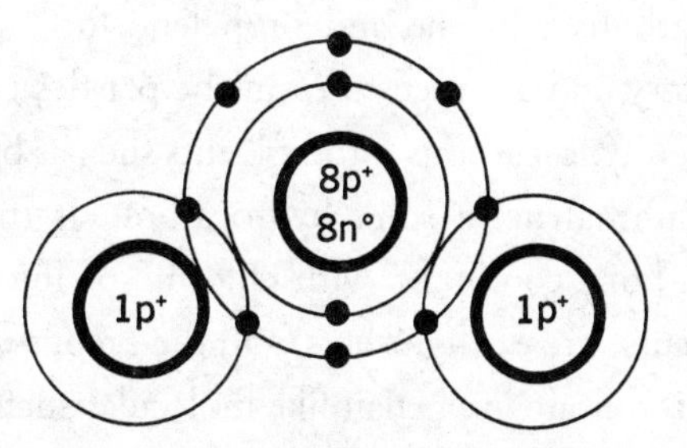

From fig. 3 in M. Toporek, *The Basic Chemistry of Life;* © 1968 by Appleton & Lange, Stamford, Conn. Reproduced by permission of the publisher.

In the compound methane (CH_4), shown below, by sharing electrons each of the hydrogens is able to make its outer shell up to 2 electrons (the maximum permitted for the inner shell) while the carbon atom is able to make its outer shell up to 8 electrons, which is again the maximum permitted.

The structure of methane.

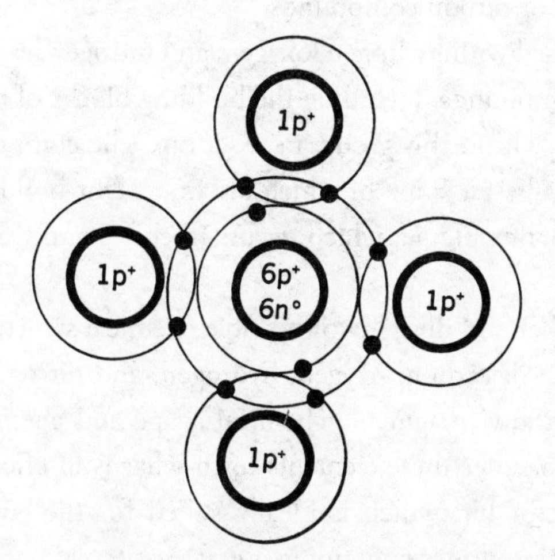

From fig. 3 in M. Toporek, *The Basic Chemistry of Life;* © 1968 by Appleton & Lange, Stamford, Conn. Reproduced by permission of the publisher.

The Diversity of Carbon Compounds

Carbon, linked with the first atom in the periodic table, hydrogen, forms the vast family of hydrocarbons, which is divided into several subdivisions. Even within this small subgroup of organic compounds, the diversity of chemical and physical properties is great. Within this group are found natural gas, liquid petroleum, gasoline, kerosene, lubricating oils, greases, and waxes. The hydrocarbons ethylene and propylene form the basis of the petrochemical industry and are used to form the plastics polyethylene and polypropylene. Other hydrocarbons form solvents such as benzene, toluene, and turpentine. Naphthalene, a solid hydrocarbon, is the insecticide in mothballs. Hydrocarbons combined with chlorine or fluorine form anesthetics, various solvents, fire extinguishers, and the Freons used in refrigeration. Some hydrocarbons are long chainlike molecules such as pentane and butene. Others contain cyclic or ringlike formations such as benzene.

In conjunction with both hydrogen (H) and oxygen (O) another vast set of carbon compounds is derived. These include the alcohols, such as ethanol and propanol, the aldehydes, the ketones, and the fatty acids. Fatty acids are composed of a long hydrocarbon chain which is attached to a carboxylic acid group at one end. Another group of compounds composed of carbon, hydrogen, and oxygen are the sugars, including glucose and fructose. Cellulose (the hard substance of wood), beeswax, vinegar, and formic acid all belong to this group of carbon compounds.

Carbon combined with hydrogen, oxygen, and nitrogen leads to a further multiplicity of compounds, including the building blocks of proteins—the amino acids. Also within this group are a set of cyclic compounds known as the nitrogenous bases, some of which are important building blocks of DNA. The stimulant caffeine, which occurs in coffee and tea, is also a nitrogenous base.

The total number and diversity of possible chemical structures that may be constructed out of carbon, oxygen, hydrogen, and nitrogen is virtually unlimited. Almost any imaginable chemical shape and chemical property can be derived. Together these elements form what is in effect a universal chemical constructor kit, which is ideally suited for the construction of complex chemical machines containing a great variety of chemical devices and calling for a vast number of different chemical components of different chemical and physical properties.

A striking aspect of this great molecular plenitude is that the atoms which comprise it—carbon, hydrogen, oxygen, and nitrogen—are among the first few atoms manufactured in the stars and also among the most abundant overall in the cosmos. And remarkably, two of these atoms, hydrogen and oxygen, form water, the matrix of carbon-based life. It is as if from the very moment of creation the biochemistry of life was already preordained in the atom-building process, as if Nature were biased to this end from the beginning.

The Mildness and Metastability of Carbon Compounds

The diversity of carbon compounds is only one of many factors which make carbon chemistry so fit for the intricate and complex metabolic processes of life. Another factor is the mildness of the chemical properties of most organic compounds. No organic acids are as violently reactive as sulfuric or nitric acid, no bases as corrosive as caustic soda. This curious mildness arises in part from the relative inertness of the carbon atom.[10] In addition to their mildness, organic compounds tend to exhibit what the great biologist J. B. S. Haldane called a characteristic "metastability," a word he defined in a symposium some time ago:

> A metastable molecule means one that can liberate free energy by a transformation, but is stable enough to last a long time unless it is activated by heat, radiation, or union with a catalyst. For example trinitrotoluene is highly metastable; a kilogram of it liberates a lot of energy. Glucose is mildly metastable, but will liberate some energy if turned into ethanol and carbon dioxide. . . . most organic molecules are metastable.[11]

This term very nicely captures an essential characteristic of organic compounds and one which makes them of great utility to living chemistry. This metastability arises because, although relatively stable, most are also quite reactive and readily undergo covalent chemical reactions with other carbon compounds under relatively mild conditions. In other words, they do not require much energy to be activated for covalent chemical reactions at ambient temperatures and conditions.

A. E. Needham commented in his *Uniqueness of Biological Materials* on the characteristic metastability which allows their manipulation without prohibitively high-energy expenditure or release: "Carbon compounds are notable for lability as much as their stability. Few remain unchanged when heated above 300°C. . . . As in so many respects carbon seems to have the best of both worlds, combining stability with lability, momentum with inertia. . . . [12] The heat of formation of carbon compounds from their immediate precursors is rarely very great, so that, once the initial steps of carbon dioxide reduction have been effected, a large variety of compounds form relatively spontaneously."[13] Henderson was also struck by the same point.[14]

However, this "metastability" which is so crucial to the utility of organic compounds is only manifest in a very narrow temperature range. As the temperature rises above 100°C, organic compounds become increasingly reactive and chemically unstable. We are all familiar with this phenomenon. Cooking nicely illustrates it. Even the gentle heating of meats or other foodstuffs, for some time, at temperatures of only about 90°C, called simmering, can cause dramatic changes in their chemical and physical characteristics. The softening of meat occurs because the collagen fibers which make up the tendons and fibrous sheets are converted into soft gelatin, which offers no resistance to a knife.[15] When sugars are heated above 100°C, they are rapidly degraded, undergoing complex chemical reactions with themselves and other biomolecules in the food. These changes are referred to as caramelization and browning in the kitchen. Many vitamins, including vitamin C, folic acid, and some of the other B vitamins—B1 and B6, for example—are rapidly broken down above 100°C.[16] Recent studies of the stability of organic compounds at 250°C, including one report in the journal *Nature,* showed that the half-life of many of the key organic compounds used by living things, including such important compounds as amino acids, used in the construction of proteins, the bases used in the construction of DNA, and the energy-rich phosphate compound known as adenosine triphosphate, or ATP (see Appendix) which plays a vital role in the energy metabolism of all living cells on earth, decompose at rates too fast to measure, or have half-lives lasting minutes or seconds at 250°C.[17]

The instability of most organic compounds, especially as temperatures rise above 100°C, was discussed by Miller and Orgel in their book *The Origins of Life on the Earth.*[18] In the case of the amino acid alanine, for example, its half-life is 20 billion years at 0°C, 3 billion years at 25°C, but only ten years at 150°C, a decrease of more than a billionfold.

While foodstuffs, made up of organic carbon compounds, undergo considerable physical and chemical changes during cooking, the metal or glass containers remain unchanged. Put a piece of glass, a piece of metal, a piece of rock, etc., in a cooker, and while foodstuffs will char, burn, or decompose at 150°C, the inorganic materials will remain unchanged. And at temperatures much above 100°C to 150°C, the reactivity and instability of organic compounds increases prohibitively and their utility for biochemistry rapidly diminishes. At low temperatures other problems emerge. Below 0°C, reaction rates are so enormously slowed that anything we would recognize as a "biochemistry" utilizing the compounds of carbon becomes impossible.

The effect of temperature on the rates of chemical reaction is quite dramatic: for each 10°C fall in temperature the rate of reaction declines by a factor of two.[19] It follows from this relationship that a fall in temperature of 100°C will slow chemical reactions nearly a thousand times! Even a temperature change of far less than 100°C causes a quite dramatic slowing of reaction times. Reactions occurring in the human body at 38°C would take place sixteen times slower at 0°C and sixty-four times slower at –20°C. As Robert E. D. Clark points out, at temperatures below –100°C all chemical reactions become vanishingly slow and at the temperature of liquid air, "only a very few reactions take place at all and these involve the exceedingly active element fluorine in its free state."[20] Even though some organic chemistry would be possible at temperatures as low as –40°C, as Wald points out, "the process that led to the origin of life within a period of perhaps a billion years upon this planet might take some 64 billion years at –40°C."[21]

The vast and unique plenitude of organic compounds can only be exploited by living systems within a temperature range of approximately –20°C to 120°C. It is only within this range that the majority of carbon compounds have their characteristic metastability, which permits the intricate and sophisticated manipulation of their constituent atoms by the chemical machinery of life.

The Temperature Range of Organic Chemistry

Now although the temperature difference between –20°C and 120°C, i.e., 140°C, appears from our ordinary perspective to be considerable, it represents in fact an unimaginably tiny fraction of the total range of all possible temperatures. Temperatures in the cosmos range from 10^{32}°C, (10 followed by 31 zeros), which was the temperature of the universe shortly after the big bang, to very close to absolute zero, which is –273.15°C.[22] The temperature inside some of the hottest stars is several billion degrees,[23] and even inside our own sun, which is not particularly hot for a star, the temperature is tens of millions of degrees and its surface temperature is 6,000°C.

The diagram below indicates the temperature range in which carbon compounds exhibit the necessary metastability to make them of utility to life. It is surely a highly suggestive coincidence that the chemical reactivity of the one great class of compounds, uniquely fit in so many other ways to serve as the building blocks of life, is of *optimal utility for the complex atomic and molecular manipulations associated with life in precisely that temperature range—0°C to 100°C—in which water, the one fluid supremely fit to serve as the matrix for carbon-based life forms, exists as a liquid at sea level on the earth.*

It is interesting to note in passing that liquid water would not exist on earth if the atmospheric pressure was less than half what it is. Which implies (since the density and pressure of a planet's atmosphere is largely determined by its size) that it is unlikely that planets much smaller than the earth would contain large quantities of liquid water for any long period of time. At pressures much higher than atmospheric, liquid water can exist at temperatures of up to several hundred degrees. Much of the water in the earth's crust is in fact much hotter than 100°C. However, water at such temperatures is of little utility for carbon-based life.

Most of the matter in the cosmos is either vastly hotter (the interior of stars) or much colder (interstellar space) than the surface of the earth. And it seems likely that one of the few environments in the cosmos where liquid water in quantity between 0°C and 100°C can exist—i.e., in the temperature range in which organic compounds are metastable and the chemical bonds which link their constituent atoms can be manipulated with "ease"—is on the surface of a planet like the earth. This reinforces the conclusion

The temperature range within which organic chemicals are metastable and of utility for biochemistry.

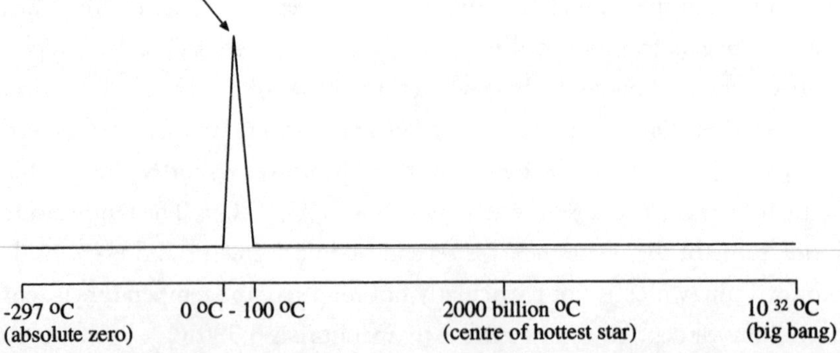

metastability of organic compounds

-297 °C (absolute zero)
0 °C - 100 °C
2000 billion °C (centre of hottest star)
10^{32} °C (big bang)

reached in the previous chapter that the surface of rocky planets like the earth provide an environment uniquely and ideally fit for carbon-based life, in not one but in several different ways.

The Weak Bonds

The covalent chemical bonds which link the atoms together in organic compounds are not the only type of chemical bond utilized in living systems. There is another class of bonds, known as weak, or noncovalent, bonds. There are several different types of weak bonds.[24]

These bonds are about twenty times weaker than covalent bonds, and they play a vital role in holding together the different parts of large organic molecules, such as proteins in complex, unique three-dimensional forms (this topic is touched on again in chapter 8). In the diagram below, the atoms (shown as black dots) of an organic molecule are linked by covalent bonds shown as continuous lines joining the individual circles.

An organic compound showing its constituent atoms (black dots) linked together by covalent bonds (black lines).

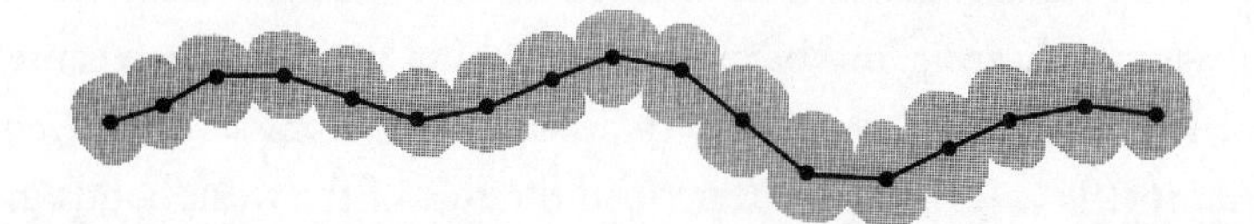

The complex, three-dimensional, functional form of large biomolecules, such as proteins and DNA, is maintained by weak bonds which hold the atoms together in different parts of the molecule. This is shown in the diagram below, where the weak bonds are represented by broken lines.

An organic compound held in a unique 3-D form by weak noncovalent bonds (dotted lines).

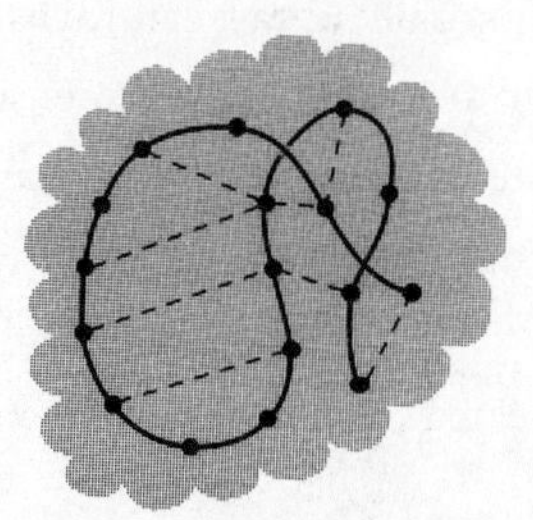

If we think of a string of beads or plastic balls held together in a three-dimensional cluster by pieces of tape, we have an analogy which conveys something of the difference between covalent and weak bonds. The string represents the strong covalent bonds, and the pieces of tape, the weak bonds. Nearly all the biological activities of virtually all the large molecules in the cell are critically dependent on their possessing very precise 3-D shapes. Nature has provided no other glue to hold together the molecular superstructure of the cell. While we cannot have carbon-based life in the cosmos without covalent bonds, as there would be no molecules, just as certainly, we cannot have carbon-based life without these weak noncovalent bonds—because the molecules would not have stable, complex 3-D shapes.

We have seen above that the covalent chemistry of carbon is of maximal utility for life in a very narrow temperature range which corresponds approximately with the range in which water is a liquid. Remarkably, these weak bonds are also of utility in approximately the same small temperature range. In fact, weak bonds are even more temperature-sensitive than covalent bonds. Most weak bonds in existing biomolecules such as proteins are disrupted by increases in temperature which leave covalent bonds intact.

The disruption of weak bonds occurs in two very familiar processes in the kitchen—in the heating and beating of egg white, both of which cause the egg white to whiten and coagulate. In a fascinating discussion in *Scientific American* in 1981, Jearl Walker described the role of the weak bonds in the making of a lemon meringue pie:

> When a cook forces a whisk through egg whites, shearing the fluid, some of the weaker bonds are ruptured and parts of the 3D structure of the proteins [present in the egg white] are destroyed. The cook does not totally disrupt the proteins because the forces [the covalent bonds] holding them in their primary structures are comparatively strong. . . . Any such altering of the structure of a protein is called denaturing.
>
> Once the proteins are partially unravelled [denatured] they begin to attach themselves to one another to form a three-dimensional mesh or gel. This interaction between the proteins is unlikely before denaturation because the proteins are relatively globular and relatively few of their sites for possible [weak] bonds are exposed.
>
> When the mixture is heated . . . the heat further denatures the proteins, un-

> ravelling them further and thus enabling the mesh to stretch . . . coagulating the whites into a firm structure.[25]

Because covalent bonds are far more robust, neither beating nor heating has any significant effect and most remain intact. In effect weak bonds are even more "metastable" than covalent bonds.

The extreme sensitivity of the weak bonds to increases in temperature limits their utility in maintaining the three-dimensional form of complex biomolecules to temperatures not far above 100°C. It seems likely that no protein could be designed to function at temperatures much above 120°C.[26] There is little experimental evidence as to the utility of weak interactions below 0°C. But it seems likely that the thermal energy at temperatures much below 0°C and certainly below –20°C would be insufficient to allow the formation and breakdown of weak bonds at speeds remotely compatible with any sort of functional biochemistry.

We can conclude that the weak bonds are only of utility for holding organic compounds into complex 3-D forms, within the temperature range of approximately 0°C to 100°C (see diagram below).

What we have, then, is another coincidence of critical significance—that the weak bonds, which are of a completely different nature from the strong covalent bonds, are also of utility in a temperature range which very nearly corresponds with that in which water exists as a liquid.

Out of the enormous range of temperatures in the cosmos, there is only one tiny temperature band in which we have (1) liquid water, (2) a great

The temperature range within which weak bonds are of utility to biological systems.

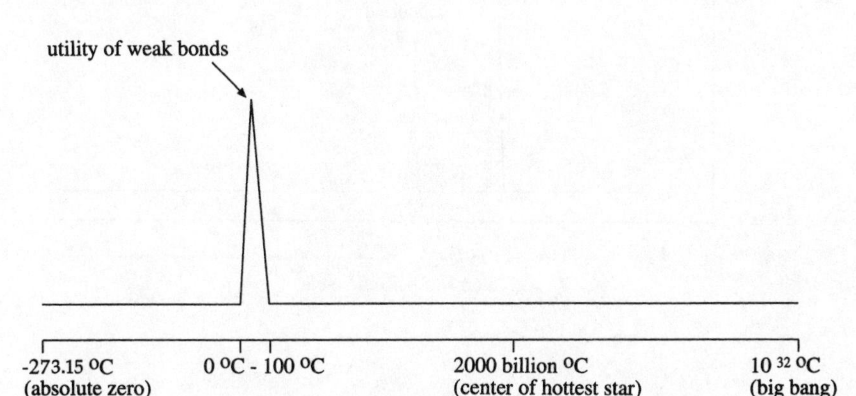

plenitude of metastable organic compounds, and (3) weak bonds for stabilizing the 3-D forms of complex molecules.

Conclusion

In short, then, the covalent compounds of carbon, and especially those containing oxygen, hydrogen, and nitrogen, the substances of life, possess just those characteristics of complexity, diversity, and metastability essential if any sort of complex chemical system is to *manipulate its atomic and molecular components in complex and intricate ways.* Moreover, this plenitude is of maximum utility in the same temperature range that water, the ideal matrix for life based on carbon chemistry, is a liquid, and where the weak bonds can be utilized to maintain the delicate three-dimensional molecular conformations upon which the functions of the cell's molecular machinery depend.

Carbon is so uniquely fit for its biological role, its various compounds so vital to the existence of life, that we may repeat the aphorism, "If carbon did not exist, it would have to be invented." The unique fitness of the carbon atom can be represented graphically as shown below.

The utility of atoms for the construction of complex chemical systems.

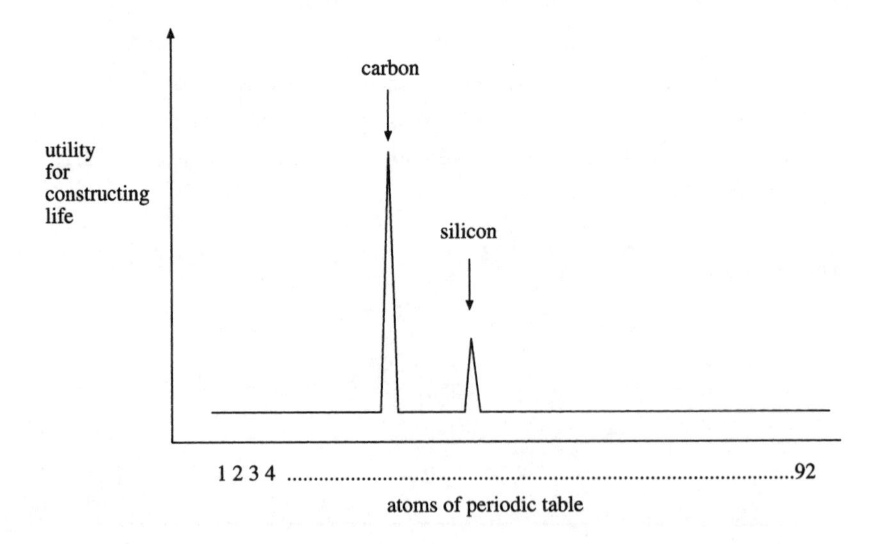

Chapter 6

The Vital Gases

In which the various adaptations which permit the use of oxidation by living systems to generate energy are examined. Oxygen is a very reactive atom and it can only be utilized by biochemical systems because of a number of adaptations, including: the attenuation of its reactivity below about 50°C; its low solubility; the fact that the transitional atoms such as iron and copper have just the right chemical characteristics to manipulate the oxygen atom; that the end product of the oxidation of carbon is carbon dioxide, an innocuous gas. Moreover, the reaction of carbon dioxide with water provides living things with a buffer—the bicarbonate buffer which has just the right characteristics to buffer organisms, especially air-breathing organisms, against increases in acidity. The chain of coincidences in the nature of things which permit higher forms of life to utilize oxygen provides further evidence of the unique fitness of nature for carbon-based life. Many of these adaptations are of special utility for large air-breathing organisms, including the fact that both oxygen and carbon dioxide are gases at ambient temperatures. Another fascinating coincidence is that only atmospheres with between 10 and 20 percent oxygen can support oxidative metabolism in a higher organism, and it is only within this range that fire—and hence metallurgy and technology—is possible.

The human lung, showing the bronchial tree.

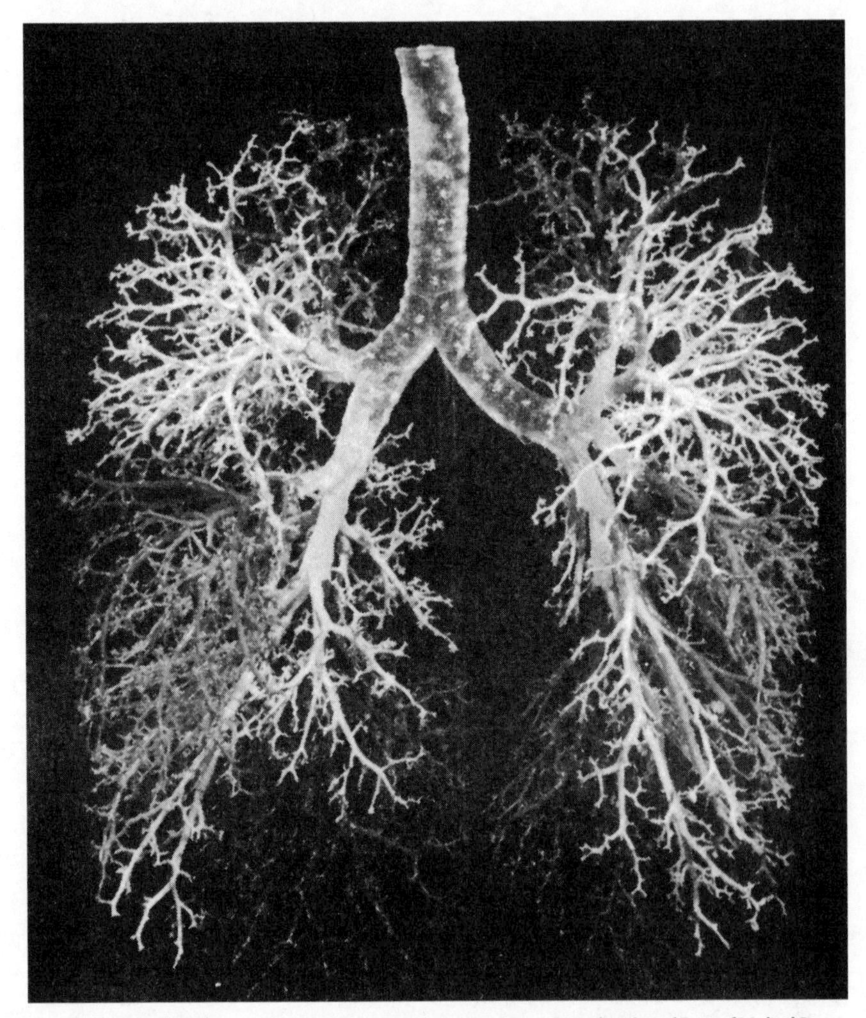

From J. B. West (1979) "Uptake and Delivery of the Respiratory Gases," in *Physiological Basis of Medical Practice,* 10th ed., J. B. Brobeck, ed. © 1979 Williams and Wilkins, Baltimore, fig 6.3. Reproduced by permission of the publisher.

And the Lord God formed man of the dust of the ground, and breathed
into his nostrils the breath of life, and man became a living soul.

—Genesis 2:7

In many science-fiction novels distant life-supporting planets resemble earth in many key respects. They often have water (if not quite as much as on earth) and complex plants and animals similar to those on earth. But the first thing humans must do on landing on an alien world is test the air to see if it is breathable. Apparently, sci-fi novelists assume that the one element of the earth's biosphere that can most easily vary is its atmosphere. Are they correct? Actually, no. As we shall see, only an atmosphere with very specific characteristics can support the life of complex, active air-breathing organisms such as mammals. And it is no accident that it also contains oxygen.

Oxidation

A major requirement for life is energy, and chemical reactions are an obvious source. Living things do, in fact, exploit a vast variety of chemical reactions to supply their energy needs.[1]

However, the great majority of the chemical reactions utilized are classed as oxidations. Oxidation is a very familiar type of chemical reaction. Two examples are burning wood and rusting iron. Life on earth utilizes a great diversity of oxidations. The bacteria at the base of the hydrothermal food chain oxidize hydrogen sulfide to sulfate and water. Other bacteria oxidize ferrous iron and hydrogen to ferric iron and water respectively. To a nonchemist it may seem curious, but not all oxidations require free atmospheric oxygen. Many can occur in an anaerobic environment, because in the absence of oxygen other compounds such as sulfates or nitrates which contain oxygen are used in place of oxygen itself.[2]

All higher organisms obtain their energy supply from one of the most important chemical reactions on earth—the complete oxidation of reduced hydrocarbons to carbon dioxide and water:

reduced carbon compounds + oxygen = water + carbon dioxide

As the oxidant in this reaction is oxygen itself, the process can only occur in an aerobic environment. This key reaction provides *many times more energy than any of the multitude of alternative energy-generating reactions.* Without it, higher active forms of life would not be possible.[3] The energy generated is used to manufacture the energy-rich molecules of ATP (adenosine triphosphate) in the mitochondria—a process called oxidative phosphorylation. (See chapter 9 and Appendix, section 3, for further details.)

Oxidation has many advantages. First, oxygen far surpasses any other chemical element except fluorine in the amount of energy liberated in the process of combining with other elements. Fluorine is, however, dangerously reactive at ambient temperatures. Also, while the chemical combination of hydrogen and oxygen results in the formation of water, when fluorine reacts with hydrogen, the product hydrofluoric acid is one of the most dangerously reactive of all acids. Moreover, fluorine has a great affinity for carbon and consequently the bonds between fluorine and carbon are very strong and can only be broken with considerable difficulty.[4]

Second, the compounds of carbon and hydrogen, which are the two most common atoms in organic compounds, are especially well qualified to be reservoirs of chemical energy liberated by oxidation, because hydrogen far exceeds any other element in the amount of energy that it yields upon oxidation and carbon is surpassed only by hydrogen and one other element, boron. Although there is less energy in compounds of hydrogen and carbon that also contain oxygen, such as sugars, proteins, and fats, a sufficient amount remains to make them highly efficient energy stores, holding far more energy than most other elements and far greater reservoirs of energy than the compounds of any other elements. Henderson was struck by the coincidence that oxygen is very nearly the most reactive atom, releasing great amounts of energy when reacting with other atoms, and that of all oxidations, those of reduced carbon compounds yield the most energy: "The very chemical changes, which for so many other reasons seem to be best fitted to become the processes of physiology, turn out to be the very ones which can divert the greatest flood of energy into the stream of life."[5]

The Reactivity of Oxygen

Could our atmosphere contain more oxygen and still support life? No! Oxygen is a very reactive element. Even the current percentage of oxygen in the atmosphere, 21 percent (partial pressure 150 mm Hg), is close to the upper limit of safety for life at ambient temperatures. The probability of a forest fire being ignited by lightning increases by as much as 70 percent for every 1 percent increase in the percentage of oxygen in the atmosphere.[6] As James Lovelock puts it in *Gaia:* "Above 25% very little of our present land vegetation could survive the raging conflagrations which would destroy tropical rain forests and arctic tundra alike. . . . The present oxygen level is at a point where risk and benefit nicely balance."[7]

A forest fire is ample witness both to the enormous energy released by the combustion or oxidation of organic compounds and testimony to the great increase in the reactivity of oxygen as the temperature rises. The greatly increased reactivity of oxygen and the consequent danger of runaway combustion as the temperature rises above about 50°C imposes an upper temperature limit on a carbon-based biosphere possessing an atmosphere like that of the earth.

Some authorities in the field of oxygen chemistry have wondered humorously why humans don't spontaneously combust, since our bodies contain so many atoms of carbon and hydrogen.[8] We do not combust because of the curious relative chemical inertness of the molecular species dioxygen (O_2), which is the molecular form of oxygen at ambient temperatures. The reason for this relative inertness lies in certain unique features of the oxygen atom. These were discussed by M. J. Green and A. O. Hill in a recent article.[9]

Curiously, the inertness of O_2, although essential if carbon-based life is to coexist with oxygen, is sufficient to pose a significant obstacle to its use in biological systems. This was already perceived to be a problem as early as the mid nineteenth century.[10] Advances over the past seventy years have revealed that the utilization of oxygen by living things is due to the catalytic action of a number of enzymes. These enzymes in turn utilize the properties of the transition metal atoms such as iron and copper, which have exactly the right atomic characteristics to carry out the activation of the O_2 molecule (see chapter 9). Activated by metal catalysts, oxygen becomes potent and highly reactive.[11]

The vigor of the reaction between carbon and oxygen is further attenuated by what Sidgwick referred to as the "characteristic inertness of carbon."[12] We have all experienced the relative unreactivity of carbon at ambient temperatures—when trying to start a coal or wood fire, particularly on a cold evening, i.e., to initiate the chemical combination between oxygen and carbon. And we have also all experienced the enormous energy released when the carbon in the coal or wood is finally coaxed to react with the oxygen and the fire starts to burn. The character of a simple wood or coal fire teaches important lessons in the chemistry of carbon and oxygen. A fire is difficult to start but, once started, releases great quantities of heat and energy and is difficult to stop. This curious unreactivity of the carbon and oxygen atoms at ambient temperatures, combined with the enormous energies inherent in their combination once achieved, is of great adaptive significance to life on earth. It is this curious combination that not only makes available to advanced life forms the vast energies of oxidation in a controlled and orderly manner but has also made possible the controlled use of fire by mankind and allowed the harnessing of the massive energies of combustion for the development of technology.

To summarize, oxygen is fit (1) because of the great amount of energy released when it combines with hydrogen and carbon, and (2) because its chemical reactivity is attenuated at ambient temperatures (below about 50°C), allowing living systems to utilize this awesome energy source in a controlled and efficient manner.

The Solubility of Oxygen

The chemical fitness of oxygen to living systems can only be exploited if additional conditions are satisfied. The solubility and rate at which oxygen diffuses in water is obviously critical. Since water is the matrix of life, if oxygen was either insoluble in water or chemically unstable in an aqueous solution, it would be incapable of playing any biological role.

The amount of a gas dissolved in a fluid depends on two factors: the partial pressure of the gas in the atmosphere, or air phase, in contact with the fluid, and the solubility of the gas in that particular fluid, which is a physical constant. For example, in the case of oxygen, the amount of oxygen that dissolves in a body of water is dependent on the solubility constant of oxygen and the partial pressure of the oxygen in the air above the water.

The earth's atmosphere contains about 21 percent oxygen, and consequently the partial pressure of oxygen is 150 mm Hg (21 percent of atmospheric pressure at sea level). Each 100 milliliters of water in contact with the air contains about 0.45 milliliters of oxygen gas at sea level.

It turns out that the solubility of oxygen is just sufficient to allow organisms, especially those with a high metabolic rate, to utilize oxidation as a means of energy generation. If it was any lower, organisms would not be able to extract oxygen from an aqueous solution at a sufficient rate to satisfy their metabolic needs. Even as it is, all actively metabolizing organisms depend on complex physiological adaptations to extract and transport sufficient quantities of oxygen to satisfy their energy needs. The delivery of sufficient oxygen to supply the metabolic requirements of the human body is critically dependent on the integrated activity of the circulatory and respiratory system and the special oxygen-carrying blood pigment hemoglobin.

The design constraints are such that it is hard to see how the oxygen-carrying capacity of the blood could be much increased.

But even if the blood's oxygen-carrying capacity could be increased several times by the use of, say, some imaginary superhemoglobin, the oxygen atoms would still have to diffuse across an aqueous layer to be "taken up by the carrier" in the lungs and diffuse across another aqueous layer when "leaving their carrier" in the tissues. As the rate of diffusion of oxygen in water is directly related to its solubility, the solubility of oxygen poses an absolute limit on the rate at which any hypothetical carrier could be loaded or unloaded with oxygen and consequently an absolute limit on the rate of delivery of oxygen to the tissues. In effect, no matter how good the capacity of the carrier system, the amount of oxygen the tissue fluids can hold is limited by the solubility constant of oxygen. If a gas is insoluble in water, then no matter how efficient a carrier molecule, the gas can never diffuse through an aqueous medium.

Clearly, if the solubility of oxygen or its rate of diffusion in water had been significantly less, then no conceivable type of circulatory or respiratory system would have been capable of delivering sufficient oxygen to support the metabolic activities of highly active, warm-blooded, air-breathing organisms in an atmosphere with a partial pressure of oxygen of 150 mm Hg.

The solubility of oxygen and hence the amount of the gas that a particular volume of water can contain falls rapidly as the temperature of water rises. The solubility at 0°C is twice that at 30°C and nearly four times that at 100°C.[13] The fact that the metabolic demand for oxygen doubles with every ten-degree increase in temperature[14] is also bound to impose an additional burden as the temperature rises above body temperature. At 58°C the demand for oxygen would theoretically be four times greater than at 38°C (the normal body temperature of many mammals), but this is just when the solubility and availability of oxygen is rapidly diminishing. *It would seem that while primitive unicellular forms of life can exist at all temperatures at which water is a liquid, higher complex multicellular life—which depends on the energy released from the complete oxidation of reduced carbon by free oxygen—is restricted to a temperature range between 0°C and 50°C.*

In passing, it is interesting to note that the specific heat of water is at its lowest between 35°C and 40°C, not far below 50°C. It is within this small temperature range, where water is most easily warmed and organisms can most easily activate their chemical machinery,[15] that most active organisms, including our own species, maintain their body temperature. It is conceivable that this same temperature range may be the optimum for the functioning of proteins and for the replication of nucleic acids and many other biochemical and physiological processes.

Because the problem of runaway combustion imposes an upper limit on the level of oxygen in the atmosphere which is close to the current partial pressure of oxygen in the earth's atmosphere of 150 mm Hg, if the solubility constant of oxygen had been significantly lower, then oxygen would be of little utility to life on earth, especially to organisms with high metabolic rates, such as mammals. It is doubtful indeed if any complex active organisms would have been possible, as no other chemical means of energy generation remotely as efficient as oxidation is available to carbon-based life forms.

On the other hand, had oxygen been more soluble than it is, this would have also produced very serious problems and would have greatly detracted from its fitness. As discussed above, oxygen is basically a very dangerous reactive substance and is highly toxic to life at levels above those normally encountered in nature. Oxygen toxicity is caused because a small proportion of oxygen atoms are continually interacting with water, producing highly reactive damaging radicals.[16] If for any reason this process increases beyond the normally low levels that occur naturally, it can be fatal. As Irwin Fridovich comments: "All respiring organisms are caught in a cruel trap. The very oxygen which supports their lives is toxic to them and they survive precariously, only by virtue of elaborate defence mechanisms."[17]

Many body cells die if directly exposed to the oxygen in the atmosphere,[18] and in fact the partial pressure of oxygen in most of the tissues is only about 50 mm Hg, which is about one third of that in the atmosphere.[19] In medicine, for example, great care must be taken in various medical procedures where oxygen is being used for therapeutic purposes. A raised partial pressure of oxygen may cause serious oxygen toxicity in a relatively short time.[20] Even slight increases in the level of oxygen dissolved in the blood can cause oxidative damage in the lungs and retina and other tissues. In fact, all organisms which utilize oxygen possess a number of enzymes specifically designed to eliminate reactive oxygen radicals. These enzymes utilize the properties of certain metal atoms, particularly the transitional metals that possess properties which are perfectly fit to handle and tame these reactive radicals. The problem of oxygen toxicity clearly imposes a limit on the maximum allowable level of oxygen in the atmosphere. Intriguingly, this is about the same value as that imposed by the problem of runaway combustion. Here we have a genuine coincidence—a case where two unrelated phenomena impose the same limit on a particular value.

It is evident, then, that oxygen's solubility (and diffusion rate) in an aqueous fluid must be very close to what it is. This is all the more remarkable considering the fact that the solubility of substances in water varies over many orders of magnitude. The solubility of many common gases varies over a range of nearly 1 million. The solubility of carbon dioxide, another gas of vital importance to life, is about twenty times greater than that of oxygen.[21]

Three authorities recently summed up the fitness of oxygen thus: "Oxygen is . . . the only element in the most appropriate physical state, with a satisfactory solubility in water and with desirable combinations of kinetic and thermodynamic properties."[22] The utilization of oxidation by living organisms is dependent on the possession by oxygen of a precise set of chemical and physical properties which are perfectly fit for its biological role in the temperature range at which complex life functions on earth. Between 0°C and 50°C in an atmosphere containing about 21 percent oxygen, sufficient oxygen dissolves in water to support oxidative metabolism. At temperatures much above 50°C the reactivity of atmospheric oxygen becomes too great, while in water the amount dissolved falls to levels which are probably increasingly insufficient to sustain active oxidative metabolism.

In summary, oxidation is fit because (1) of the enormous energies released when oxygen combines with other atoms, (2) the activity of oxygen is attenuated at ambient temperatures, and (3) oxygen has the appropriate solubility in water. Further, as we shall see in chapter 9 the utility of oxygen for life can only be exploited because the transitional metals have just the correct electronic properties to transport and handle it in an aqueous solution. These same metals also play a key role in the mitochondria in the generation of the energy-rich molecules of ATP.

Large complex, metabolically active life forms such as ourselves are entirely dependent on the energy released from the complete oxidation of reduced carbon:

reduced carbon + oxygen = water + carbon dioxide

And this reaction can only be exploited because oxygen has the precise properties it has. Two scholars recently noted: "For those who find it a meaningful pastime to speculate on the existence of intelligent life elsewhere in the universe, they might heed the caveat that the evolution of large complex forms of life on Earth was only possible due to the advent of atmospheric oxygen and the subsequent evolution of oxidative phosphorylation. This requirement significantly reduces the probability of the evolution of complex life forms on some remote planet."[23]

Curiously, the very many simple microbial species that utilize reactions which do not require the presence of free oxygen in the atmosphere are probably essential to aerobic life in a number of ways. For example, many may be involved in the cycling of the elements through the hydrosphere, and it may be that the origin of life occurred in an anoxic environment; and save for the capacity of some primitive unicellular organisms to thrive without oxygen, it may never have occurred.

Atmospheric Pressure

It is not only the chemical and physical properties of oxygen that must be precisely as they are if oxidation is to be exploited by carbon-based life forms on a planet like the earth; the overall composition and general character of the atmosphere—its density, viscosity, and pressure, etc.—must be very similar to what it is particularly for air-breathing organisms.

In addition to the 21 percent oxygen, the earth's atmosphere also contains about 78 percent nitrogen and has a pressure of 760 mm Hg and a density at sea level of about 1 gram per liter. Its viscosity at sea level is about one-fiftieth that of water. Current research suggests that the atmospheric pressure of the earth's atmosphere has been between about 500 mm and 1,000 mm Hg throughout most of the history of the earth.[24]

The total pressure of the earth's atmosphere is critical to life, particularly to highly active aerobic organisms like mammals, which depend on a complex respiratory system to deliver the oxygen in the air to the blood in the lungs. Recall first that respiration in vertebrates involves drawing air into the lungs (inspiration) via a system of branching tubes into tiny air sacs, or alveoli, where the oxygen in the air is absorbed by the blood, and then its expulsion (expiration) via the same set of tubes. Again, it is hard to imagine how the respiratory system in higher vertebrates could be much improved. In the adult human, gaseous exchange occurs across a special respiratory membrane lining the lungs which consists of 300 million alveoli. The total surface area available for oxygen absorption is 50 to 100 square meters, about the area of a tennis court. The process of drawing air into and expelling it from the lungs is critically dependent on the fact that both the viscosity and the density of the air which contains the vital oxygen are very low—about one-fiftieth and one-thousandth respectively of that of water. Even with these low values the total work of breathing is considerable. Although at rest this takes up only a small fraction of the total consumption of oxygen, during voluntary hyperventilation up to 30 percent of the oxygen consumption of the body is devoted to the work of breathing.[25] The very low viscosity and density are particularly critical because a significant proportion of the work of breathing is involved in overcoming what is termed "airway resistance," and this is determined directly by the density and viscosity of the air.[26]

However, as the atmospheric pressure is raised, the density also increases. Breathing becomes much more difficult. At about three times atmospheric pressure, extended periods of strenuous work become impossible because the effort involved in moving the air takes up a prohibitive proportion of the total energy available. When the pressure is increased to several times atmospheric pressure, this resistance becomes prohibitive.[27] It is clear that if either the viscosity or the density of air were much greater, the airway resistance would be prohibitive and no conceivable redesign of the respiratory system would be capable of delivering sufficient oxygen to a metabolically active air-breathing organism. If the atmospheric pressure were ten times greater, the work of respiration would be prohibitive. If it were about ten times less, the body fluids would vaporize at 38°C.[28]

By plotting (as shown opposite) all possible atmospheric pressures against all possible oxygen contents, it becomes clear that there is only one unique tiny area, A, where all the various conditions for life are satisfied.

It is surely a coincidence of enormous significance that several essential conditions are satisfied in this one tiny region in the space of all possible atmospheres. Fire is possible, but runaway combustion is avoided, oxygen toxicity is relatively low, the solubility of oxygen is sufficient to support oxidative metabolism, and the density is sufficiently low so that the work of breathing during strenuous exercise is not prohibitive.

And what is perhaps even more remarkable is that long-term atmospheric stability on a planet the size of the earth may only be possible in this same unique region of the atmospheric space. James Lovelock has recently specu-

The region of atmospheric space fit for life.

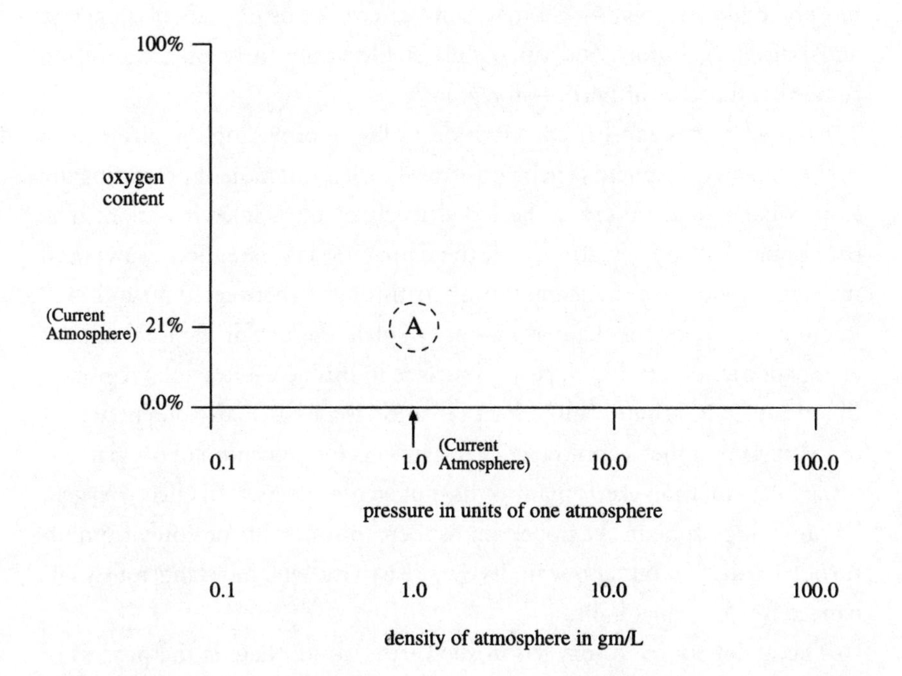

lated along these lines.[29] Of course, because climatic modeling is still in its infancy, it is impossible to know to what extent this may be true. However, it would seem relatively likely that if the atmospheric pressure was, say, only one-fifth as great as it is, the seas might eventually vaporize and the increased water vapor in the atmosphere might cause a massive and runaway greenhouse effect. On the other hand, if the atmosphere was several times more dense, this might reduce the amount of water vapor in the atmosphere and the continents might be converted to arid wastelands.

The current atmosphere of the earth is the natural result of planetary evolution resulting from a complex set of feedback controls involving complex interactions between the hydrosphere, the biosphere, and the material making up the crust of the earth. That such a process should have generated an atmosphere with just the appropriate amount of oxygen and of the appropriate pressure to support active aerobic metabolism and maintained it for perhaps half a billion years is again perfectly in keeping with the biocentric hypothesis. It may be that all planets of the size and composition of the earth the same distance from their sun—i.e., those capable of sustaining a biosphere like our own—will inevitably, after billions of years of planetary and crustal evolution, end up with a stable atmosphere of composition somewhat like that of Earth—in region A.

Oxygen makes another contribution to life in providing the ozone layer in the upper atmosphere which performs the vital function of protecting life from what would otherwise be lethal levels of ultraviolet radiation. The ozone shield effectively absorbs all the damaging UV radiation below 0.30 microns. Although a small amount filters through—between 0.30 and 0.40 microns, as we saw in chapter 3—only a small amount of the total radiant energy of the sun reaches the earth's surface in this near-ultraviolet region of the spectrum. Curiously, ultraviolet radiation is a particularly potent activator of oxygen in the near ultraviolet and it is via the activation of oxygen that ultraviolet radiation exerts many of its mutagenic effects.[30] In effect, oxygen, by providing ozone in the upper atmosphere, protects life not only from ultraviolet radiation but also from its own reactivity, which is dangerously enhanced by ultraviolet light.

The major source of oxygen in the earth's atmosphere is the process of photosynthesis, which involves the building up of sugars from water and carbon dioxide, using energy provided by sunlight. It is fortunate that the oxygen thus formed creates the ozone shield to prevent lethal UV from reaching the earth's surface, because photosynthesis itself obviously necessitates direct exposure to the sun's radiation, which, in the case of many terrestrial plants, would be impossible without the protection of the ozone layer. Thus, oxygen, via a strange sort of self-referential loop, provides the crucial protecting shield for the delicate machinery that generates it in the first place, attenuating both the flow of UV as well as its own photo activation by UV light.

We are so familiar with the act of breathing that it mostly goes unnoticed. But it is far, very far, from ordinary. Our ability to breathe and to utilize the vital properties of the oxygen atom depends on a long and deep chain of coincidences in the nature of things. There is in the end nothing contingent about the choice of oxidation as the major source of energy for life on earth. Without the energy inherent in the chemistry of oxidation, life would have remained frozen forever at the primitive unicellular stage it reached on earth long before the Cambrian explosion and the development of complex multicellular life. It is not that life adapted to oxygen or to the atmospheric conditions on the earth, but rather that long ago, long before the first organisms, long before the formation of the earth, the design of oxidative metabolism and the general character of the atmosphere of our planet was already built into the order of the cosmos.

Carbon Dioxide

Despite all the energy that oxidation supplies to life, unless the end products of oxidative metabolism were innocuous and harmless and easy to dispose of, oxidative energy would not be available to life. In fact, the final two products of the oxidative breakdown of organic compounds are water and carbon dioxide.

reduced carbon + oxygen = water + carbon dioxide

Water is not only harmless to life, it is the very matrix of life. And we have already seen just how wonderfully and in so many ways water is adapted to life. Organisms have at their disposal a great number of means by which to rid themselves of excess water produced in the course of metabolism: via kidneys, via evaporation, via contractile vacuoles, and so forth.

The other end product of oxidative breakdown of organic compounds, carbon dioxide (CO_2), possesses a number of physical and chemical properties which are critical to life on earth. If carbon dioxide had been a toxic substance, if it had been a liquid insoluble in water, if it had been a solid, if it had dissolved in water forming a strong acid, the complete oxidation of carbon to carbon dioxide would have been impossible and complex carbon-based life would in all probability never have evolved. However, carbon dioxide is none of these things.

Excretion

In fact, carbon dioxide is a relatively unreactive compound and a gas at ambient temperatures. That it is a gas should, as Needham points out, "be emphasised since it is one of the very few gaseous oxides at ordinary temperatures (water vaporises more than most others)."[31] Moreover, that carbon dioxide is an innocuous soluble gas which can be readily excreted from the body of terrestrial organisms via respiration is of enormous utility. As Henderson says:

> In the course of a day a man of average size produces as a result of his active metabolism, nearly two pounds of carbon dioxide. All this must be rapidly removed from the body. It is difficult to imagine by what elaborate chemical and physical devices the body could rid itself of such enormous quantities of material were it not for the fact that . . . in the lungs . . . [carbon dioxide] can escape into air which is charged with little of the gas. Were carbon dioxide not gaseous, its excretion would be the greatest of physiological tasks; were it not freely soluble, a host of the most universal physiological processes would be impossible.[32]

The Regulation of Acidity

The gaseous nature of carbon dioxide is not only of great utility for excretion in the case of terrestrial organisms. It is also of great utility for the regulation of the level of acidity in the body.

Most of the carbon dioxide transported from the site of its production in the tissues to the lungs is not merely a simple solution of the gas. As every medical student learns, from estimates of the total amount of carbon dioxide dissolved in the blood and from estimates of the difference in the amount of dissolved carbon dioxide in arterial and venous blood, it can be shown that most of the 200 milliliters of carbon dioxide produced per minute in an average adult human cannot be transported in simple physical solution to the lungs. When carbon dioxide is dissolved in water, it gradually interacts with water molecules to form carbonic acid (H_2CO_3), which then ionizes to produce hydrogen ions (H^+) and the base bicarbonate (HCO_3^-):

carbon dioxide + water = hydrogen ions + bicarbonate base

Now the fact that the gas CO_2 reacts reversibly with water to produce the base bicarbonate has physiological consequences of very great significance because it provides the organism with a wonderfully elegant means of protecting or buffering itself against fluctuation in the level of acidity of the body.

As can be seen from the above equation, when acid (hydrogen ions, H^+) accumulates in the body, the hydrogen ions combine with bicarbonate (HCO_3^-) to produce water and carbon dioxide (CO_2), which can be excreted readily from the body via the lungs. In effect, the excess acid (hydrogen ions) is simply breathed out of the body.

Hence, an organism's defense against the accumulation of acid, which is an inevitable consequence of the oxidative metabolism of organic compounds, is provided by one of the major end products of this same metabolic process, carbon dioxide and its hydration product bicarbonate. The other major product is water, itself the ideal medium for life. Amazingly, it is water, the other final product of oxidative metabolism, which dissolves the CO_2 and carries it to the lungs, and it is water which reacts with carbon dioxide to generate bicarbonate, "the perfect buffer."

Many authors have commented on the fitness of the CO_2– bicarbonate buffering system for the maintenance of acid-base balance. Henderson comments: "there is I believe, except in celestial mechanics, *no other case of such accuracy in a natural regulation of the environment.* Moreover the chemist has discovered no means of rivalling the efficiency and delicacy of adjustment of the process."[33] Like Henderson, the protein chemist John Edsall was also struck by the remarkable nature of the system: "The combination of the acidity and buffering power of H_2CO_3 with the volatility of CO_2 provides a mechanism of unrivalled efficiency for maintaining constancy of pH in systems which are constantly being supplied as living organisms are with acidic products of metabolism."[34]

The bicarbonate–carbon dioxide buffer system is a paragon of elegance and efficiency which solves two basic and very different physiological problems—the ridding of the body of the end product of oxidative metabolism, and the maintenance of acid-base homeostasis in the same basic equation:

$$\text{carbon dioxide} + \text{water} = \text{hydrogen ions}^{+} + \text{bicarbonate}^{-}$$

Moreover, this remarkable system is particularly fit for air-breathing terrestrial organisms like ourselves. And there are additional subtle aspects of the CO_2– bicarbonate buffering system which contribute further to its fitness for acid-base homeostasis. (For a more detailed description of the nature of buffers and of the bicarbonate buffer system, see Appendix, sections 6 and 7.)

It turns out, then, that both the maintenance of acid-base balance in the body and the excretion of the end product of oxidative metabolism, CO_2, depend crucially on the chemical and physical properties of CO_2 itself and its hydration product, bicarbonate. *Thus both the problem of excretion of the end product of carbon metabolism and the problem of acid-base balance are both elegantly solved in the properties of the same remarkable compound—carbon dioxide. It is a solution of breathtaking elegance and parsimony based on another set of mutual adaptations in life's constituents.*

The diagram below summarizes the basic interactions which occur during oxidative metabolism.

Oxidative metabolism.

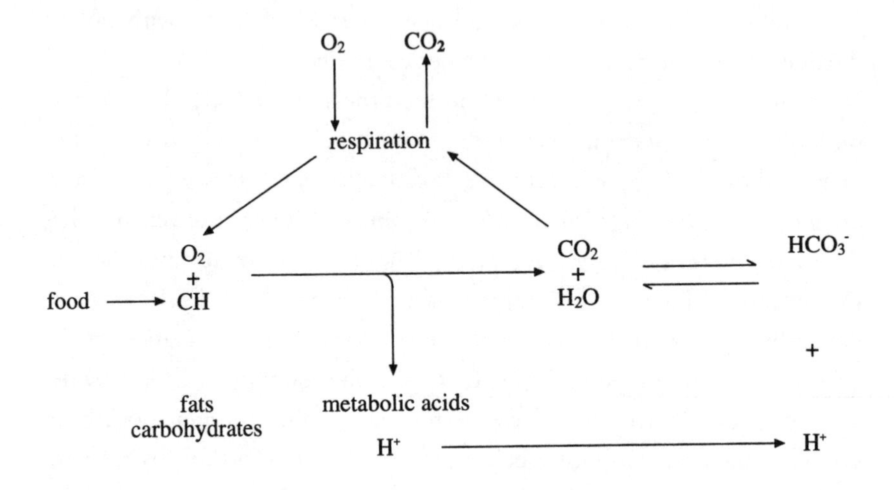

Although the above scheme is well known, it remains a paragon of elegance and parsimony. The elegance of the fact that both O_2 and CO_2 enter and leave the body via the same route, which depends on their both being gases at ambient temperatures. The fact that CO_2 is both the major excretory product of oxidation and at the same time, by reacting with water, the source of bicarbonate, which mops up excess acid produced inevitably during periods of relative oxygen lack. The fact that both CO_2 and bicarbonate are innocuous compounds, that both are soluble in water. The fact that oxygen is soluble in water and its reactivity is attenuated at ambient temperatures. The fact that oxidation is the most efficient means of energy generation and so on. The mutual fitness of the chemical actors for their respective roles in this the central metabolic drama of life is simply astounding. If these compounds did not possess precisely the chemical and physical properties they do, the drama would be impossible, and it is exceedingly difficult to envisage an alternative.

Curiously, despite the fact that these key components have precisely the properties required for their biological roles, which together make the execution of oxidative metabolism a marvel of parsimony and efficiency, we are seldom struck by the immensity of the teleology which underlies it all. Failing to see the wood for the trees, we take the entire system and all the coincidences for granted, assuming erroneously that things could not have been otherwise. This same point was made by Richard Bentley in his "Confutation of Atheism," which is one of the classics of seventeenth-century natural theology and was cited in chapter 1. It is worth repeating:[35]

> What we have always seen to be done in one constant and uniform manner; we are apt to imagine there was but that one way of doing it, and that it could not be otherwise. This is a great error and impediment in a disquisition of this nature: to remedy which, we ought to consider every thing as not yet in Being; and then diligently examine if it must needs have been at all, or what other ways it might have been as possibly as the present.

For Bentley's point is well taken. Things could have been otherwise as far as we can tell. For example, if O_2 and CO_2 were not gases, the design of large terrestrial carbon-based organisms obtaining energy by oxidative metabolism would in all probability be impossible. Carbon-based life forms such as mammals are critically dependent not only on the fact that O_2 and CO_2 are gases but also on the low viscosity of water which makes possible (as we saw in the previous chapter) a circulatory system which is itself essential if the gaseous properties of O_2 and CO_2 are to be exploited. Water is not only a key chemical player in the metabolic scheme of oxidative metabolism, but also through its low viscosity it provides the physical means, i.e., the circulatory system, by which the various chemical and physical properties of the other players, particularly the gases O_2 and CO_2, may be utilized in the case of large terrestrial life forms.

Every detail of the chemistry of carbon dioxide we examine seems to reveal additional aspects to its fitness. For example, take the actual process of hydration itself. Because of certain molecular characteristics of the carbon dioxide molecule, the process of hydration occurs relatively slowly. Now this apparently esoteric point, the slowness of the hydration of CO_2, happens to be of great physiological importance. If hydration was instantaneous, this would mean that whenever the metabolism of carbon was increased, the increased quantities of CO_2 generated would immediately hydrate, producing carbonic acid which would then dissociate, releasing H ions and subjecting the cell to sudden violent fluctuations in acidity that might well be lethal in higher organisms. Undoubtedly, the slowness of the hydration of CO_2 contributes further to the biological fitness of carbon dioxide. In the body the speed of hydration is increased by the presence of the enzyme carbonic anhydrase, which catalyzes the hydration reaction in the red blood cell and in the lining of the lungs.[36]

The unique utility of the bicarbonate buffer system for carbon-based life forms can be represented again in graphic form as shown below.

It should be noted that the CO_2-bicarbonate system is not just the main buffering mechanism responsible for the maintenance of the neutrality of the body fluids; it also plays the same role on a global scale, preserving the neutrality of the oceans and all bodies of water on the earth's surface. Weak solutions of carbonic acid also play a role, probably quite vital, in the process of weathering by greatly increasing the rate at which minerals are washed out of the rock. As Henderson comments, "it is the united action of water and carbonic acid . . . which sets free the inorganic constituents of the earth's crust and turns them into the stream of metabolism."[37]

The fact that CO_2 is very soluble in water means that wherever there is water on the planet, CO_2 will be also present to provide a ready source of carbon atoms for the synthesis of biological materials.

Another feature related to its solubility, which ensures the universal distribution of CO_2 throughout the hydrosphere, is the fact that its absorption coefficient in water is close to one. This means that whenever water is in

The fitness of buffers for pH homeostasis.

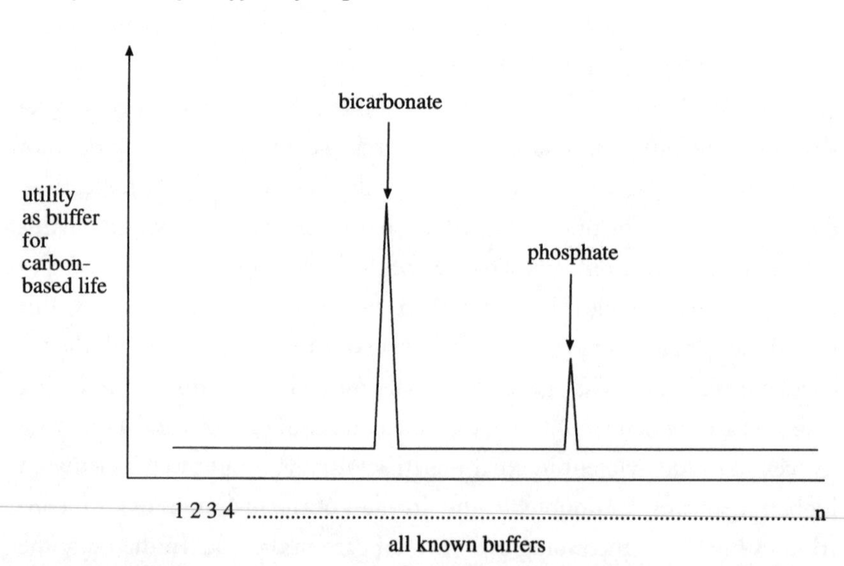

contact with air, and equilibrium has been established, the amount of CO_2 per liter of water will be equal to the amount per liter in air. Thus, as Henderson remarks, "the waters can never wash [CO_2] out of the air, nor keep it from the waters. It is the one substance which thus, in considerable quantities relative to its total amount, everywhere accompanies water."[38] Not only does the gaseous nature of carbon dioxide greatly facilitate the excretion of the carbon from the body of large organisms; this same gaseous nature and its solubility and absorption properties provide what would seem to be the perfect means of distributing the carbon atom to every part of the hydrosphere in the atmosphere and in the rivers, lakes, and seas.

Photosynthesis

Photosynthesis is almost the exact reverse of oxidation. During oxidation, reduced carbon compounds such as sugars and fats are converted to carbon dioxide and water. In photosynthesis light energy is used to convert water and carbon dioxide to oxygen and reduced carbon compounds. The essence of the process can be represented by the reaction:

water + carbon dioxide = oxygen + reduced carbon

Photosynthesis is another absolutely vital biological process. Nearly all complex plant and animal life on earth depends upon it. It generates all the fuel—the reduced carbon compounds such as the sugars and fats, etc.—which energize complex life on earth. Like respiration, it is so familiar that its remarkable nature fails to strike us. And like respiration, it is very hard to imagine any other process which could replace it to sustain complex life. Simple forms of life can obtain energy from sources other than sunlight, but for a rich, complex world on the surface of a planet there is no alternative to photosynthesis. And just like respiration, photosynthesis is possible only because each of the key players in the process—water, carbon dioxide, and oxygen—have precisely those properties they have. The fact that both CO_2 and O_2 are gases which can be readily taken up or excreted by the plant cells is crucial; also important are the facts that carbon dioxide is distributed universally throughout the hydrosphere, that water is ubiquitous, and that the solubilities of carbon dioxide and oxygen are as they are. Moreover, we have already seen that in addition to all this, the sun's light is perfectly fit for photochemistry, and the transparency of the atmosphere and liquid water are perfectly fit for sunlight's penetration to the surface of the earth. Photosynthesis also depends on the unique light-absorbing characteristics of the magnesium atom in chlorophyll, which is touched on again in chapter 9.

Conclusion

The three basic chemical reactions upon which all higher life depends are:

1. Oxidation.

   reduced carbon + oxygen = water + carbon dioxide

2. The regulation of acidity.

   carbon dioxide + water = hydrogen ions + bicarbonate base

3. Photosynthesis.

   water + carbon dioxide = oxygen + reduced carbon

They can only be exploited by living organisms because of a unique fitness in the nature of things. It is solely because of this unique fitness that higher organisms can obtain, via photosynthesis, an almost unlimited supply of food in the form of reduced carbons, can carry out the oxidation of reduced carbon compounds to provide energy, can rid themselves of the end products of oxidative metabolism, and can maintain acid-base homeostasis. The treatment here is by no means exhaustive. Many other important compounds involved in oxidative metabolism such as the sugars, the sugar-storing molecule, glycogen, and the phosphates also appear uniquely fit for the key roles they play in the process. (See Appendix, sections 1 and 3.)

And so the coincidences lengthen further. In case after case, the constituents of life—water, the carbon atom, the oxygen atom, carbon dioxide gas, the bicarbonate base—turn out to be uniquely and ideally fit in so many diverse and complex ways for their respective biological roles.

Henderson's conclusion in *The Fitness of the Environment* has certainly stood the test of time:

> Accordingly, we may finally conclude that the fitness of water, carbonic acid, and the three elements make up a unique ensemble of fitness for the organic mechanism. . . . There is nothing about these substances which is . . . inferior to the same thing in any other substance . . . not a single disability of the primary constituents . . . has come to light.[39]

> The fitness . . . [of these compounds constitutes] a series of maxima—unique or nearly unique properties of water, carbon dioxide, the compounds of carbon, hydrogen and oxygen and the ocean—so numerous, so varied, so complete among all things which are concerned in the problem that together they form certainly the greatest possible fitness.[40]

Conclusion

From the evidence presented so far, we are now in a position to play the "life game" advocated by Robert Clark in chapter 8 of his book *The Universe: Plan or Accident?* (cited at the beginning of chapter 5) where he suggests we imagine ourselves in the position of Plato's Demiurge, setting out to create life from scratch, being free to choose at every stage of the process the most ideal materials and components available and being constrained only by the laws of physics. Playing the game is instructive, for it highlights one of the main arguments of the book, namely that the laws of nature are fit for only one specific type of life—that which exists on earth. To start the game we must choose an atom out of which to create life. And, in Clark's words, "we soon find that carbon is the most promising . . . as we continue we find, with increasing astonishment, that it is not a case of carbon will do, but that carbon atoms have all the properties we could desire."

Having decided to use carbon, we face our next problem: we must find a medium in which to base our carbon-based life. We examine several liquids and soon find one—water—which seems to have some of the properties we need. Again, our astonishment grows as we find that water has not only one or two useful properties but seems mutually fit for carbon-based life in almost all of its characteristics. Next we face the problem—how shall we ensure that the carbon atom is always readily available? "One possible solution," Clark suggests, "might be to place the carbon in the air, since air is the only material disseminated over the whole earth. For this a gaseous compound would be necessary."

However, as Clark points out, such a "compound would also have to fulfil rather exacting conditions." To begin with, it would be preferable if it were a highly oxidized compound if we are planning to utilize the energy of oxidation to drive our chemical machine, for only a highly oxidized substance can coexist with atmospheric oxygen. The best possible solution would be an oxide containing the maximum amount of oxygen. Moreover, as Clark continues, our carbon-containing oxide must be soluble in water but not so soluble that it is washed out of the atmosphere. "Carbon dioxide might be the compound we require. If this will not do, it seems that nothing can take its place. It is the oxide of carbon richest in oxygen and so stable in an oxygen-containing environment." But does it have, asks Clark, all those properties we desire? "We find with delight, that the gas dissolves in about an equal volume of water—just what we want."

However, oxides are often troublesome compounds. In water they usually produce powerful acids or bases. "What we need is a very weak acid, . . . one that will not interfere with the valuable properties of water. . . . The strengths of acids vary by factors of billions. We consult our tables of these strengths—hardly two are close together and none are close to that of water. Anxiously we measure the strength of the acid formed when our gas dissolves in water. . . . The incredible happens again! It comes out right!"

And continuing our life game, next we must find a means of obtaining energy. Clearly, the oxidation of reduced carbon compounds is an ideal candidate. When reduced carbon compounds are oxidized, we find that the process generates water, our chosen matrix for life, and carbon dioxide, our chosen means of universally distributing the carbon atom over the earth. So oxidation will do wonderfully. But what about the solubility of oxygen and its reactivity? Amazingly, its solubility is right and its reactivity attenuated to just the right extent.

For life we need the carbon atom and water, and for complex life we need oxygen, we need carbon dioxide, we need bicarbonate, we need the transitional metals, we need an atmosphere like that on earth, and we need all their chemical and physical properties precisely as they are. And for life anywhere in the cosmos it will be the same. For there is no alternative. So if there is in some distant galaxy another carbon-based biosphere as rich as our own, containing large active terrestrial organisms, they will, like us, inhale oxygen and exhale carbon dioxide and use the bicarbonate buffer system. No matter how many times we play the Demiurge, we will always be led via the same chain of mutual adaptations to the same unique solutions.

Chapter 7

The Double Helix

In which evidence is presented which supports the notion that DNA and RNA may be uniquely fit for their respective biological roles—DNA as the genetic data bank and RNA as the temporary information carrier—in complex carbon-based life forms. DNA is fit for its biological role in a number of ways: it is chemically stable in an aqueous medium, its structure allows for highly accurate and rapid duplication, it possesses conformational plasticity which enhances its informational capacity and facilitates DNA-protein interactions, and it has an enormous capacity for compaction because of its supercoiling ability.

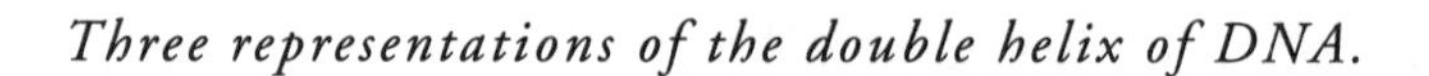

Three representations of the double helix of DNA.

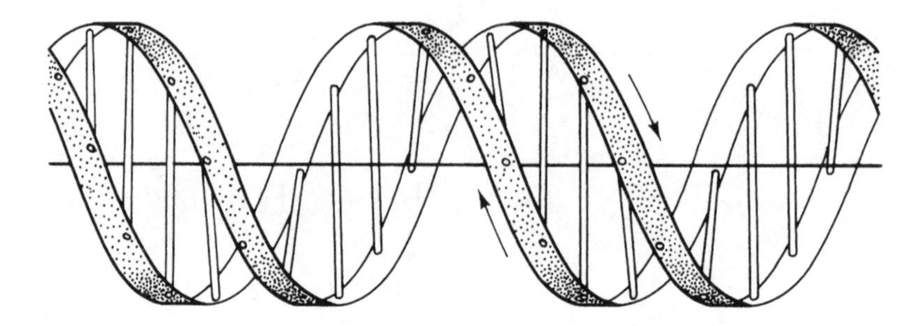

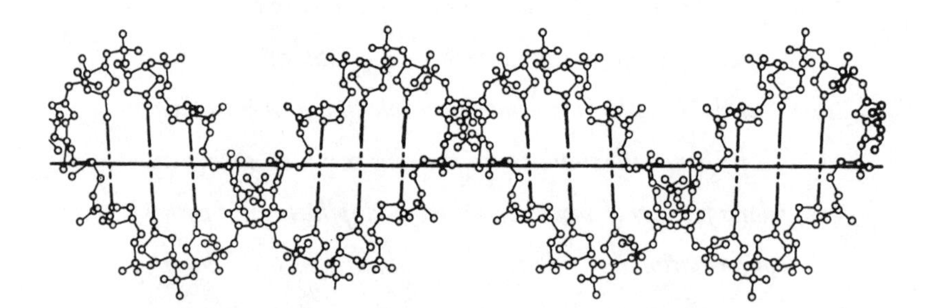

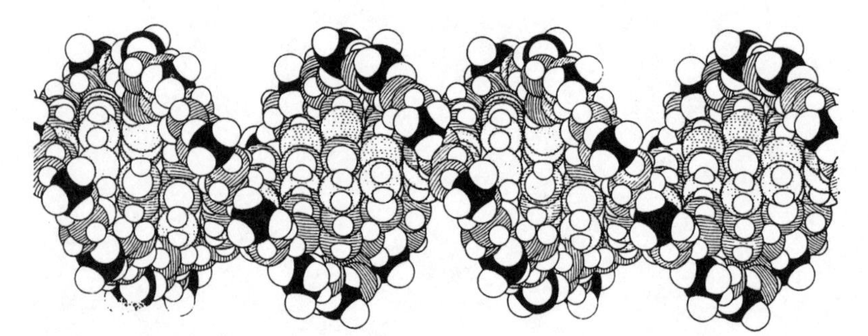

From S. Miller and L. Orgel (1974) *The Origins of Life on the Earth* (Englewood Cliffs, N.J.: Prentice-Hall), fig. 6.8. Reproduced by permission of Leslie Orgel.

That morning Watson and Crick knew . . . the entire structure: it had emerged from the shadow of billions of years, absolutely pure and simple, and there was seen and understood for the first time.

—H. F. Judson, *The Eighth Day of Creation,* 1979

I would rather stress that the structure made Watson and Crick. After all, I was totally unknown at the time and Watson was regarded as too bright to be really sound. But what I think is overlooked in such arguments is the intrinsic beauty of the DNA helix. It is a molecule which has style, quite as much as the scientists.

—Francis Crick, *Nature* magazine, 1974

Much of our current knowledge of the biochemical basis of life has resulted from a succession of discoveries from the late 1940s to the early 1960s which completely transformed the biological sciences. These discoveries, collectively known as the molecular biological revolution, revealed for the first time the molecular structure and biological function of some of the most important macromolecular constituents of the cell, including DNA, RNA, and proteins, and revealed for the first time the mechanism by which living things achieve the miracle of self-replication. It has turned out to be one of the most remarkable stories in twentieth-century science.

Curiously, over the same period of time, while biology was being transformed, dramatic developments were also occurring in the fields of information theory, cybernetics, and artificial intelligence, developments which subsequently led to the information revolution and the computer age of today. These parallel developments led to a rather strange historical coincidence. While biologists were busy working out the actual basis of biological replication, information theorists were working on the theoretical possibilities of constructing artificial systems capable of self-replication, machines that could make themselves, or, in other words, "artificial life."

One of the greatest of these figures was the mathematician John von Neumann, who developed his model of a self-replicating machine in the early 1950s before the actual self-replicating machinery of the cell had been worked out.[1] Von Neumann, as was mentioned in chapter 3, visualized the surface of a vast body of fluid covered with infinitely many copies of each kind of element required for the construction of the automaton, distributed in random fashion over the surface of the lake. The automaton, like an animated erector set, floats on the surface of the hypothetical lake and, by picking up elements from the fluid and assembling them together, eventually constructs a copy of itself. The automaton consists of two components: an information bank and a mechanical assembly unit capable of manipulative robotic activities—what von Neumann called the "constructor." The information bank provided all the information and instructions necessary to direct the constructor to assemble a copy of itself. When the constructor had finished constructing a copy of itself, it then made a copy of the information bank and inserted this new copy into the newly assembled offspring constructor. Thus, the automaton makes a complete copy of itself.

Artificial Life

The subject of artificial life has recently become a scientific discipline in its own right. Recently, a symposium entitled "Artificial Life" and subtitled "The Proceedings of an Interdisciplinary Workshop on the Synthesis and Simulation of Living Systems," devoted entirely to the subject, was held at the Santa Fe Institute in New Mexico. The organizer of the symposium, Christopher Langton, explained:

> Artificial Life is the study of man-made systems that exhibit life-like behaviour characteristic of natural living systems. . . . By extending the empirical foundations upon which biology is based beyond the carbon-chain life that has evolved on earth, Artificial Life can contribute to theoretical biology by locating *life as we know it* within the larger picture of *life as it could be.*[2] [My emphasis.]

The idea that self-replicating machines will eventually be built that will equal or even surpass life in sophistication and complexity in the not too distant future has become almost a defining characteristic of late-twentieth-century science fiction. Indeed, much of the literature in this field is concerned with the increasing difficulty of distinguishing between life and machinery. And there is no doubt that technology today has advanced to levels that were undreamt of, simply unimaginable, even as recently as fifty years ago. At an ever-accelerating rate one technological advance has followed another. We have built machines in which we have flown like a bird and others in which we have traveled to the bottom of the ocean. With radio telescopes we have listened to the murmurs of the most distant galaxies. We have trod in moon dust and we have sniffed the air of Mars. We have machines that can calculate a billion times faster than a man.

The gap between living things and machines seems to have narrowed with every advance in technology. This trend has become particularly obvious over the past few decades. Today there seems hardly any feature of living systems that does not have some machine analogue. Machines use artificial languages and memory banks for information storage and retrieval. Advanced machinery utilizes elegant control systems regulating the assembly of parts and components. Fail-safe devices and proofreading systems are utilized for quality control; assembly processes utilize the principle of prefabrication. All these phenomena have their parallel in living systems. In fact, so deep and so persuasive is the analogy that much of the terminology we use to describe the fascinating world of the cell is borrowed from the world of late-twentieth-century technology.

And the achievements to date may be vastly exceeded if the development of a new field of microminiaturized technology—nanotechnology—is successful. Conrad Schneiker recently reviewed the history of this fascinating field. He relates that it was the science-fiction author Robert Heinlein who first envisaged the possibilities of nanomachines: "the extensive use of teleoperator hands . . . complete with sensory feedback for full remote-controlled telepresence . . . for building and operating a series of ever smaller sets of such mechanical hands . . . the smallest of which were hardly an eighth of an inch across, used to manipulate living nerve tissue."[3] But it was physicist Richard Feynman who authored the definitive source paper on the subject:

> [Consider] the final question as to whether, ultimately . . . *we can arrange the atoms the way we want, the very atoms all the way down!* . . . [when] we have some control of the arrangement of things on a small scale we will get an enormously greater range of possible properties that substances can have, and different things that we can do. . . . if we go down far enough, all our devices can be mass produced so that they are absolutely perfect copies of one another.[4]

More recently, various workers in the field of nanotechnology have expanded in more detail the revolutionary concept of machinery in the nanometer range, envisaging gears, bearings, and motors scaled down to the atomic level, and even the construction of a nanocomputer, much smaller than a bacterial cell, based on logic operations mediated entirely by molecular rods made of carbine, sliding into ON or OFF positions in a complex maze of channels embedded in a three-dimensional matrix composed of atoms near carbon in the periodic table. One of the gurus in the field, Eric Drexler, has envisaged tiny factories, smaller than a grain of sand, containing molecular assemblers, tiny atomic machines that could assemble, atom by atom, bearings, rods, rollers, etc.[5] He has envisaged rows of these miniature assemblers programmed to assemble all manner of specific machines. Drexler has even envisaged a microminiaturized submarine smaller than a red blood cell which could be programmed to hunt out and destroy invading bacteria or cancer cells even in the smallest capillaries. According to an article in a recent issue of the *Scientific American:*

> Such products, depending on design and purpose, might roam through the human body, invading cancerous cells and rearranging their DNA. Other ma-

> chines might swarm as a barely visible metallic sheen over an outdoor construction site. In a few days an elegant building would take shape. . . . Every hour entire factories no larger than a grain of sand might generate billions of machines that would look like a mass of dust streaming steadily from the factory doors—or like a cloudy solution suspended in water.[6]

The sheer genius of modern technology and its achievements encourages the belief that, however complex life's design, it must eventually be equaled in a machine. Possessing a technology so sophisticated that we can contemplate the design and construction of a submarine as small as a red blood cell, a computer smaller than a bacterium, objects which are every bit as complex in terms of number of components per unit volume as living systems, encourages the belief that machines will one day be built which are capable of self-replication, and that artificial life based on a completely different design to that on earth will finally be achieved.

The Magic of Self-Replication

Yet despite the dreams of artificial life and the gurus of nanotechnology, the undeniable fact remains that many characteristics of living organisms are still without any significant analogue in any machine which has yet been constructed. Every living system replicates itself, yet no machine yet possesses this capacity even to the slightest degree. Nearly half a century after von Neumann, Claude Shannon, Norbert Wiener, and their circle dreamed of self-replicating machines, the dream is nowhere near realization. Nor does there exist even a well-developed, detailed blueprint in the most advanced area of nanotechnology for a machine that could carry out such a stupendous act. In the case of von Neumann's model, for example, no serious consideration was given to the fuel and energy supply problem. Von Neumann assumed conveniently that his automata would have unlimited energy!

The challenge is enormous. A self-replicating machine requires a data storage system which must be accessible or comprehensible to the constructor device. It requires that the constructor be assembled from a very small number of readily available substances. It requires a means of energy generation, storage, and distribution to its working components and so forth. None of these problems has been solved. Yet every second, countless trillions of living systems from bacterial cells to elephants replicate themselves on the surface of our planet. And since life's origin, as the earth has circled thousands of millions of times around the sun, endless life forms have effortlessly copied themselves on unimaginable numbers of occasions.

And it is not just the act of self-replication which has not been copied in our technology. Even the far less ambitious end of component self-assembly which is utilized by every living cell on earth, exhibited in processes as diverse as the assembly of viral capsules to the assembly of cell organelles such as the ribosome, a process whereby tens or hundreds of unique and complex elements combine together, directed entirely by their own intrinsic properties without any external intelligent guidance or control, is an achievement without any analogue in modern technology.

The well-known self-reorganizing, self-regenerating capacities of living things have been a source of wonderment since classical times—phenomena such as the growth of a complete tree from a small twig, the regeneration of the limb of a newt, the growth of a complete polyp, or a complex protozoan from tiny fragments of the intact animal. These are all phenomena without analogue in the realm of the mechanical. Imagine a space ship, a computer, or indeed any machine ever conceived, from the fantastic star ships of science fiction to the equally fantastic speculations of nanotechnology, being chopped up randomly into small fragments. Imagine every one of the fragments so produced (no two fragments will ever be the same) assembling itself into a perfect but miniaturized copy of the machine from which it originated. Nature does this constantly. It is an achievement of transcending brilliance which goes beyond the wildest dreams of even the most ardent proponents of artificial life. I doubt there is anyone who has witnessed the regeneration of a protozoan through a microscope who has not been struck with an almost metaphysical awe at the wonder of the process.[7]

The contrast between the apparent ease with which life forms assemble and replicate themselves and the absolute failure to simulate this effortless activity in any sort of nonliving artificial system is very striking. While engineers have been dreaming about the possibilities of artificial self-replicating automata over the past fifty years, advances in biology since the early fifties have gradually revealed how the miracle of self-replication is actually realized in living things.

The Structure of DNA

One of the crucial requirements for self-replication is a means of storing information in some sort of data bank. In the case of living organisms this critical function, that of the information bearer, is of course carried out by the molecule now referred to universally by the three-letter acronym DNA, for dioxyribonucleic acid. The blueprint for every organism on earth—for humans, redwoods, flies, and mushrooms—is encoded in a linear script in this remarkable polymer.

Every human body cell contains a one-meter-long string of DNA coiled into a tiny ball about 5 microns (five-thousandths of a millimeter) in diameter in the cell nucleus. The DNA polymer itself is made up of four subunits called nucleotides (see below). Each nucleotide consists of a phosphate (P), a ribose sugar (the pentagon), and one of four bases: guanine (G), cytosine (C), thymine (T), or adenine (A).

The DNA in the cell is composed of two strands—i.e., is double stranded

A short section of DNA.

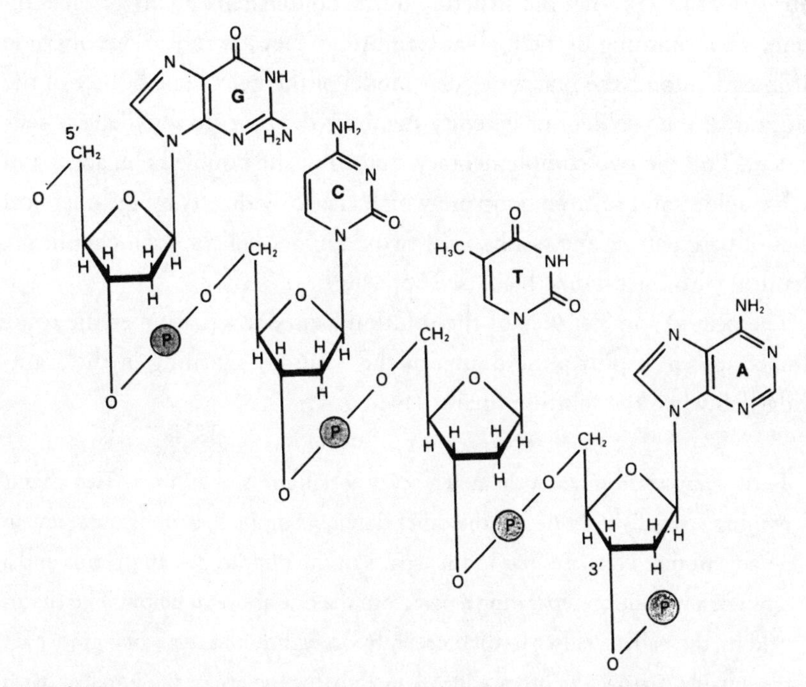

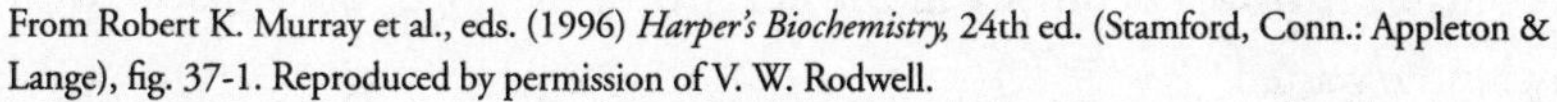
From Robert K. Murray et al., eds. (1996) *Harper's Biochemistry,* 24th ed. (Stamford, Conn.: Appleton & Lange), fig. 37-1. Reproduced by permission of V. W. Rodwell.

(see figure on page 142), and the two strands are twisted around one another to form the celebrated double helix. The genetic messages are encoded in the sequence of the bases in the molecule in exactly the same way a sequence of letters forms a word in a human language. From the initial fertilization of the egg cell every single one of the unimaginable infinity of biochemical and developmental events which shapes the growing mass of embryonic cells into a human form is under the control of the DNA master tape. It is this remarkable information bearer that has carried the human blueprint down through time, through all the generations since the birth of the human race. And it is the DNA that will carry the human blueprint forward into the distant future.

Now on any design hypothesis one would expect to find such a key component to be ideally fashioned for the biological end it serves, and in this regard DNA certainly does not disappoint.

The elucidation of the double helical structure of DNA in 1953 was undoubtedly the single most important advance in biological knowledge in the twentieth century. "Double helix" soon became a household term, as did the names of its codiscoverers, Watson and Crick. It was on Saturday morning, February 28, 1953, that the structure of the double helix finally dawned on them. That morning in their lab at Cambridge they assembled out of crude cardboard cutouts the first molecular model of the helix. The beauty of the solution to the problem of heredity manifested in the double helix is self-evident. Pull the two complementary strands of the double helix apart and each single strand forms a template which elegantly directs by the chemical rules of base pairing the synthesis of two daughter helices, each chemically identical with the original helix (see opposite).

The beauty and elegance of the solution caused a sensation at the time. Horace Judson captures the drama of the Saturday morning in the Cambridge lab when the solution finally dawned:

> Twenty angstrom units in diameter, seventy billionths of an inch. Two chains twining coaxially . . . one up, the other down, a complete turn of the screw in 34 angstroms. The bases flat in their pairs in the middle, 3.4 angstroms and a tenth of a revolution separating a pair from the one above or below. The chains held by the pairing closer to each other, by an eighth of a turn, one groove up the outside narrow, the other wide. A melody for the eye of the intellect, with not a note wasted.[8]

The replication of DNA.

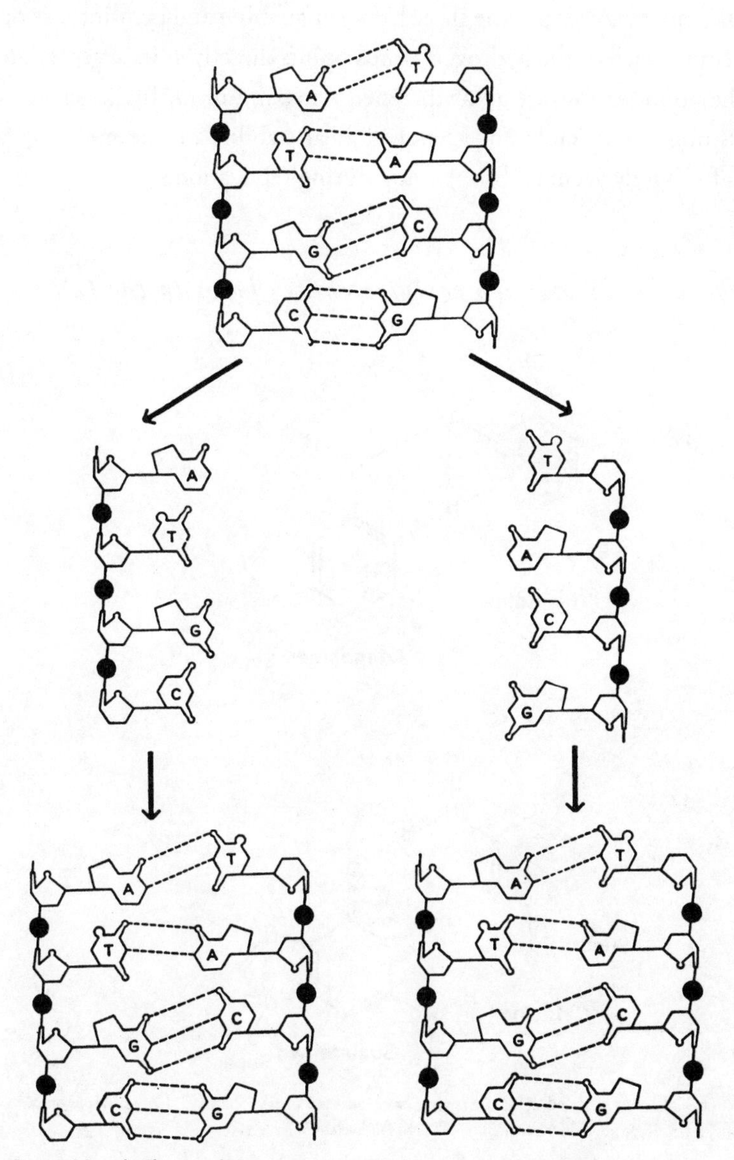

From Jacques Monod, *Chance and Necessity,* trans. A. Wainhouse; © 1972 by Alfred A. Knopf. Reproduced by permission of the publisher.

The structure of DNA was "flawlessly beautiful" according to Judson in *The Eighth Day of Creation.*[9] It had style and intrinsic beauty, according to Crick.[10]

The geometric perfection of the molecule is particularly evident in the fact that the strength of each of the five hydrogen bonds—the two between adenine and thymine and the three between guanine and cytosine—is optimal because each of the hydrogen atoms points directly at its acceptor atom and the bond lengths are all at the energy maximum for hydrogen bonds. This is most remarkable, for it confers great stability on the molecule and makes for highly accurate base pairing during replication.

Hydrogen bonds between complementary bases in the DNA.

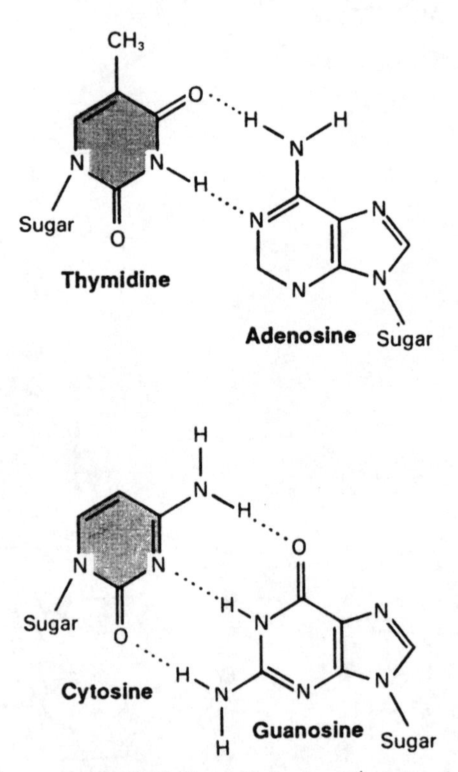

From Robert K. Murray et al., eds. (1996) *Harper's Biochemistry,* 24th ed. (Stamford, Conn.: Appleton & Lange), fig. 37-3. Reproduced by permission of V. W. Rodwell.

Metastability

Since 1953 a great deal has been learned of this remarkable molecule, and it has become increasingly apparent that it possesses additional properties, not suspected in 1953, which contribute further to its fitness for its biological role. To begin with, as befits the information bearer, DNA has turned out to be a remarkably stable molecule. Most researchers in molecular biology are familiar with the stability of DNA; it is more stable than the great majority of laboratory biochemicals. Unlike many biochemicals, it remains relatively stable in a solution, even at room temperature for months. The stability of DNA in an aqueous solution is due in part to the negatively charged phosphate groups on the backbone, which tend to retard hydrolytic breakdown of the molecule. The negatively charged phosphate groups are also necessary to maintain the solubility of DNA, which is essential for function in an aqueous environment. Recently, DNA has been extracted from the bones of Neanderthals, and some workers have claimed to have extracted it from fossil insects and leaves up to 100 million years old. The movie *Jurassic Park* may have been science fiction, but the very great stability of DNA is well established.

Although the two strands of the helix bind strongly, their affinity is not so great that they cannot be pulled apart and manipulated by the biochemical machinery of the cell. Rapid dissociation of the two strands is essential during replication. The flexibility of DNA and its ability to adopt a variety of different conformations also plays an essential role in gene expression (see below). The negatively charged phosphate groups in the backbone may again play a role here, causing a degree of repulsion between the two backbones which weakens the overall affinity of the two strands. Obviously, there is a conflict between the need for stability—essential for the genetic repository—and the need for flexibility and conformational variety—essential for gene expression and replication. It is likely that the strength of binding of the two strands in the DNA helix must be very close to what it is for biological function. If it were any stronger, both strands would be frozen into an immobile, lifeless embrace. But if it were weaker, the molecule would fall apart. The biological function of DNA is therefore dependent on the molecule possessing a certain metastability—a term, which, as we saw earlier, was used by Haldane to describe the basic character of all organic molecules.

Compaction

Another feature of DNA which contributes to its fitness is its remarkable compacting capacity. The amount of information carried in the chromosomes of higher organisms like man is very great, and it requires an enormous length of DNA to encode it. In man, for example, this requires a length of DNA 1 meter long. Yet, as mentioned above, this 1-meter-long molecule is compacted into a tiny ball less than 5-thousandths of a millimeter in diameter.[11] How is this achieved? Anyone who has worked with DNA will be aware that DNA solutions of approximately 1 milligram of DNA per milliliter of solution are very viscous and difficult to manipulate. Yet in the cell nucleus the concentration of DNA is about a thousand times greater. This super packing capacity is possible because one of the hidden talents of DNA helices is their capacity to twist and bend into superhelices, and these superhelices can be bent into higher-order helices, and so on, thereby permitting the highly dense packaging that is actually observed in the cell nucleus.

Recently, the structure of one of the fundamental compacting units in the nucleus—the nucleosome—was worked out.[12] It consists of a coil of DNA wrapped around a cluster of proteins known as histones. Interestingly, the negatively charged phosphate backbone and the small groove of the helix are both critically involved in binding the helix to the histone core. So it seems that the negative charge on the phosphates has other roles in addition to protecting the DNA from hydrolysis and lowering the binding affinities of the two strands to a level commensurate with biological function.

The ability of DNA to store information is so efficient that all the information needed to specify an organism as complex as a man weighs less than a few trillionths of a gram. The information necessary to specify the design of all the organisms which have ever existed on the planet, a number, according to G. G. Simpson,[13] of approximately 1 billion, could be easily compacted into an object the size of a grain of salt!

The compacting ability of DNA is critical to its biological role. Many processes in which DNA is involved, such as meiosis and mitosis, would be impractical if the DNA molecule could not be tightly compacted. If DNA did not have the compacting capacity it does, the cell system would have to be radically redesigned. The cell, for example, might have to be much larger to accommodate a vast tangle of disordered DNA fibers. However, there are diffusional constraints on cell size. Cells cannot be much larger, as they are dependent on diffusion for their supply of oxygen and nutrients. As we have seen in chapter 2, diffusion is only efficient over distances not much greater than the average diameter of the cell. The diffusional constraint on cell size suggests that the compactness of DNA makes a vital contribution to its biological fitness.

In addition to its transmission from one generation to another, the information in any molecule playing the genetic role would also have to be retrievable or accessible.

How Information Is Retrieved

In the case of DNA, this means that components of the cell must be able to recognize specific regions of the helix. These components are of course the proteins. Over the last decade it has become clear that specific proteins recognize particular sections of the helix by feeling for the unique electrostatic shape of target sequences within the major and minor grooves of the helix. Just how proteins recognize sections of DNA is a fascinating story in itself, which, as we will see in the next chapter, depends on a set of coadaptations between proteins and DNA.

When the problem of protein-DNA recognition was first considered in the early 1960s, it was a matter of some speculation as to how it worked. The trouble was that the helix was believed to be so regular that it was hard to see how one part would differ sufficiently from another to allow recognition to occur. We now know that because of its basic metastability (discussed above) different regions of the helix exhibit a vast number of unique variations in structure and conformation.

To begin with, there is a considerable degree of electrostatic variability in the major groove which proteins can "feel" when searching for particular base sequences. Moreover, all along the helix there are minor structural variations caused by differences in the actual base sequence itself. Groove width, local twist, displacement of average base pair from the helical axis—all vary in different sections of the molecule. Moreover, DNA, because of its metastability, is also more flexible than was originally assumed and can adopt a number of different conformations. These various departures from "helical perfection" confer unique stereochemical properties to different sections of the DNA helix which can be recognized by proteins, thus providing an additional crucial element to the fitness of DNA for its biological role—making the information stored in its base sequence more readily decipherable.[14]

So DNA is not only fit for self-replication and for the transmission of information, but it also lends itself to information retrieval because of the many subtle minor distortions which it can undergo and which can be recognized by proteins that interact with specific sections of the DNA.

It is evident, then, that DNA has not one, but many properties which are wonderfully fit for its role as the genetic molecule: (1) the essential double helical structure, which is fit for self-replication and for the transmission of genetic information, (2) great chemical stability in water, (3) a metastable character consequent on the relatively low binding affinity of the two strands, which assists the machinery of the cell in pulling apart the helix—for example during replication—and which confers flexibility on the molecule permitting it to adopt a variety of alternative shapes which are critical to gene expression, (4) the tiny distortions along the length of the helix—another consequence of its metastability—which greatly facilitates information retrieval by proteins, (5) the ability to be superfolded and -compressed into highly compact structures, which allows the storage of massive amounts of information in very tiny volumes (an essential requirement if it is to perform the genetic role in complex multicellular forms of life).

RNA

RNA is also a nucleic acid polymer that is very closely related chemically to DNA. In the cell, RNA carries the genetic message from the DNA in the nucleus to the cytoplasm, where the message is translated.

A short section of RNA.

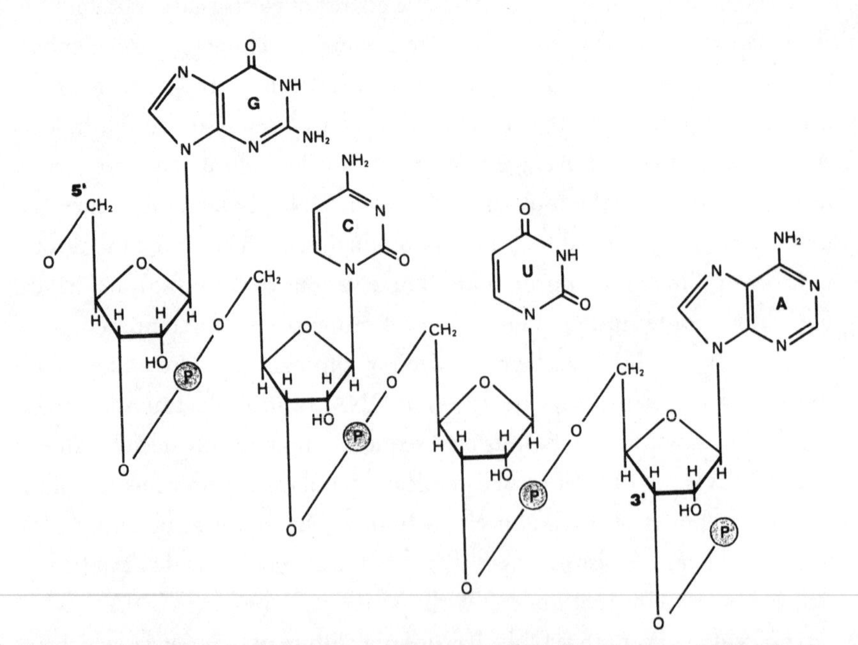

From Robert K. Murray et al., eds. (1996) *Harper's Biochemistry,* 24th ed. (Stamford, Conn.: Appleton & Lange), fig. 37-6. Reproduced by permission of V. W. Rodwell.

The flow of information from DNA via RNA to protein.

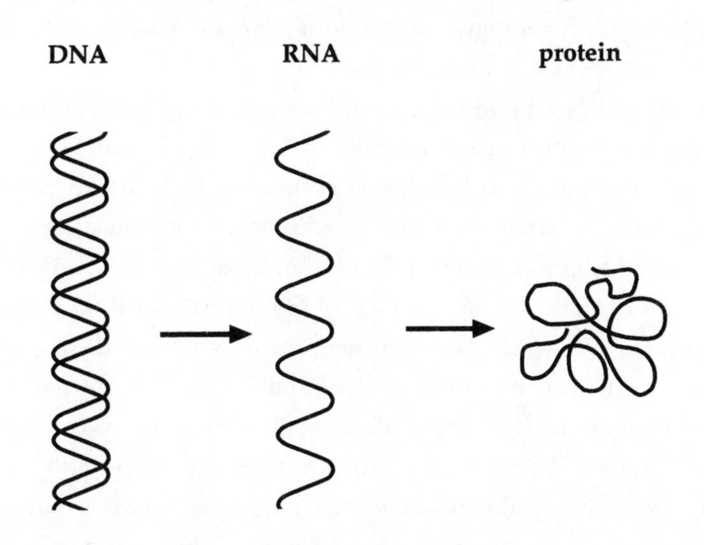

Although the chemical differences between DNA and RNA are very small, they have a significant influence on their chemical properties. Double-stranded RNA polymers exhibit fewer of the minor structural distortions which make the DNA helix so ideal for protein recognition.[15] Also, the double-stranded RNA helix has a very deep major groove, which is less accessible to proteins than in the case of DNA.[16] This means that the base sequence in a section of double-stranded RNA cannot be recognized as easily by proteins as it can in DNA. Consequently, information encoded in sequences of double-stranded RNA is less accessible or decipherable than in DNA. And as the double-stranded structure is the most chemically stable form of either DNA or RNA, then on this count DNA is clearly fitter than RNA to serve as the ultimate repository of genetic information, which must be both stable and accessible.

However, protein recognition of particular RNA molecules, or sections of particular RNA molecules, is vital if RNA is to fulfill its biological role in the cell. It is proteins which handle and process the RNA and assist in the transport of RNA molecules from nucleus to cytoplasm. But where RNA is less fit than DNA for protein recognition in the superstable double helical conformation, in the cell much of the *RNA is single-stranded* and proteins therefore do not have to recognize sections of RNA molecules in the hard to recognize, double-helical form. Single stranded RNA folds into complex 3-D conformations and it is mainly these shapes which are recognized by proteins. The RNA molecule is not only a "tape" but a "tape" which folds into a "shape." In this combination of information and conformation RNA is perfectly adapted for its biological role.

Moreover, RNA, by virtue of its additional hydroxyl group, is more reactive than DNA and therefore less stable and again less fit than DNA to serve as the ultimate repository of genetic information. Anyone who has had any experience in preparing DNA and RNA in a laboratory will attest to the relative instability of RNA compared to DNA. Because of the ability of single-stranded molecules to adopt complex 3-D conformations, RNA is capable of catalytic activities which cannot be performed by the stiff double-stranded DNA helix. Remarkably, these catalytic abilities enable RNA molecules to change their chemical structure, allowing them to convert themselves, during processing in the cell nucleus, from the large initial copy of the DNA sequence to the far smaller messenger RNA sequence, which is translated by the ribosome into the amino acid sequence of a protein (see below).

As more is learned of these two molecules, it is becoming increasingly apparent how their many subtle structural differences are adapted for their respective biological roles—DNA as the permanent accessible stable information store of the cell and RNA as the transitory carrier of the genetic information from the DNA to the translational machinery.[17]

The formation of mRNA from the larger initial transcript of the DNA.

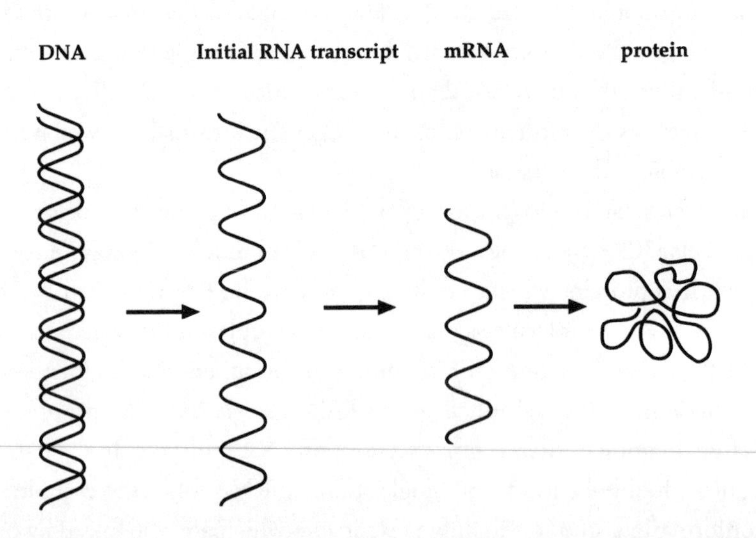

Possible Alternatives

Although it is clear that both DNA and RNA are wonderfully fit for their respective biological roles, the question remains, are they uniquely fit? Or might there be other candidate information carriers even fitter than either DNA or RNA? Is every chemical detail of the two molecules essential for biological function? Could different bases be used? Could different sugars be used? Could the phosphate be replaced by some other joining molecule? None of these questions can be answered definitely, but the evidence suggests that any change would be detrimental and no other polymers are known which possess precisely the chemical and physical properties of DNA and RNA.

DNA and RNA are two members of a vast class of closely related nucleic acid polymers, some of which have been recently synthesized. Some have hexoses (a six-carbon sugar) instead of riboses (a five-carbon sugar) in the backbone, others such as the peptide nucleic acids (PNA) have a peptide backbone linked together by amide bonds.[18] The study of these "alternatives" is preliminary. However, they have already provided evidence that the fidelity of base pairing in DNA and RNA is probably dependent on the sugar being a ribose rather than a hexose. And in the case of one recently synthesized "alternative RNA" in which a slightly different ribose is incorporated, the strands are less flexible than in the case of natural RNA, and this may reduce its fitness for biological function.[19] In the PNAs the binding affinities of the two strands may be too strong for biological function. As mentioned above, it is likely that the precise binding strength and flexibility of the strands in DNA and RNA are critical for biological function, conferring on the molecules a crucial "metastability" which is not possessed by any of the alternative closely related nucleic acid polymers.

One fascinating aspect of this work is the possibility that some of the various alternative helical replicators, such as the PNAs, may have been used by very primitive life before the current DNA-RNA system evolved. They may even have played a crucial role in the origin of life. If indeed some alternative DNA-like replicators were utilized by early life on earth, then on purely selectionist grounds it is hard to see how they could have been superior and yet selected against and thus have lost out in the evolutionary race.

Different Bases

Some years ago one of the world authorities on DNA structure pointed out: "It is sufficient to note that there are *very few* additional complementary pairs which can fit into the DNA."[20] Recently, chemists have incorporated some of these additional base pairs into DNA. However, although the evidence is not conclusive, it seems unlikely that DNA constructed out of any of the unnatural base pairs would be as biologically fit. Studies of alternative base-pairing schemes involving various unnatural bases in DNA helices are very preliminary but tend to support the view that DNA constructed out of these unnatural bases would probably not be as chemically stable nor as faithfully copied as DNA constructed out of the 4 natural bases.[21] In the view of Wolfram Saenger, one of the world's leading authorities on nucleic acid chemistry, the 4 natural bases are superior in many critical ways compared with the various unnatural alternatives.[22] Theoretical consideration of the chemical characteristics of the natural and alternative unnatural bases in Saenger's words makes it "clear why nature has selected the four natural bases as primary tools."[23] That there may be only these 4 bases available has important implications related to the encoding of information for the specification of proteins and for the retrieval of genetic information by proteins from DNA.

Although the current evidence is insufficient to establish absolutely the unique fitness of DNA and RNA for their respective biological roles, all the available evidence is at least consistent with this position. According to Wolfram Saenger, the available evidence strongly suggests that the natural nucleic acids are optimally fit for their respective biological roles and any chemical change to the phosphate radical, the riboses, or to any of the 4 bases used would certainly decrease greatly their biological fitness and may even abolish it. (See endnote for Saenger's summary of the fitness of the natural nucleic acids.)[24]

The Fitness of the Code

Every high-school student knows that the linear nucleotide sequence of a gene contains a message specifying the amino acid sequence of a protein, and that this message is decoded via a system of rules, the genetic code, which relate particular DNA sequences 3 bases long, known as codons, to one of the 20 amino acids used in the construction of proteins (the structure and functioning of proteins is dealt with in the next chapter).

Is the genetic code uniquely and ideally adapted for its biological role? Is it, in other words, the best possible coding system we can envisage? Or might an alternative coding system have been possible using more or less than the 4 bases in DNA or using a different set of amino acids? This is a problematical area and no clear answer can be given at present. However, some of the possible alternative arrangements would appear to be less fit.

The genetic code.

Codon	Amino acid	Codon	Amino acid	Codon	Amino acid	Codon	Amino acid
UUU		UCU		UAU		UGU	
UUC	Phenylalanine	UCC	Serine	UAC	Tyrosine	UGC	Cysteine
UUA		UCA		UAA	Stop	UGA	Stop
UUG		UCG		UAG		UGG	Tryptophan
CUU	Leucine	CCU		CAU	Histidine	CGU	
CUC		CCC	Proline	CAC		CGC	Arginine
CUA		CCA		CAA	Glutamine	CGA	
CUG		CCG		CAG		CGG	
AUU		ACU		AAU	Asparagine	AGU	Serine
AUC	Isoleucine	ACC	Threonine	AAC		AGC	
AUA		ACA		AAA	Lysine	AGA	Arginine
AUG	Methionine or start	ACG		AAG		AGG	
GUU	Valine	GCU		GAU	Aspartic acid	GGU	
GUC		GCC	Alanine	GAC		GGC	Glycine
GUA	Valine or start	GCA		GAA	Glutamic acid	GGA	
GUG		GCG		GAG		GGG	

Note that because of its primary base-paired structure DNA can only be constructed with pairs of complementary bases. Therefore, the number of bases used in its construction has to be an even number. The consequences of using only 2 bases (i.e., 1 base pair) rather than the 4 actually used in natural DNA was raised briefly by Alexander Rich: "We may ask why nucleic acids have four units in them at the present time. . . . In order to contain information, it is obvious that two bases would be enough; for example, simply adenine and thymine. A primitive organism whose nucleic acid contained only two complementary bases could still develop a similar type of biochemical system for information transfer."[25]

Consider a DNA molecule made up of only 2 bases, adenine and thymine. It would be formed of (A-T) base pairs as shown below:

DNA constructed out of two bases: A and T.

A--T
A--T
T--A
A--T
T--A
T--A
T--A
A--T
A--T

The table below shows the number of codons available, either 2, 3, 4, 5, or 6 nucleotides long, in a coding system using only 2 bases:

Genetic coding systems utilizing two bases.

Coding system using 2 nucleotide bases					
Number of nucleotides in codon	2	3	4	5	6
Number of codons	4	8	16	32	64

From the above table we can see that to code for 20 amino acids the codons would have to be at least 5 bases long. A potential problem would be that with only 32 different codons such a system would not provide many alternative codons for the same amino acid. Most amino acids would be specified by only one codon. The actual 4-base coding system involves a relatively high level of redundancy, so that most of the amino acids are specified by more than 1 codon.

The redundancy of the code was a considerable puzzle in the early 1960s when the details of the genetic code were first worked out. Since then, as knowledge of gene expression has increased, it has become increasingly clear that without the flexibility the redundancy confers, the embedding of additional information in protein coding sequences related to sophisticated types of gene regulation would be impossible.

To achieve the same level of redundancy with a 2-base system, we would have to increase the length of the codons to 6 bases long. This would provide the same level of redundancy as in the existing code. However, the genes and messenger RNA molecules in such a system would be twice as long and energy required for protein synthesis would be doubled. The transfer RNA molecules (the molecules which recognize particular codons and transfer the appropriate amino acid to the growing polypeptide chain) would in all probability have to be larger, as might the entire protein synthetic apparatus. The process of protein synthesis would in all probability be slowed down considerably. In short, such a system would be more complicated than the existing system, cost twice as much in terms of materials and energy, and provide no obvious advantage.

What would the consequences be of using a coding system utilizing more than 4 bases—6, 8, or 10, perhaps? A hypothetical double-stranded DNA molecule constructed out of 3 different base pairs, including the 2 base pairs A-T and G-C used in natural DNA and a third hypothetical base pair X-Y, is shown below:

DNA constructed out of six bases: A, T, G, C, X, and Y.

C--G
T--A
X--Y
G--C
C--G
T--A
Y--X
T--A
C--G

Although it is possible to imagine a coding system using 6 bases, it is difficult to see how it would be advantageous compared with the existing 4-base system. The table below shows the number of codons available, either 2, 3, or 4 nucleotides long, in a coding system using 6 bases:

Genetic coding system utilizing six bases.

Coding system using 6 nucleotide bases			
Number of nucleotides in codon	2	3	4
Number of codons	36	216	1296

The use of 6 bases would clearly involve certain problems. If the codons in a 6-base system were 2 bases long, this would provide 36 different codons sufficient to specify 20 amino acids. However, 32 codons may not provide the necessary element of redundancy. Moreover, the fidelity of a theoretical decoding mechanism based on matching only 2 base pairs would probably be lower than the existing system.

On the other hand, if three nucleotides were used in a 6-base system, this would mean that 216 codons would be available for specifying amino acids. If all these 216 codons were to be used, the complexity of the transitional system would be considerably increased, involving four times the number of transfer RNA molecules. But if a proportion of codons were not used to specify amino acids and were in other words nonsense codons, this would vastly increase the chance of a mutation destroying the meaning of the genetic message. The use of 6 different nucleotides would also involve additional metabolic pathways for their synthesis, as well as complicating the mechanism of DNA synthesis.

An imaginary 6-base system is in any case only hypothetical because, as was mentioned above, on the evidence available, nature has not provided an extra base pair capable of quite the same perfect base pairing as that which occurs between the A-T and G-C base pairs in natural DNA.

We could continue the game by considering other possible coding systems. However, given our current understanding of protein chemistry, it is increasingly hard to escape the conclusion that we need no more than 20 amino acids to specify for proteins with a diverse range of functional properties. To specify for 20 amino acids in a DNA sequence, the current system based on 4 nucleotides, consisting of 64 codons made up of 3 nucleotides each, with a considerable degree of redundancy, is just about the most elegant solution possible.

It is fortunate, then, that nature provides 2 unique base pairs for the construction of the DNA helix. If there was only 1 base pair available, then the utility of DNA for encoding information for the specification of proteins made out of 20 amino acids would be greatly diminished. That there are in all probability no additional base pairs available with the same ideal qualities as the 2 actually utilized in the structure of DNA is no drawback. Because unless more than 20 amino acids were required for some reason—and it is difficult to imagine what this might be, given the fantastic functional diversity of natural proteins—no obvious advantage would accrue by increasing the coding potential of DNA by incorporating additional base pairs.

If the coding system can be flawed in any way, it is in the curious variation in the number of codons specifying different amino acids. The amino acid serine, for example, has 6 codons, while methionine and tryptophan have only one. Yet serine is not used six times more frequently in proteins than tryptophan or methionine. Most other amino acids are specified by 2 or 4 codons. But again, there does not seem to be any exact relationship between number of codons and frequency of occurrence of an amino acid in proteins. Does this mean that the code is less than ideal, or is it possible in this case that there may be a number of equally "ideal" alternatives so that no particular system is preferable? A graph plotting all possible coding arrangements against their utility would contain multiple peaks.

However, from the evidence currently available we cannot be sure that the codon assignments are not in fact optimized for a variety of functional reasons which have not yet been clearly defined. The apparently excessive number of codons for serine and threonine may be related to the fact that these amino acids are favored for phosphorylation, which involves the addition of a phosphate group to a protein. Phosphorylation changes the activity of a protein and is used widely by cells as a regulatory device. The fact that proline has 4 codons may relate to its helix-breaking properties (a probable reason for its original choice as one of the 20 amino acids in the first place—see the discussion in the next chapter). Before concluding that the code is not maximally fit, we should remember also that the origin of the coding system and its early evolution are still mysterious. It may be that there is a reason for these apparent anomalies rooted in as yet undiscovered necessities associated with the evolution of the code.

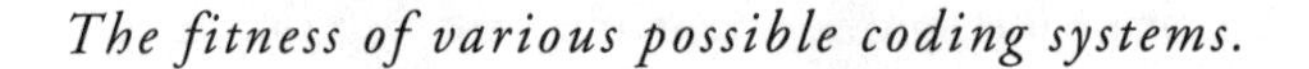

The fitness of various possible coding systems.

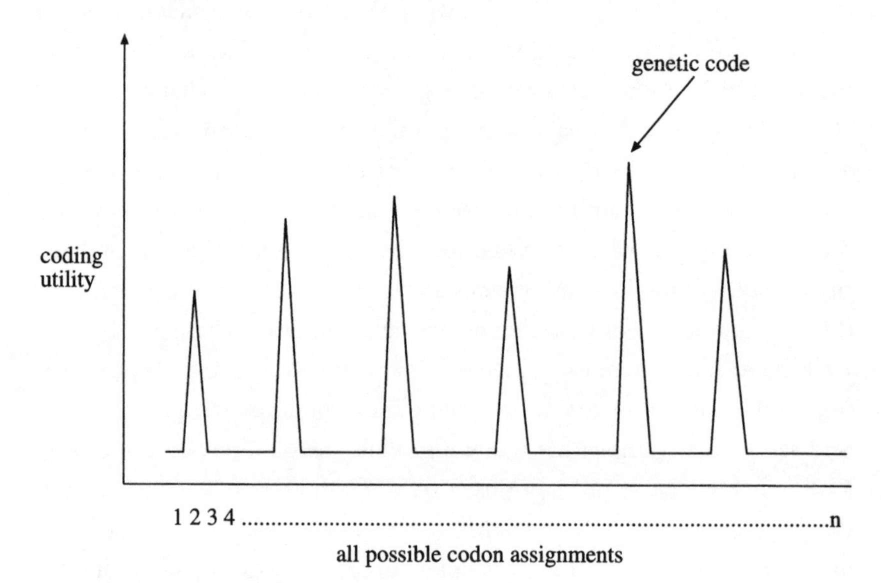

Conclusion

If life is the result of design, then every component must be perfectly fit for the end it serves. There can be no exceptions. If the genetic code is indeed less than optimum, then the entire teleological worldview collapses. Fortunately, in the case of the code we still have insufficient knowledge of protein structure and function to judge the code as clearly "nonoptimal." Our knowledge of evolution is also incomplete. In the case of DNA and RNA, we are on safer ground; nearly everything we have learned since 1953 is at least consistent with the possibility that DNA and RNA are both ideally and uniquely fit.

And DNA may be fit for its biological role in other ways of which at present we have only the haziest notion. For example, DNA can form many other conformations in addition to the double helix. It can also form what are called cruciform structures and the so-called triple helices. Triplex DNA is particularly intriguing, as there are a number of possible biological processes in which it could function, such as in recombination and in regulating gene expression.[26] Then there is the equally intriguing possibility raised recently by a paper in *Science* that DNA may provide the basis for a subcellular computing system.[27]

Although as we have seen in this chapter, DNA molecules are wonderfully fit to perform the function of information carrier, such long, relatively stiff polymers are not fit to manipulate individual atoms or molecules by making and breaking specific chemical bonds. DNA cannot in itself carry out all the various sophisticated manipulative and structural functions which must inevitably be performed in any self-replicating system in the process of copying itself. It cannot in itself perform the task of "constructor device."

This essential need for automated constructor devices or robots to carry out the instructions encoded in the memory bank is one of the major unsolved problems in the design of artificial self-replicating machines. It is certainly very difficult to imagine how any sort of self-replicating machine would be feasible and function unless its component "constructor or manipulator devices" (capable of reading and decoding the message in the data bank and carrying out its instructions) were self-assembling to a very large degree. Clearly, the subproblem of designing self-assembling devices will have to be solved before anything approaching genuine artificial self-replication can be attempted. The daunting nature of this subproblem was nicely captured by Ted Kaeler, a contributor to a recent symposium on nanotechnology, when he described the current predicament as an "impasse or wall": "I view the problem of developing a proto assembler as a *wall.* We are on one side of the wall and on the other side of the wall there is an assembler that can make other assemblers."[28]

The biological solution to the "constructor device problem," the way through the wall, is of course to be found in the characteristics and properties of a remarkable class of self-assembling biopolymers—the proteins. As I shall try to explain in the following chapter, as the universal nanoconstructors in a self-replicating automaton, they have no peer. They represent a solution of surpassing brilliance to von Neumann's problem of the universal constructor.

Chapter 8

The Nanomanipulators

In which it is argued that no other class of polymers are known which are as fit as proteins for the central biological role they play in living systems. The functional and structural properties of proteins are astoundingly diverse, and in addition, proteins are capable of self-assembly. Because of their ability to adopt alternative shapes, the biological activities of proteins can be finely regulated—a phenomenon known as allostery. The mutual adaptations of proteins and DNA are also examined, including the fit of the α helix into the large groove of the DNA and the fact that the a helix can "feel" about 4 bases in the DNA. It is concluded that the evidence is consistent with the possibility that the DNA-protein partnership is uniquely fit for its biochemical role.

The atomic structure of the protein cytochrome

Showing all the atoms in the protein except for hydrogen. The protein is made up of a long chain of 104 building blocks known as amino acids. Embedded in the center of the protein is one heme molecule, which is a planar cyclic compound made up of nitrogen and carbon atoms and containing at the center of the ring one iron atom (Fe). Altogether, the protein contains about 1,000 atoms and the complexity of their spatial arrangement is apparent. In the cell, cytochrome *c* is found in the mitochondrion and forms part of what is known as the respiratory assembly—a set of proteins involved in generating energy from oxidative metabolism. Its major function is to shuttle electrons across the membrane surrounding the mitochondrion. The protein measures about 5 nanometers, or five-millionths of a millimeter, across.

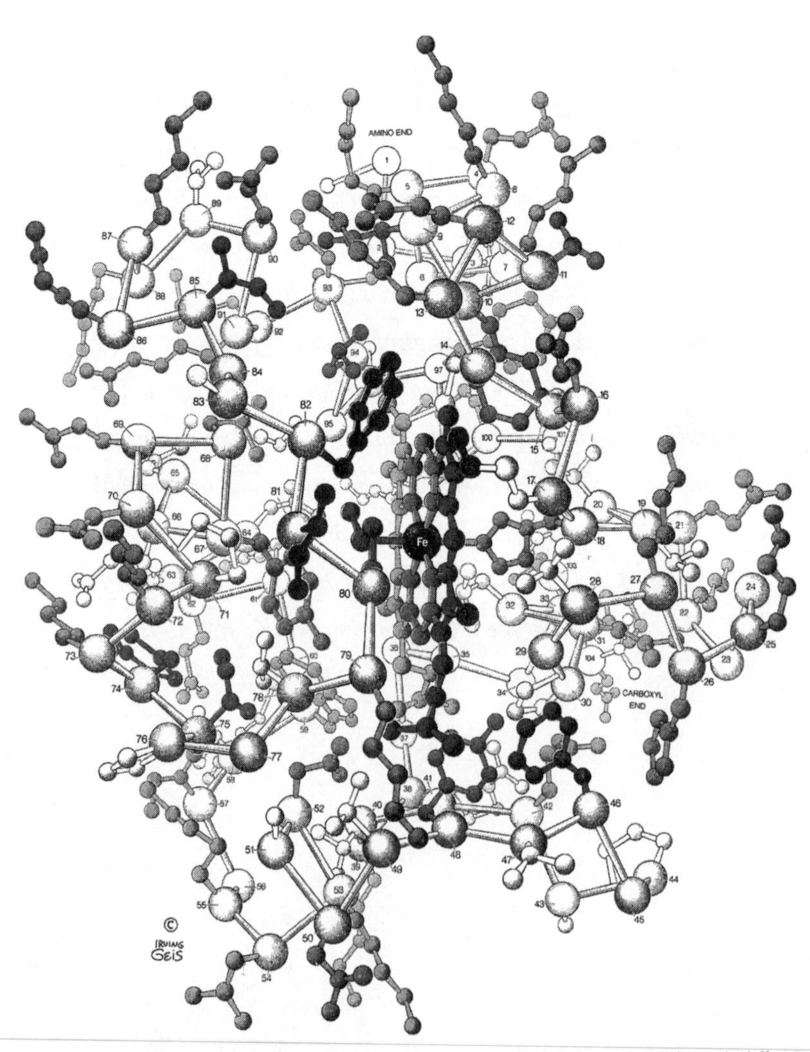

From Geoffrey, Zubay, *Biochemistry,* 2nd ed. © 1989, p. 59; reprinted by Prentice-Hall, Englewood Cliffs, N.J., p. 59. Reproduced by permission of the publisher.

We would like to think ourselves necessary, inevitable, ordained from all eternity. All religions, nearly all philosophies, and even a part of science testify to the unwearying, heroic effort of mankind desperately denying its own contingency.

—Jacques Monod, *Chance and Necessity,* 1972

It struck me recently that one should really consider the sequence of a protein molecule, about to fold into a precise geometric form, as a line of melody written in canon form and so designed by Nature to fold back upon itself, creating harmonic chords of interaction consistent with biological function. One might carry the analogy further by suggesting that the kinds of chords formed in a protein (incorrectly folded) . . . are dissonant, but that, by giving an opportunity for rearrangement . . . they modulate to give the pleasing harmonics of the native molecule. Whether or not some conclusion can be drawn about the greater thermodynamic stability of Mozart's over Schoenberg's music is something I will leave to the philosophers of the audience.

—Christian Anfinsen, *New Perspectives in Biology,* 1964

Where DNA is the data bank of life, the ultimate repository of all biological information, the proteins are life's animated actors, the universal constructor devices, the nanomanipulators which translate the one-dimensional DNA dream into the vital three-dimensional reality of the cell. By reading and following the instructions in the DNA, the proteins manipulate the atoms and molecules of life into the trillions of unique and specific conformations upon which the miracle of self-replication and self-assembly depends.

As the constructor devices of the cell, they represent what is in effect the realization of von Neumann's dream. As much as DNA, they are the secret of life. Without what Jacques Monod called their "demoniacal functions,"[1] we would certainly not be here to marvel at the sheer brilliance of the solution they represent.

For over 4 billion years, these tiny nanoconstructors have been reading DNA scripts. Building atom by atom, they assembled over eons of time every living structure that ever existed on earth. They built the first cell; they built the human brain; they erected the dinosaurs and all past life on earth. All the vital chemical functions of every cell on earth are all dependent on the activities of these tiny nanomachines. The living kingdom, in a very real sense, has its being and origin in the infinitely rich, ever-changing patterns woven from the interactions of these tiny fragile collections of dancing atoms, a billion times smaller than the tiniest visible speck of dust.

It we think of the cell as being analogous to a factory, then the proteins can be thought of as analogous to the machines on the factory floor that carry out individually or in groups all the essential activities on which the life of the cell depends, apart from the transmission of genetic information, which as we have seen is the role of the DNA. Proteins are also the basic structural building blocks of the cell, for it is largely by the association of different proteins that all the forms and structures of living things are generated. In terms of the factory, proteins form not only the machines but the walls, roof, floor, stairwell, and doors as well.

Protein Diversity and Versatility

The diversity of the structural and functional properties of proteins is astounding. Some proteins form the hard Teflon-like materials which make up hair, nails, and feathers. Others form tough nylonlike materials which make up the tendons that attach muscles to bone and the fibrous sheaths which encase the various compartments and organs in the body. Some form the rubberlike elastic materials that surround the major arteries in the body, while others maintain the smooth elasticity of skin. Still others make up the transparent materials which form the lens of the eye.

In addition to their structural roles, proteins play an infinite variety of functional roles in the cell. They can act as catalysts speeding up the rates of chemical reactions billions of times. Working closely together in teams, proteins can build up all the chemical components of the cell, including complex lipids and carbohydrates. As well as building things up, proteins can utilize their catalytic powers to break down the cells' macromolecular constituents back into simple organic compounds. It is through the catalytic activity of proteins that cells derive their energy. It is through these same catalytic powers that the energy of the sun, trapped by the chloroplasts, is converted into reduced carbon fuels, the basic energy source of life.

It is proteins that form the essential components of the contractile assemblies in the muscles, providing organisms with the capability of movement. It is proteins which comprise the basic building blocks of the tubule system of the cell, providing the scaffold which determines the shape of the cell. These same tubules also provide the cell with a system of interconnecting conduits for the movement of the cells' constituents in an orderly fashion around the cell. In addition to forming the structural basis of the cells' transport system, it is proteins which also perform the role of transporters, selectively binding to certain chemicals at particular stations in the cell and then releasing them at other sites. Proteins also play the role of chemical messengers, being manufactured in one location and then traveling to another site where they bind to some other molecule to cause some appropriate chemical response. It is proteins which are also the receptors of chemical messages, selectively binding to "messenger molecules" (which are often other proteins) and responding to the arrival of the "message" by bringing about, generally via specific interactions with other proteins, changes to the functional state of the cell. Proteins also form the gates and pumps that control the passage of innumerable different types of chemicals through the cell membrane, either by opening and closing chemical channels or by actively pumping chemicals from one side to another. The list of structural and functional properties of proteins is virtually endless.

Although some of the properties of proteins are equaled by particular types of polymers or other classes of compounds. For example, in its elasticity and strength, collagen resembles nylon; the toughness of keratin, the protein which makes up nails and hair, is rivaled by the carbohydrate polymer chitin; the transparency of the crystallines of the eye is like that of the plastic Perspex; and so forth. However, no other individual class of molecules is known which possesses even remotely such a diversity of properties. Moreover, their catalytic properties are effectively unparalleled by any other type of molecule. So are their powers of molecular discrimination whereby one particular protein molecule is able to interact with unerring specificity with another specific molecule in the cell.

The very great structural and functional diversity of proteins is one of the key characteristics of these remarkable molecules which contributes to their unique fitness as the molecular constructor devices of the cell. To appreciate more fully some of the other characteristics of proteins that tailor them so superbly for their biological role as the working components of life, it is necessary to digress here and review some basic aspects of their chemical design for those readers unfamiliar with this area of biochemistry.

Protein Structure

It is immediately obvious even to someone without any previous experience in molecular biology that the arrangement of the atoms in a protein (see figure 8.1) is unlike any ordinary machine built or conceived by man, or indeed unlike any object of common experience with which we are familiar. On superficial observation one is immediately struck by the apparent illogic of the design and the lack of any obvious modularity or regularity. The sheer chaos of the arrangement of the atoms conveys an almost eerie, otherworldly impression.

A similar feeling of the strangeness and chaos of the arrangement of atoms in a protein struck the researchers at Cambridge University after the molecular structure of the first protein, myoglobin, had been determined in 1957. As one researcher put it at the time:

> Perhaps the most remarkable features of the molecule are its complexity and its lack of symmetry. The arrangement seems to be almost totally lacking in the kind of regularities which one instinctively anticipates, and it is more complicated than had been predicted by any theory of protein structure.[2]

Ten years later another member of the original Cambridge team confessed that the diversity and complexity of the 3-D atomic configuration of proteins is "so baffling that it stops protein crystallographers remembering the conformation of any protein but their own."[3]

Despite the complexity of proteins, their basic chemical structure is relatively simple. All are polymers built up from twenty different small organic molecules called amino acids. The figure below shows the chemical structures of four amino acids. Each amino acid contains an amino group (NH_2) and a carboxyl acid group (COOH) linked by a carbon atom (the α carbon atom), as well as a unique side chain.

Because of their different side chains each amino acid has different chemical properties. Some amino acids, like alanine in the diagram below, have uncharged side chains so are relatively insoluble in water. These are the hydrophobic, or water-avoiding, amino acids. Other amino acids have charged side chains and are soluble and hydrophilic. In a protein the amino acids are

Symbolic diagrams of four amino acids.

Note that the molecular skeleton enclosed within the dotted frame is the same in all. The attached groups, or radicals, outside the chain give each amino acid its unique chemical characteristics. Glycine, the simplest amino acid, has one hydrogen atom outside the frame. In alanine, the hydrogen atom is replaced by a CH_3 group. In cysteine, a sulfur atom has been incorporated with the CH_3 group. In tyrosine, a benzene ring has been incorporated.

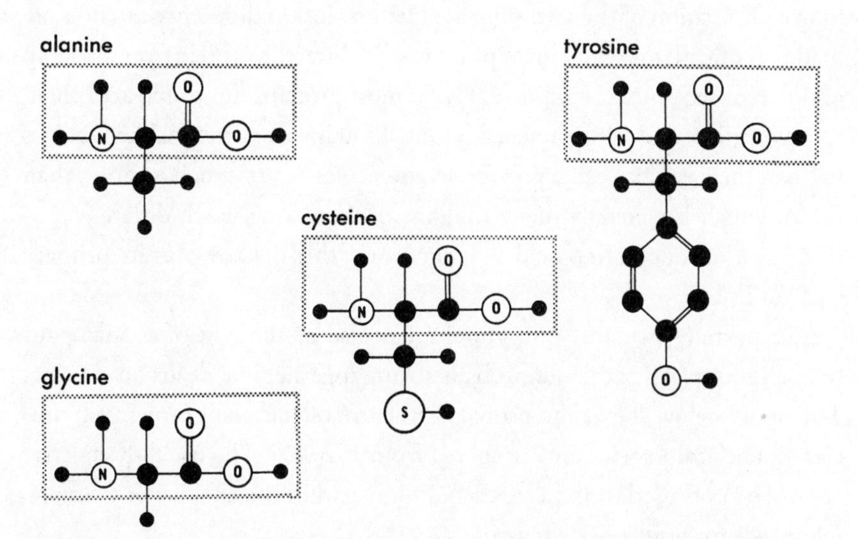

From N. J. Berrill (1996) *Biology in Action* (New York: Dodd, Mead), fig. 3.10. Reproduced by permission of N. J. Berrill.

linked by their amino and carboxyl acid groups to form a long linear polymer which is known as the primary structure of the protein. The figure below shows a short section of a polypeptide chain.

A short section of the amino acid backbone of a protein. Showing the linked carbon (large filled circles) and nitrogen atoms (N) forming the backbone or chain and the attached hydrogen (small filled circles) oxygen (O) and amino acid radicals (R).

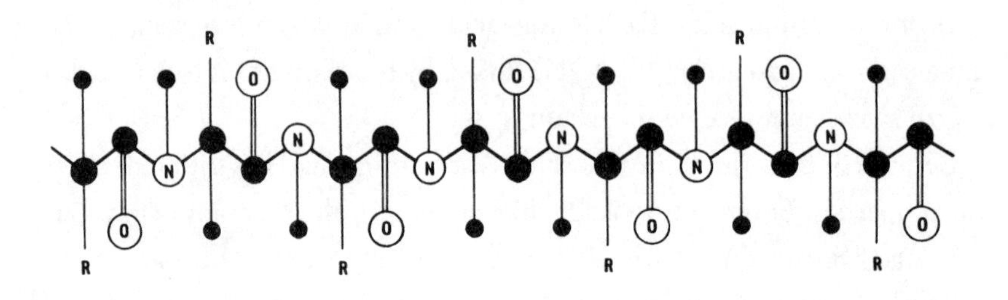

From N. J. Berrill (1996) *Biology in Action* (New York: Dodd, Mead), fig. 3.11. Reproduced by permission of N. J. Berrill.

The backbone, which is formed by linkage of the amino acid and carboxylic acid groups, is identical throughout the molecule. It is the unique side groups that jut out from the backbone which confer different chemical properties to different regions of the amino acid chain. The linear sequence of amino acids in a protein can be thought of as a sentence made up from a long combination of the 20 amino acid letters. Just as different sentences are made up of different sequences of letters, so different proteins are made up of different sequences of amino acids. In most proteins the amino acid chain is between 100 and 500 amino acids long. Proteins are therefore like DNA—in that they are basically polymeric molecules—but much shorter than DNA, which is generally many millions of units long. Each different protein has a unique amino acid sequence, and this is known as its primary structure.

The primary structure of a protein may also be thought of as analogous to a series of plastic table-tennis balls strung together like beads on a string. The figure below shows the primary structure of the small protein ribonuclease. The amino acids are numbered from 1 to 124. The usual three-letter abbreviation is used to indicate each amino acid in the chain. For example, Lys stands for lysine; Ser, for serine, etc.

The primary structure of the protein ribonuclease.

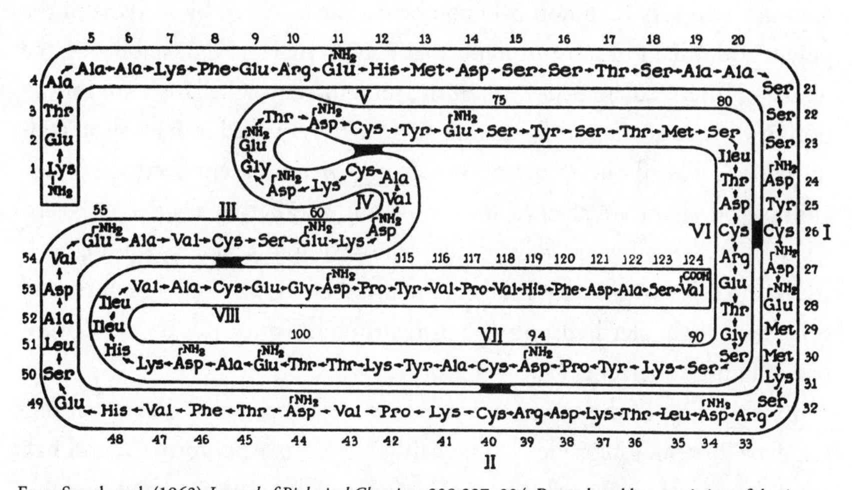

From Smyth et al. (1963) *Journal of Biological Chemistry* 238:227–234. Reproduced by permission of the American Society for Biochemistry and Molecular Biology.

Could Nature Have Chosen Other Amino Acids?

At this point, let us stop to consider an obvious question. Over two hundred amino acids occur in nature. The number of amino acids which are theoretically possible is enormous. Even in the human body many amino acids that do not occur in proteins are utilized for a variety of metabolic functions. If the laws of nature are programmed for life as we know it, there should be something special about the choice of these particular twenty amino acids. As discussed in the previous chapter, in the case of DNA it is relatively easy to rationalize the choice of the four bases on grounds of their superior fitness. However, in the case of the twenty amino acids, it is very difficult to judge whether the structure or function of proteins would have been better if a different selection of amino acids had been chosen for their construction.

From our current understanding of the structure and function of proteins, it is possible to list at least some of the criteria that must be satisfied in the choice of amino acids used if an amino acid sequence is to fold into a stable form. There is clearly a requirement that a proportion of the amino acids possess hydrophobic side chains and that these chains be not too large or they would be impossible to pack neatly together into the protein interior. There is a requirement that a fraction of the amino acids have side chains which specifically favor what are known as α helix and β sheet formations, two very common 3-D conformations adopted by sections of the polypeptide chain in most proteins that play a crucial role in stabilizing the final and intermediate 3-D form of the protein. The α helix is a conformation in which a section of the amino acid chain is twisted into a helical conformation. The β sheet formation occurs when adjacent sections of the amino acid chain are stacked on top of one another in a series of layers. There is the requirement that some of the hydrophilic side chains be positively and others negatively charged. Finally, there is the requirement that all side chains, whether hydrophobic or hydrophilic, must not be too disruptive of either α helix or β sheet conformations.

Examination of the twenty amino acids used in proteins shows that several have hydrophobic side chains, half are good α helix formers, and half are β sheet formers. The choice of side chains looks to be just about right. It is less easy to rationalize the choice of individual amino acids. However, the inclusion of the smallest and simplest amino acid, glycine, may be necessary because of its essential role in one of the most important of all proteins—collagen—which forms the fibrous matrix that binds and holds together all the cells and tissues of multicellular organisms. Every third amino acid in the coiled polypeptide chains of collagen is glycine, and its small size plays an essential role in the design of the collagen molecule because it allows the coiled chains to wrap tightly together, conferring on the collagen fibers far greater tensile strength than would be possible if glycine were replaced by any other amino acid with a longer side chain. Without glycine, collagen fibers might not possess the requisite strength to glue together the cells of multicellular life forms. The sulfur-containing amino acid cysteine may also be essential because of the transitory role played by disulfide bonds in stabilizing transient intermediate conformations via which many proteins fold into their final native form.[4] The choice of serine, threonine, and tyrosine may have been dictated because their hydroxyl groups lend themselves to phosphorylation, one of the key mechanisms whereby the cell regulates the activities of protein. Proline may be included because it is a helix breaker. Moreover proline is also used in the construction of collagen in which it may play, like glycine, an essential role.

Finally, in this context it is interesting to note that the amino acids in proteins are often modified chemically in a variety of ways after the amino acid chain has folded into its native form. In effect, because of these chemical modifications, nature has available far more amino acids than the twenty actually used as the basic building blocks of proteins.

Protein Folding

The chemical bonds which link the successive carbon atoms in the backbone of the protein are known as covalent bonds. As discussed in chapter 5, covalent bonds are formed when atoms share electrons to complete electron shells. And as explained, nearly all the atoms in the organic compounds utilized in living organisms—sugars, amino acids, fats, the nucleotide bases in DNA, etc.—are linked together by covalent bonds. There are, however, as also mentioned in chapter 5, another class of chemical bonds which do not involve sharing electrons but arise out of weaker electrostatic forces between adjacent atoms, and these are known as noncovalent bonds, or weak chemical bonds. The linear chain of amino acids folds automatically under the influence of these weak electrochemical forces into the complex three-dimensional aggregate of atoms.

In terms of the "string of plastic balls" analogy, we can think of the folded three-dimensional conformation of a protein as being analogous to the string of balls tightly folded together to form a globular mass of balls in which each ball is stuck loosely to one or several of its neighbors with lengths of tape. The tape represents the weak chemical bonds that hold together the various parts of the protein structure.

Folding occurs in such a way so as to bring about the maximum number of favorable atomic interactions between the various constituent amino acids. During the folding, which takes a fraction of a second, negatively charged groups tend to associate with positively charged groups, and hydrophobic side chains tend to stack at the center of the molecule while hydrophilic side chains tend to arrange themselves on the surface in contact with water. The final three-dimensional arrangement of the atoms is dictated directly by the primary amino acid sequence. The figure below shows the folded chains of a protein in its native conformation.

Because the unfolded amino acid chain is capable of folding into its native form in vitro without assistance from any other component of the cell in which it is synthesized, this must mean that the information or direction for the folding process is contained in the amino acid sequence. This is somewhat analogous to the situation in origami, where all the instructions

The weak chemical bonds which hold a protein in its native 3-D conformation.

The hydrophobic core, shaded, consists mainly of hydrophobic amino acids.

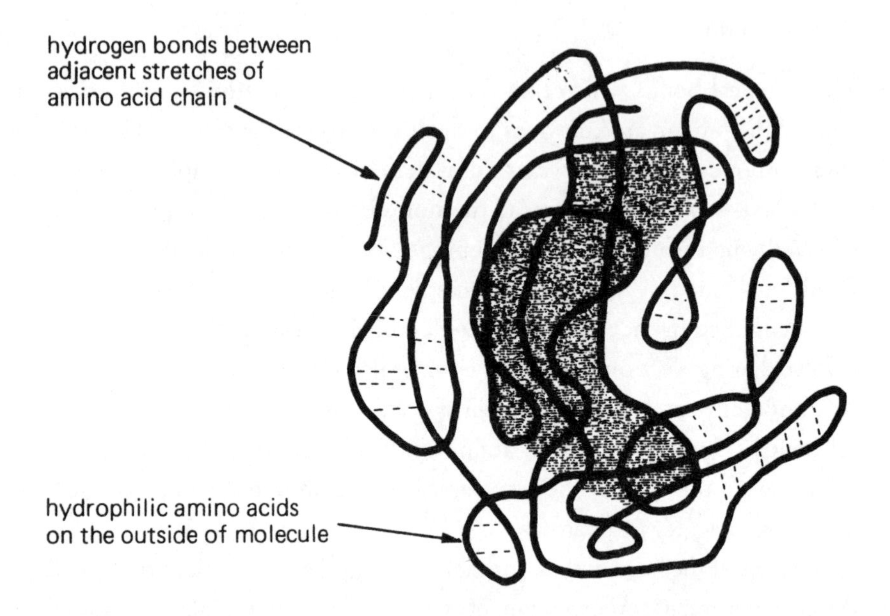

necessary to reach the final folded 3-D form are contained in a series of lines drawn on the initial 2-D piece of flat paper.

The mechanism by which proteins achieve the extraordinary end of self-assembly is a subject of intense research. Already it is clear that the folding process is complex and depends on global cooperative interactions, which involve a high proportion of all the amino acids scattered all along the sequence of the chain. Recent evidence suggests that the folding may occur (again, as in origami) through highly specific or even unique pathways involving partially folded intermediates which exhibit (as in origami) nonnative conformations, i.e., conformations which are not present in the final native folded form. This must mean that in many biological proteins the amino acid sequence not only specifies the native conformation but also specifies a folding pathway which includes in some proteins the specification of partially folded nonnative intermediate conformations; and these, like stepping-stones, lead the folding process along an energetically favorable route to a magic disclosure, the exquisite beauty of the three-dimensional native form of the protein.

The ability of proteins to assemble themselves automatically is a key capability which is essential to their biological role as the atomic manipulators or constructor devices of the cell. No sort of self-replicating machine could function unless its component machinery was self-assembling. The inability to specify completely feasible self-assembling robots able to assemble themselves *without assistance from any external agent* lies at the heart of the failure to achieve self-replication in artificial systems.

Proteins are not the only polymers which can fold themselves into specific three-dimensional forms. Such a capacity is inherent in many complex organic molecules. Recently, organic chemists have created a variety of polymers, known as "foldamers," which can mimic proteins in folding themselves into specific 3-D forms. Moreover, it seems likely that artificial enzymes and foldamers mimicking some of the structural roles of proteins may be created in the next ten years. But even if artificial proteinlike foldamers are achieved, it seems likely that they will be very heterogeneous in terms of subunits and chemical design. The study of these alternative polymers is still in its infancy, but there are no grounds for believing that any one single generic class of foldamers will exhibit the complexity, the range of chemical and structural abilities, and the precise degree of metastability essential for allostery (a subject discussed below) of natural proteins. It seems likely that, as in the case of the study of alternative DNA polymers, study of these foldamers will merely underline the unique fitness of proteins for their biological role.

Within the context of current scientific knowledge proteins are, as far as we know, the only available molecular constructor devices possessing, first, the capacity to carry out a vast diversity of structural and functional chemical roles, involving every imaginable type of specific atomic and molecular manipulations and, second, the capacity to assemble themselves automatically without the help of an external agent.

There is a fascinating aspect to the story of protein folding. The hydrophobic force not only folds the protein into its native conformation, but by packing together the hydrocarbon side chains of the hydrophobic amino acids, it creates a nonaqueous microenvironment in the center of the protein. This nonaqueous core has enormous functional significance. It provides a chamber where various chemical reactions may be carried out which would be impossible or very difficult in an aqueous medium (including the syntheses of the hydrophobic amino acids and lipids). In effect, those amino acids that are only sparingly soluble in water are forced by water into a tight water-avoiding ball. This ball provides the environment in which synthetic reactions, particularly condensation reactions that involve the removal of a molecule of water and which are almost impossible to carry out in an aqueous medium, can be easily performed. The two defects of water mentioned in chapter 2—that it is unable to dissolve lipids and compounds containing hydrocarbon chains and that it is a poor medium for carrying out organic syntheses—are both overcome in the center of a protein. One defect—the insolubility of hydrocarbons—is, as it were, utilized to overcome the other!

The Relative Strength of Weak and Strong Bonds

As the constructor devices of the cell, it is the proteins that carry out all the atomic manipulations upon which life depends. To carry out these nanomanipulations proteins must necessarily associate intimately with other molecules in the cell. The molecule that a protein associates with, whether it is a small molecule like an amino acid, or a large molecule like another protein, is termed a "ligand."

Nearly all these associations between a protein and its ligand are formed by the weak chemical bonds. As we have seen (in chapter 5 and in the discussion above), it is these bonds which hold biomolecules, including proteins, in their characteristic or native three-dimensional form. Because of the weakness of these bonds a stable interaction between a protein and a ligand can only occur if this involves multiple weak interactions.

The strength of the weak bonds is obviously critical to the ability of the proteins to interact selectively with other ligands in the cell. If the weak

The ligand binding site of a protein.

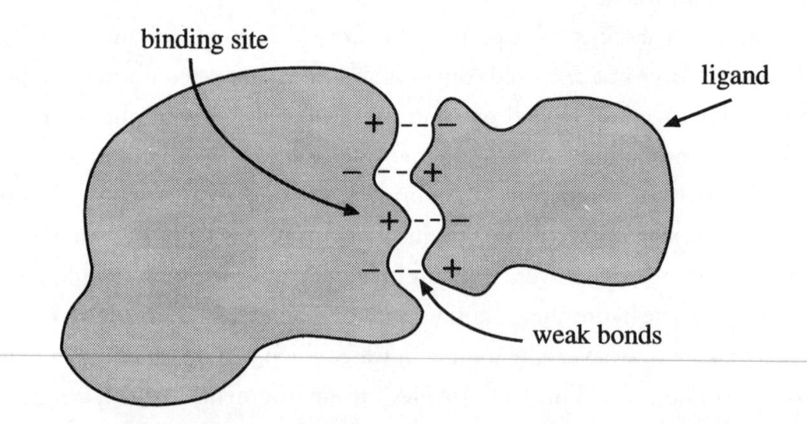

bonds were weaker, then no protein would be able to bind specifically to any other molecule in the cell.

If on the other hand these bonds were stronger, then proteins and their ligands would be bound so strongly that, once in place, they could never be removed. Consequently, the rapid association and dissociation of the protein and ligand on which so many protein functions, such as enzymic functions, depend would be impossible. In effect, proteins and all the constituents of the cell would be frozen into rigid immobile structures. The low diffusion rates would be incompatible with cellular existence. As Watson points out in *Molecular Biology of the Gene,* the low energy levels of these bonds are precisely what is required for enzymic function:

> Enzyme-substrate complexes can be both made and broken apart rapidly as a result of random thermal movement. This fact explains why enzymes can function so quickly, sometimes as often as 10^6 times per second. If enzymes were bound to their substrates by more powerful bonds they would act more slowly.[5]

Just as the strong and weak atomic forces must be exactly as they are, so both the absolute and the relative strength of the weak and strong chemical bonds must be very close to what they are or biochemistry would be impossible. In fact, the weak bonds are about twenty times weaker than the strong. This 20-to-1 ratio is vital because it means that for the majority of substrates the combined energy of binding to the catalytic site is close to that of an individual covalent bond. Which means that enzyme-substrate interactions have sufficient energy to strain and break individual covalent bonds within the substrate, thus making it possible for enzymes to carry out the various specific atomic manipulations upon which the life of the cell depends.

If we imagine a world in which the ratio between the strength of covalent and weak bonds was, say, 2 to 1 rather than 20 to 1, then the weak interactions would tend to rip the strong bonds apart. Pulling off and reattaching our tape would break the very string holding our plastic balls together.

On the other hand, if the ratio between the strength of covalent and weak bonds was increased to 200 to 1, combinations of weak bonds would be incapable of achieving the necessary energy levels to strain, break, and form specific covalent bonds within the substrate. The atomic manipulation of matter by proteins would be impossible.

The Miracle of Self-Regulation: Allostery

It is because the energy levels of the weak bonds, which hold protein molecules together in their characteristic three-dimensional form, are so low that proteins exhibit their characteristic "metastability," which is manifest in their fragility and instability in the face of very minor physical or chemical challenges. Increase their temperature a few degrees, and they unfold. Change the chemical character of the medium they are in ever so slightly, and they, again, unravel. Attach another molecule to their surface, and they change shape. Proteins are stable, but only just. They are delicately balanced, on the threshold of chaos.

Precisely because of their metastability and the weakness of the interactions that hold them on the edge of chaos, the conformation of a protein can easily be altered if it binds to another molecule. Any such interaction will cause molecular distortions which will be transmitted throughout the molecule. These discrete conformational changes effect the functioning of the protein. In the case of an enzyme, they often result in significant changes in its activity.

These reversible conformational changes which are transmitted through a protein when it binds to another molecule are the basis of allostery. Via such changes, the function of a protein may be modulated by association or dissociation with other molecules in the cell. An important aspect of this phenomenon is that the interactions are between the protein and each ligand separately and not between the ligands themselves. Consequently, the protein is able to integrate information from several different chemical inputs, each being determined by the concentration in the cell of a particular chemical. Thus allostery confers on proteins a remarkable dual ability—to carry out unique chemical reactions while at the same time integrating information about the concentration of various chemicals in the cell, and to respond intelligently to this information by turning up or down their own enzymic activity. In this, allosteric enzymes function like electronic relays.

Jacques Monod, who discovered the phenomenon, was not exaggerating when he called it "the second secret of life." For it is hard to exaggerate the significance of the *dual functionality* this phenomenon confers on these remarkable microminiaturized machines. Proteins are not only capable of carrying out a specific chemical reaction but are also able to integrate and intelligently respond to changes in their chemical environment. In other words, the functional units that carry out the basic chemical processes are also regulatory units. It is this dual functionality which is crucial to the coherent functioning of the cell, allowing it to avoid the chemical chaos that would ensue were enzymic activity not precisely tuned to the ever-changing requirements of the cell.

In this remarkable dual ability, to combine the role of microprocessor and functional machine in the same object, proteins are far in advance of any artificial device. Invariably, in artificial machines even the most advanced modules which regulate and integrate the function of the machine are separate from the working parts. In an oven there is a thermostat (regulator) and a heating device (functional unit). In a protein they are one and the same.

By sensing the concentration of molecules one or several chemical steps removed from the reaction catalyzed by the protein itself, and then responding to this information intelligently by increasing or decreasing its own activity, allosteric proteins are able to control the flow of metabolites along a metabolic pathway. It is a vast integrated network of such proteins in the cell that in Jacques Monod's words "guarantees the functional coherence of the intracellular chemical machinery."[6]

If it were not for this remarkable phenomenon, the control and regulation of the cell's metabolic activities would seem to present almost insuperable problems. It would appear that the self-regulatory capacity of proteins is not a gratuitous characteristic but rather a necessity. For at present it is impossible to envisage any alternative regulatory system which could maintain the homeostatic state of cellular metabolism at such a peak of efficiency and coherence. If proteins were incapable of self-regulation, we would have to envisage a vast, almost infinite regress of molecular control devices external to and separate from the individual enzymes which carry out the actual chemical work of the cell. Even if one were to postulate such a cybernetic system, it is hard to see how the actual activity of individual enzymes could be modulated other than by the conformational transitions actually utilized in allosteric enzymes to turn up or down their catalytic activities.

Could There Be Other Proteins

On any design hypothesis the specific proteins utilized by living organisms should be the fittest available candidates for their specific biological roles. Hemoglobin, for example, should be the fittest oxygen-transporting protein. Collagen should be the fittest structural protein for binding together the cells and tissues of multicellular organisms. While there could perhaps be "alternative" proteins of very different amino acid sequence and perhaps even basic design which are functionally equivalent to hemoglobin or collagen, if the teleological position is correct, then no "alternatives" should be fitter than the natural products. Unfortunately, protein chemistry is not sufficiently advanced to provide any clear answers. Consequently, the teleological position cannot be subjected to a vigorous test.

Recent advances in protein chemistry have revealed that the millions of different functional proteins appear to fall into about a thousand major families which share basic structural characteristics. But it is still uncertain whether these families represent a major fraction of all possible structural forms or just a small subset. Again, our knowledge is too limited.

RNA

During the 1980s the unexpected finding was made by Thomas Cech that RNA molecules could act like enzymes and catalyze chemical reactions. Since then a considerable number of RNA-catalyzed reactions have been documented.[7] However, despite their many catalytic capabilities, it seems very unlikely that RNA molecules could carry out the vast diversity of biological functions carried out by proteins. For example, many synthetic reactions catalyzed by proteins are carried out in hydrophobic niches where water is excluded and RNA molecules are unable to form large nonpolar niches. The chemical properties of RNA molecules are also less diverse, being constructed out of only four bases, while proteins are made up of twenty different amino acids. Also, as the authors of a recent review point out, RNA is far less fit for allosteric regulation than proteins.[8] RNA strands are also far less amenable to being folded into complex, compact molecular structures. Although we cannot completely exclude the possibility, it seems extremely unlikely RNA could substitute for proteins in the great majority of diverse devices and materials used in living systems.

Although RNA molecules cannot compete with modern proteins, RNA, because it can both carry information and function as an enzyme, may have served the function of both DNA and proteins in the most primitive cells shortly after the origin of life. It is an intriguing fact that RNA molecules are particularly suited for carrying out manipulations of RNA molecules—self-manipulations—and in a world where the cell was largely a collection of RNA molecules, this may have been an ability of critical significance.

Could Nanotechnology Provide an Alternative to Proteins

The uniqueness of proteins as atom-manipulating assemblers was also highlighted in a recent *Scientific American* review of the latest state of nanotechnology and the immense technical difficulties that will have to be overcome before any artificial device remotely as capable of manipulating atoms and molecules could be constructed.[9] The review showed just how far nanotechnology is at present from creating artificial nanoassemblers capable of carrying out the sorts of atomic or molecular manipulating tasks carried out with such effortless efficiency by biological proteins. According to George M. Whitesides, an authority on molecular self-assembly who was cited in the article, the nanoists' dream of self-replication is "at the moment . . . pretty much science fiction. . . . Even after a fair amount of thought, there is no way that one could see of connecting this idea to what we know [now] or can even project into the foreseeable future."[10] The article also cites the chemist David E. H. Jones, best known perhaps as the author of the "Daedalus" column in *Nature* who has "provided a pointed critique of the idea that individual atoms could serve as constructor elements in the ultimate erector set."[11] Jones points out just how challenging are the many difficulties that need to be overcome before an artificial nanoerector could be built: "Single atoms of the more structurally useful elements at or near room temperature are amazingly mobile and reactive . . . they will combine instantly with ambient air, water, each other, the fluid supporting the assemblers or the assemblers themselves."[12] Jones believes the advocates of nanotechnology fail to take into account many critical questions, such as: How would the assemblers obtain information about which atom is where in order to manipulate it? How would the assemblers know where they are in order to navigate from the atom supply point to the correct position in which to place the atom? As Jones concludes: "Until these questions are properly formulated and answered, nanotechnology need not be taken seriously." [13] As the author of the article concludes, in the face of so many seemingly intractable problems, the nanotechnology dream of constructing an artificial self-assembling atom-manipulating device currently resembles a form of postmodern alchemy: "just another latter day cargo cult."[14]

Peerless Molecular Machines

From this very brief review of the properties of these remarkable nanomanipulators, we can conclude that, as candidates for the basic constructor role in a self-replicating nanomachine, proteins are fit, first, because of their functional and structural diversity (a remarkable enough characteristic given that all proteins are short polymers composed of a string of a hundred or so amino acids), second, because of their ability to assemble themselves, and, third, because they can perform the additional miracle, through the phenomenon of allostery, of regulating their own activities as well as integrating their individual activities with other proteins and enzymes to create a cybernetic control network of unparalleled elegance and efficiency.

In the entire realm of science no class of molecule is currently known which can remotely compete with proteins. *It seems increasingly unlikely that the abilities of proteins could be realized to the same degree in any other material form.* Proteins are not only unique, but give every impression of being ideally adapted for their role as the universal constructor devices of the cell. Again, we have an example in which the only feasible candidate for a particular biological role gives every impression of being supremely fit for that role.

And there is yet another, final aspect to the remarkable fitness of proteins.

The Fitness of Proteins for DNA Recognition

While proteins are wonderfully fit for the role of the constructor devices, performing all the various structural and functional activities which must be performed during the replicative cycle of the cell, they are not suited to performing the genetic role. It is only because, first, the information necessary to specify a protein can be encoded in the DNA sequence, and second, the information stored in the DNA can be readily retrieved and decoded by proteins, that the replication of living systems can proceed. Neither of these tasks could be achieved were it not for the fact that both molecules exhibit a set of remarkable adaptations which marvelously tailor them to function together.

To begin with, DNA and proteins are both linear polymers made up of a limited number of subunits, which means that the sequence in one can be readily translated into the sequence of the other. If proteins had been more complex structures, say, branched polymers like polysaccharides, they could not have been readily encoded in the DNA or in any other linear type of molecule. No matter how fit as atomic manipulators, they would not be fit for encoding in DNA.

From first principles, proteins, because of their limitless variety, functional diversity and inherent flexibility, and their ability to undergo allosteric transitions, etc., are obvious candidates for the crucial function of recognizing and binding to a particular section of a DNA molecule. But on top of their general properties, which tailor them so superbly for almost any conceivable biochemical task, there is a fascinating and highly specific aspect of protein structure that appears to fit them precisely for DNA recognition: the fact that the α helix of a protein, one of the most common conformations found in proteins, fits almost perfectly into the major groove of the DNA helix.[15]

As mentioned above, the α helix conformation is a region of the amino acid chain of a protein which is twisted into a helical conformation. There are about 3.6 amino acids per turn of the helix—about 18 amino acids in five turns. The figure below shows a short stretch of an α helix, showing the carbon and nitrogen atoms of the amino acid backbone. The carbon atoms

The α helix.

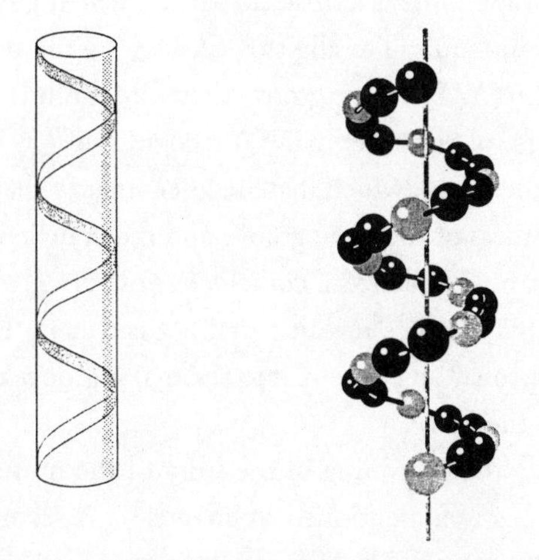

From figure 2.38 in W. T. Keeton (1976) *Biological Science,* 3rd ed. (New York: Norton). Reproduced by permission of the publisher.

are shown in black. Note the atoms of the backbone follow a helical path. The side groups of the amino acids (not shown in the figure) project outward from the central backbone.

The fact that one of the most fundamental protein conformations fits very neatly into the large groove of the DNA obviously greatly facilitates protein DNA recognition because it allows the protein to have intimate access to the DNA sequence. Of course, there are many other conditions which must be met in addition to this coincidence if proteins are to be able to recognize particular sequences in the DNA.

To begin with, if a protein is "to read" a particular base sequence in a particular region of the DNA, the protein must be able to distinguish between the different base pairs along the helix. Of course, the protein cannot actually "see" but must feel the sequence of the DNA like a blind person reading braille until it finds (feels) the sequence it is looking for. It turns out, and this is surely another coincidence of great significance, that of the two grooves in the DNA, the major and the minor, *it is the major groove*—the one into which the α helix happens to fit so perfectly—that provides hydrogen bond patterns which are distinctive for each of the four base pairs and can therefore be felt by the α helix. (The minor-groove hydrogen bond patterns are not so distinctive.) So the actual base sequence of the DNA can be "felt" most readily by a protein feeling the sequence in the major groove. In effect, the major groove is fit for protein recognition not only because its dimensions match that of the α helix but also because in the large groove each base pair presents a unique electrostatic pattern which greatly facilitates sequence recognition—quite literally the "feeling" by the α helix of the base sequence of the DNA. The large groove therefore exhibits two independent adaptations for its role in protein-DNA recognition—its electrostatic variability and its dimensions, which match closely that of the α helix.

The mutual fitness of the large groove and the α helix for DNA-protein recognition must be considered a coincidence of very great significance, as recent work in this area has revealed that a great many DNA-recognizing proteins insert a protruding α helix into the major groove of the DNA helix when binding to the DNA.

But this is only the beginning of the story of the mutual fitness of proteins and DNA. As was mentioned in chapter 7, it seems that nature has provided only 4 bases capable of forming geometrically perfect base pairs and possessing the required chemical stability to function in the genetic tape. Another coincidence is that an α helix can "feel" no more than about 4 contiguous bases in the large groove of the DNA.[16]

Recent work has confirmed that in the case of most DNA recognition proteins there are generally 4 or less contiguous bases which are critical in DNA recognition by an α helix, except for one important set of DNA-recognizing proteins that use a special type of helix, known as a zinc finger. Zinc fingers are used very widely as DNA-recognizing motifs by proteins in higher organisms. As two authorities comment in *Science*: "The helix of each zinc finger fits directly into the major groove of the DNA, and the side chains from the amino-terminal portion of the helix contact the edges of the base pairs. . . . Each finger makes its primary contacts along a section of the large groove three base pairs long."[17] From detailed study of DNA-protein interactions over the past ten years, it appears, as a rule, that either three or four contiguous base pairs are involved in direct binding of α helices to their DNA target sequences.

The four bases in the DNA can be arranged in 4^4, or 256, different base sequences 4 bases long (quadruplets)—i.e., ATAT, GTAT, TGCT, TCTT, etc. This raises the interesting possibility that perhaps each different DNA sequence 4 bases long, or quadruplets, might be recognized by a specific α helix—in other words, that there might be a code relating each unique quadruplet of bases in the DNA to a particular α helix. However recent work suggest this is unlikely to be the case.[18]

The fact that DNA is made up of 4 bases has further implications. From simple combinatorial considerations it is easy to show that using 4 bases, unique recognition sequences in the genomes of higher organisms must be about 15 bases long. Remarkably, as Mark Ptashne shows in his book *A Genetic Switch,* specific reversible binding of a protein to a particular DNA sequence requires about 10 to 18 weak interactions, and it turns out that this is just about the number of weak bonds that occur when alpha helices of a recognition protein associate with a DNA sequence about 15 bases in length. Using 2 bases, recognition sequences would have to be 20 to 30 bases long and the number of bonds involved would be approximately doubled. This would mean that, given the strength of the bonds, the binding by proteins to such extra-long recognition sites would be so strong that it would be in effect irreversible. Using 6 bases, if nature had provided them, simple calculation shows that it would be possible, in higher organisms, to have unique recognition sequences only 8 bases long. This gains no advantage, however, because a sequence 8 bases long is too short to provide the requisite number of weak interactions to tie a recognition protein to a specific sequence in the DNA.[19]

From this brief review of protein-DNA interactions and the discussion of the code in the previous chapter, we are led to an intriguing conclusion—that the use of 4 bases for the construction of the DNA double helix is fit for two completely independent biological functions: First, for a coding system (the genetic code) to specify proteins composed of twenty amino acids, and second, for DNA-protein recognition, involving the recognition by proteins (using mainly α helices) of unique target sequences in the genome.

The protein-DNA recognition system contains a particularly intriguing play on the number four. In the largest genomes unique combinations of 4 bases are about 15 bases long. As we have seen above, given the existing energy levels of the weak chemical bonds involved in protein-DNA binding, protein recognition complexes can bind reversibly to DNA sequences up to this length but not to lengths much greater. We have also seen that because of the natural twist in the DNA double helix, protein recognition motifs such as the α helix can only feel along about 4 bases in the DNA double helix. It has often been said that God is a mathematician; on the evidence of molecular biology we might add that He is keen on the number *four.*

Conclusion

Everything that has been learned about the chemical and physical properties of DNA and protein since the early 1950s increasingly confirms the wonderful fitness of these two remarkable molecules for their respective biological roles in the replicative cycle. The number and complexity of their mutual adaptations is growing continually as biological knowledge advances. But already the picture is impressive enough: the mutual fit of the α helix into the large groove of the DNA; the fact that both DNA and proteins are linear polymers so that the information in a DNA sequence can be translated via a coding system into the amino acid sequence of a protein; the fact that the four bases confer geometric perfection and great chemical stability on the DNA helix; the fact that four bases seems to be the ideal number for two different coding systems—the well-known genetic code specifying for the 20 amino acids in proteins and the DNA-protein recognition system whereby proteins are able to recognize unique DNA sequences long enough to function as unique target sequences in the genome; the fantastic diversity of proteins and their ability to regulate their own activities; the fact that the energy levels of the weak interactions are at precisely the level needed to confer on proteins their metastable character and to bind reversibly to unique DNA sequences and thereby to retrieve the information in the genes.

We have seen that, in the case of water, the carbon atom, the process of oxidation, the light of main sequence stars, the earth's hydrosphere, etc., the evidence suggests strongly that each is uniquely and optimally fit for its particular biological role. If the teleological position is correct, the DNA-protein system should also be uniquely and maximally fit for the advanced type of cellular life that exists on earth today. Note that the teleological position does not imply that *all* self-replicating chemical systems will necessarily utilize or depend on this particular partnership, that self-replication can only be achieved using DNA and protein. The early evolution of life, for example, may have proceeded via a series of simpler replicating systems—which contained neither DNA nor proteins—including some based entirely on RNA or RNA analogues. Teleology only implies that the partnership should be uniquely fit for the self-replication of a biochemical system as sophisticated and complex as the current cell system. And the evidence is certainly consistent with such a conclusion. Considering the bewildering suite of mutual adaptations—discussed above and in the previous chapter—it seems hardly conceivable that there could be any other two molecules as mutually fit, or more perfectly adapted to play the fundamental roles of "information bearer" and "constructor device" in a self-replicating automaton as complex and intricate as the cell.

Chapter 9

The Fitness of the Metals

In which the unique chemical properties of the metals are examined. In keeping with the concept that the cosmos is uniquely fit for carbon-based life, living things utilize the properties of metals from each of the main subgroups of the periodic table, and even particular metals, such as iron, calcium, copper, molybdenum, and magnesium, appear to be adapted for specific biological processes of a critical significance without which no world of life remotely as rich as ours would be possible. Iron and copper are essential for the manipulation of oxygen, molybdenum for nitrogen fixation, etc. It is concluded that there could be no biology or biosphere without metals.

The Hot Springs of the Waikato in New Zealand

From Oliver Goldsmith (1876) *A History of the Earth and Animated Nature.* Hot springs, geysers, and volcanoes provide dramatic evidence of the heat and turbulence within the crustal rocks of the earth. The heat in the earth's core and crustal rocks is derived primarily from two sources—from the continual radioactive decay of uranium and from energy liberated shortly after the earth's formation as iron was drawn by gravity to the center of the earth to form its molten core. Without the two metals, uranium and iron, the earth would be cold and dead and there would be no tectonic turnover of the earth's crust to ensure the vital chemical constancy of the earth's surficial layers.

Chlorophyll, for example, contains magnesium and it is thought that the process of reduction in the leaf may depend upon the characteristic of this element. . . . In like manner, haemoglobin contains iron and the capacity of haemoglobin to unite with oxygen and as oxyhaemoglobin to carry it from the lungs to the tissues is unquestionably due to the chemical behaviour of that metal.

—Lawrence Henderson, *The Fitness of the Environment,* 1913

Why grass is greene, or why blood is red. Are mysteries which none have reach'd unto. In this low forme, poore soule, what wilt thou doe?

—John Donne, *Of the Progresse of the Soule,* 1633

Of all the metals there is none more essential to life than iron. It is the accumulation of iron in the center of a star which triggers a supernova explosion and the subsequent scattering of the vital atoms of life throughout the cosmos. It was the drawing by gravity of iron atoms to the center of the primeval earth that generated the heat which caused the initial chemical differentiation of the earth, the outgassing of the early atmosphere, and ultimately the formation of the hydrosphere. It is molten iron in the center of the earth which, acting like a gigantic dynamo, generates the earth's magnetic field, which in turn creates the Van Allen radiation belts that shield the earth's surface from destructive high-energy-penetrating cosmic radiation and preserve the crucial ozone layer from cosmic ray destruction. And it is iron which by its delicate association with oxygen in the hemoglobin in human blood is able to convey in subdued form this most ferociously reactive of atoms, the precious giver of energy, to the respiratory machinery of the cell, where oxygen's energies are utilized to fuel the activities of life.

Without the iron atom, there would be no carbon-based life in the cosmos; no supernovae, no heating of the primitive earth, no atmosphere or hydrosphere. There would be no protective magnetic field, no Van Allen radiation belts, no ozone layer, no metal to make hemoglobin, no metal to tame the reactivity of oxygen, and no oxidative metabolism.

The intriguing and intimate relationship between life and iron, between the red color of blood and the dying of some distant star, not only indicates the relevance of metals to biology but also the biocentricity of the cosmos and why we are indeed, as Sagan so succinctly expressed it, "in the most profound sense children of the Cosmos."[1]

Our understanding of the important role of metals in biology was virtually nonexistent in Henderson's day. Even as recently as a few decades back, knowledge in this field was so limited that Sir Hans Krebs was able to comment "that for all he and I knew most metal ions found in biology could be damaging impurities and therefore had been sequestered or rejected."[2] We now know that Krebs was quite wrong and that metals play so vital a role in living systems that one of the experts in this field, Professor Robert J. P. Williams of Oxford University, in a fascinating review entitled "The Symbiosis of Metals and Protein Function," which summarized current knowledge in this area, concludes:

> In this essay I have not wanted just to repeat the message that metal ions are incorporated and used in biology in a particular way. Rather I wish to assert that

> biology without metal ions does not exist any more than biology exists without DNA or proteins. Metal ions are . . . an essential part of energy and dynamics. . . . No matter what we know about DNA and RNA and even of sugars the nature of the machinery of life rests with these two components, metal ions and proteins. . . . The all pervading influence of metal ions in biological systems is such that I now declare that in my mind there is *no biology without metal ions.*[3] [My emphasis.]

More than half of the most abundant atoms in the cosmos are metals, including sodium (Na), potassium (K), calcium (Ca), magnesium, (Mg), iron (Fe), and Copper (Cu) (see chapter 4). Iron, for example, is nearly as common as carbon. Given their abundance, in any biocentric view, one would expect that the metals would be of considerable utility for life. And it has turned out that many of the metals do indeed play a vital role in some of the most fundamental biological processes and the evidence increasingly suggests that many of these processes are dependent on the precise chemical and physical properties of particular metal atoms. Close to one-third of all enzymes involve a metal ion as an essential participant.[4] An excellent review of this topic is given by Fraústo da Silva and Williams in their *Biological Chemistry of the Elements.*[5]

Electron Conductors

One key role plays by metals in the cell is the formation of electronic circuits, and one area where these play a vital role is in energy metabolism. Moreover, it is only the transitional metal atoms, particularly iron and copper, which possess precisely the properties required to form an electronic circuit. No other atoms will do. Only the transitional metals, having far more complex electron shells with many more energy states than the simpler atoms such as sodium (Na), calcium (Ca), carbon (C), nitrogen (N), etc., possess the appropriate electrochemical characteristics to trap and channel electronic energy. The unique electric conducting properties of the transitional metals are also utilized in human technology to make wire conductors. As Fraústo da Silva and Williams comment, "man makes his wires from metals such as copper; biology makes hop conductors from metal ions embedded in protein."[6]

It is only by utilizing the conducting properties of the transitional metals that the cell is able to channel the electron flow through discrete energy steps and utilize each energy drop to perform useful chemical work. No organic compounds can substitute for the transitional metals in this regard. It is fortuitous indeed that the transitional metals possess precisely those unique chemical characteristics essential for stepwise electron transport in the respiratory assembly and in the photosynthetic apparatus. If no atoms in the periodic table were specifically fit for this highly specialized role, then the controlled and efficient utilization of the energy of oxidation could not be achieved. Advanced life forms would in all probability be impossible. If we are to have electronic circuits in living organisms, these will be made of transitional metal wires. But the transitional metal atoms not only provide the electronic circuits of the cell upon which the efficient exploitation of oxidative energy is critically dependent, they also possess precisely the required complement of chemical and physical properties which permit organisms to manipulate the oxygen atom, without which oxidative metabolism would be impossible.

Oxygen Transport

The capacity of the transitional metal atoms to handle oxygen is illustrated by the oxygen-carrying molecule, hemoglobin, which transports oxygen in the blood of higher vertebrates, including humans. Hemoglobin is made up of the protein globin, the small planar cyclic compound heme, and the iron atom which is chelated with the heme. The figure below shows the structure of heme.

The structure of a heme.

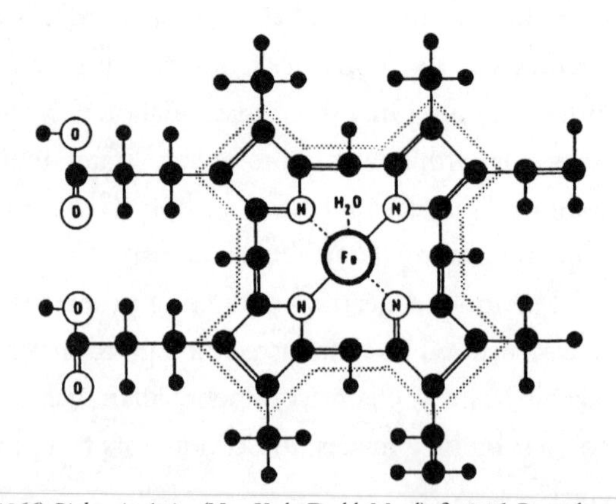

From N. J. Berrill (1966) *Biology in Action* (New York: Dodd, Mead), fig 3.16. Reproduced by permission of N. J. Berrill.

The oxygen-carrying capacity of mammalian blood is about 25 milliliters of oxygen per 100 milliliters of blood. This is fifty times the amount that can be dissolved in ordinary solution! However, an efficient respiratory pigment must satisfy additional criteria beyond the mere ability to carry large quantities of oxygen. As Ernest Baldwin points out in his *Introduction to Comparative Biochemistry*: "It must take up oxygen where the partial pressure is high and give it up again equally readily to the tissues in which the pressure is low. . . . in other words, the compound of the . . . respiratory pigment with oxygen must be such that it readily dissociates."[7]

The reversible binding between iron and oxygen in hemoglobin depends on three critical properties: on the hydrophobic nature of the interior of proteins, on the unique characteristics of the iron atom complexed with the heme moiety in the globin, and on the unique characteristics of the dioxygen molecule that are complementary to those of the iron atom and allow association and dissociation at precisely the range of oxygen concentrations that permit uptake in the lungs and release in the tissues.

Consideration of the detailed events which occur during reversible binding suggests that no other metal atom could exactly mimic the properties of iron in heme.[8] None of the other transitional metal atoms closely related to iron will substitute for iron in hemoglobin, because none are of precisely the same size, nor do any possess precisely the same chemical characteristics allowing them to undergo the same subtle changes on associating with oxygen.

As the efficient transport of oxygen is essential to the viability of any large active organism with a high metabolic rate, a molecule with the properties of hemoglobin would seem to be essential. Might there be any alternatives to hemoglobin? None of the many other oxygen-carrying molecules which occur in the blood of invertebrates, such as the copper-containing proteins of the molluscs, come close to the efficiency of hemoglobin in transporting oxygen in blood. As Ernest Baldwin commented, "Mammalian haemoglobin is far and away the most successful of the respiratory pigments from this point of view," and Joseph Barcroft has written of it that "but for its existence, man might never have achieved any activity which the lobster does not possess."[9]

The question arises as to whether a respiratory pigment designed on radically different principles to hemoglobin might be possible. The question was raised by Earl Frieden when he asked, "Why has no other essential metal . . . or other type of respiratory protein developed to satisfy this important function?" Because, he continues, such a pigment "needs to be able to form a stable dissociable complex with the highly reactive molecule, O_2, and, as he points out: "Transition metals excel in this capacity; few other chemical groups can do this. In fact all efforts to devise other physiologically compatible, model oxygen carriers have failed to date. . . . The compounds that come closest to emulating the oxygen-binding properties of haemoglobin . . . *contain transitional metal ions.*"[10] (My emphasis.)

It would seem that in designing an oxygen-transporting molecule from first principles we are led inevitably to a molecule very like hemoglobin and to the choice of iron or at least one of the transitional metals to carry out the key oxygen-binding role. Water would also almost certainly have to be excluded from the binding site, and this would lead inevitably to something like the hydrophobic heme cleft in hemoglobin. The evidence is consistent with the possibility that hemoglobin is the ideal and unique respiratory pigment for metabolically active air-breathing organisms such as ourselves, and that its unique abilities depend in turn not only on the unique properties of the transitional metal atoms but on the specific properties of one of these atoms—iron.

The elegance of the way the hemoglobin system functions is simply astounding, and a source of wonder to everyone who is familiar with its intricate ingenuity. And as with the elegance of so much of the biochemical machinery of life, this elegance is only possible, as in so many other instances, because of the provision by nature of atoms and molecular structures perfectly fit for particular vital biological functions.

The fitness of the iron atom for reversible binding to oxygen is of course only one of many mutual adaptations in the nature of things which make possible the delivery of oxygen to the metabolically active tissues in a large organism like a mammal. There is also the fact that oxygen is soluble in water; that the viscosity of water is sufficiently low to make the design of a circulatory system possible; that the viscosity of a non-Newtonian fluid—i.e., one containing a suspension of particles—decreases as the pressure increases, a phenomenon which greatly facilitates the propulsion of the blood through the tissues in times of high metabolic activity; that carbon dioxide is a gas, and so on.

Manipulating Oxygen

The unique oxygen-manipulating capabilities of the transitional metals are also utilized in a variety of enzymes that defend the cell from the destructive effects of oxygen. One such "protector" enzyme, which occurs in all aerobic organisms and which utilizes the properties of the transitional metal copper, is known as superoxide dismutase. The reaction it catalyzes furnishes a means of disposing of a highly reactive (and highly destructive) oxygen radical, the superoxide ion O_2^-.[11] And as Frieden points out, the protection offered by the activities of this enzyme "is a *prerequisite for the adaptation of all living cells to the utilisation of oxygen.*"[12] (My emphasis.)

One of the most important of all enzymes that utilize the oxygen-handling capabilities of the transitional metals is cytochrome *c* oxidase, which is the terminal member of the respiratory assembly and performs the critical final reaction of oxidative metabolism. This involves uniting the electrons flowing through the respiratory assembly to atoms of oxygen and hydrogen. It sits astride one of the bilayer lipid membranes in the mitochondrion (the organelle concerned with energy generation).

On its significance, Frieden comments: "If a biochemist is asked to identify the one enzyme which is most vital to all forms of life, he would proba-

Cytochrome oxidase.

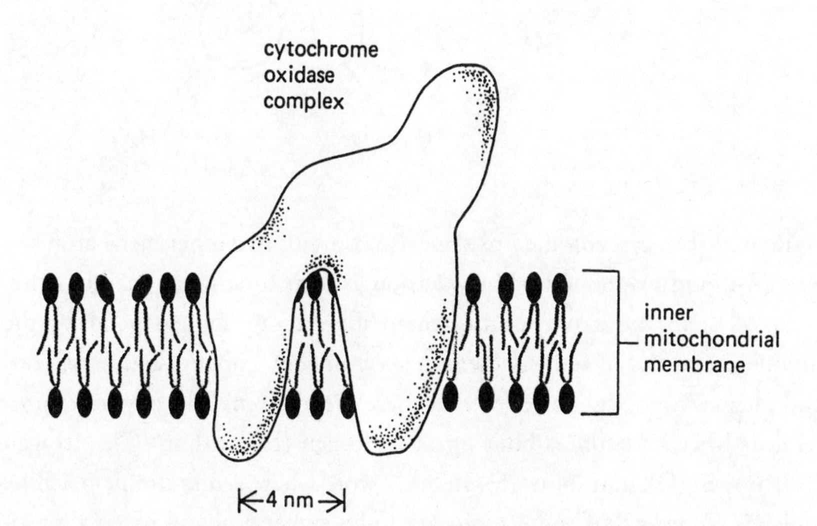

From B. Alberts et al. (1983) *The Molecular Biology of the Cell* (New York: Garland Publishing) fig. 9. 31. Reproduced by permission of the publisher.

bly name cytochrome *c* oxidase. This is the enzyme, found in all aerobic cells, which introduces oxygen into the oxidative machinery that produces the energy we need for physical activity and biochemical synthesis. . . . This enzyme may be regarded as the ultimate in the integration of the function of iron with copper in biological systems. Here in a single molecule, we combine the talents of iron and copper ions to bind oxygen, reduce it with electrons from the other cytochromes in the hydrogen electron transport chain and, finally, to convert the reduced oxygen to water."[13] The diagram below, redrawn from a recent *Science* article,[14] shows the electrons flowing within the molecule, through a "transitional metal wire" composed of a succession of iron and copper atoms which conducts them to the final catalytic center where they cause the reduction of oxygen to form water.

Cytochrome c *oxidase.*

Showing the electron flow and chemical reactions.

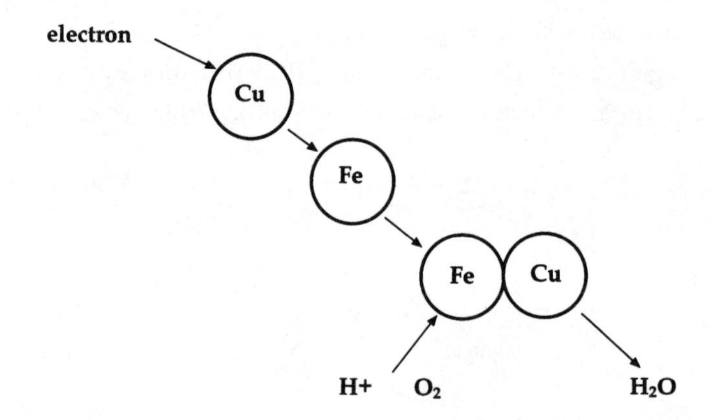

Remarkably, cytochrome *c* oxidase also contains two other metal atoms—zinc (Zn) and magnesium (Mg)—although their function is mainly structural. All in all, the activity of this remarkable nanomachine depends on the unique properties of four metal atoms—iron (Fe), copper (Cu), zinc (Zn) and magnesium (Mg) and as well as these metal atoms, like any other protein, its basic structure is built up of hydrogen (H), carbon (C), nitrogen (N), oxygen (O), and sulfur (S). In other words, here is one atomic machine which is composed of and exploits the unique chemical and physical properties of nine of the ninety-two naturally occurring elements.

Molybdenum

Another transitional metal atom which appears to possess unique properties essential to life is molybdenum (Mo). Molybdenum is an essential component in the two enzymes involved in nitrogen fixation: nitrogenase, which catalyzes the reduction of nitrogen to ammonia and nitrate reductase, which catalyzes the reduction of nitrate to nitrite. These two activities are absolutely critical to life on earth. All the nitrogen utilized by living things is initially captured via these two reactions. As two authorities point out: "The molybdenum atom possesses a number of unique characteristics which can account for its role in these two critical reactions. No other atom, with the possible exception of vanadium (V), could possibly substitute for molybdenum."[15] And they raise the fascinating question "Is the requirement for molybdenum an absolute necessity or could the chemistry have been done in some other way and is what we observe an accident?"[16]

It is hard to believe that the choice of molybdenum is accidental. While molybdenum is one of the commonest metals in seawater, its availability on land is patchy. Because the ability to fix atmospheric nitrogen and utilize it for organic syntheses is obviously of great selective advantage, if there was any alternative to molybdenum, particularly in those environments where molybdenum is rare or unavailable, then surely some microorganism would have discovered it at some stage during the 4 billion years of evolution. The absence of any alternative suggests strongly that the vital processes by which nitrogen enters the biosphere are absolutely dependent on the special characteristics of the molybdenum atom. Again, we have what appears to be another case in which life on earth is critically dependent on the specific chemical properties of a unique constituent.

Rapid Information Transmission

Being far smaller than even the simplest organic compounds such as sugars and nucleotides, the metal ions of sodium, potassium, magnesium, and calcium diffuse far more quickly in aqueous solutions and are therefore ideal for rapid information transmission.

The capacity for rapid movement, however, is not the only criteria which must be satisfied by a chemical messenger. A chemical messenger must not only be able to move quickly but also to associate reversibly with a specific target in the cell. The necessity for specific high-affinity binding largely excludes the monovalent metal ions sodium and potassium, which are too "simple" to be conveniently utilized in this role. Of the two divalent ions calcium and magnesium, the binding of calcium to most sites is a thousand times tighter and hence the most fit for this role.

In biological systems, it is calcium which is preeminently used where chemical information must be transmitted at great speed, as in the triggering of muscle contraction, transmission of nerve impulses across the synapse, triggering hormone release, the changes following fertilization, etc. As Williams points out in his review, "Amongst the metal ions available to biology *only calcium* can be high in concentration, can diffuse rapidly, can bind and dissociate strongly."[17] (My emphasis.)

Of particular relevance to its role as the "mercury of the cell" is the fact that the chemical characteristics of the calcium ion are perfectly fit for specific association with proteins—the key functional biomolecules in the cell and as such the most likely components of all the cell's nanomachinery to be receivers and transducers of chemical information. And for protein binding, calcium is far superior to magnesium. This is because, first, magnesium requires a more regular geometrical binding site than calcium and such sites are difficult to arrange in a protein because of the basic irregularity of its structure and, second, because of the particular affinity of calcium ions for oxygen atoms, which are readily provided by the amino acids of proteins. Proteins in their molecular irregularity and in their possession of readily accessible oxygen atoms provide an ideal molecular matrix for the design of calcium binding sites.

Another fascinating aspect of the calcium-protein relationship is the capacity of the α helix, which as we have seen is one of the basic structural subunits used in the building of proteins, to react with great rapidity to the stimulus of calcium binding. As one author, commenting on the suitability of the helical structures in proteins to respond rapidly to the stimulus of calcium, remarks: "The proteins which are in the muscle or the internal filamentous units of cells must have activity matching that of calcium. . . . These proteins are largely based on helices. In a general sense a helical rod is useful in that its movement economically connects activities at either end through rotational-transitional movements like that of a screw or worm gear. It is fast since helix-helix movements need not break hydrogen bonds. We see in the helix the potential for matching the dynamics of the calcium ion."[18]

Magnesium

Magnesium is calcium's sister atom in the periodic table and is similar in many of its properties to calcium, but it binds to proteins less quickly and less tightly than calcium. It is certainly less fit for the role of chemical messenger than calcium, but its gentler affinity for proteins is utilized by the cell in the more subtle molecular rearrangements which accompany enzymic activities. In present day life forms its involvement is vital to many crucial enzymic processes.

Of particular interest is the role of magnesium in photosynthesis. The chlorophyll molecule, the green pigment of plants, which is the key component in the photosynthetic apparatus, contains one centrally positioned magnesium atom. Although the magnesium can be replaced by copper, nickel, cobalt, iron, and zinc, none of these metals can mimic the light-absorbing capacity of magnesium. Compared to iron, for example, the light absorption of magnesium is several thousand times greater. Precisely what it is about the magnesium atom that confers on chlorophyll its magic light-absorbing capacity is not known. But there must be, as Melvin Calvin comments, "something very special about the electronic structure of the magnesium"[19] and about the way the chlorophyll molecules are packed together in the chloroplast. Whatever the basis for the unique properties magnesium confers on the chlorophyll, it is clear that the capture of light energy is remarkably efficient.

An intriguing aspect of the light-absorbing properties of chlorophyll, which seems to be an exception to the general rule that the constituents of life seem maximally adapted to their biological roles, is the curious point that the light-absorption properties of chlorophyll are maximum in the violet and near ultraviolet and in the red and infrared regions of the spectrum. These regions do not coincide with the regions of the spectrum which contain most of the sun's radiant energy at the earth's surface, which is in the blue-green range. George Wald, who elucidated the biochemistry of vision, raised this question in 1959 in a *Scientific American* article entitled "Life and Light": "What properties do the chlorophylls have that are so profoundly advantageous for photosynthesis as to override their disadvantageous absorption spectra."[20] Chlorophyll would appear on the face of it to be less than maximally fit for its biological role; maximal fitness would appear to demand that it absorb light in the blue-green range of the spectrum.

However, in the case of other apparent "defects" in the fitness of the basic ingredients of life, such as water's inability to dissolve hydrocarbons and oxygen's low solubility in water, it often turns out that, with increasing knowledge, such anomalies are revealed to be highly beneficial in some respect that was overlooked at the time. Another example may be the apparently anomalously high absorption by chlorophyll of light in the violet and near-ultraviolet regions of the spectrum. As mentioned in chapter 6, this region of the spectrum is damaging to life largely through the production of free radicals of oxygen. It is possible that the anomalous absorption of radiant energy in the violet and near-ultraviolet region of the spectrum may confer on chlorophyll an element of biological fitness previously unsuspected, that of attenuating the flux of ultraviolet radiation. Sunburn is less severe on grass than on sand.

There are several other metal atoms—vanadium (V), chromium (Cr), manganese (Mn), cobalt (Co), nickel (Ni), copper (Cu), and zinc (Zn)—which are also essential to life and where the unique property of the individual atom appears to be exploited in some vital biological process. An excellent summary of the biological role of these metals is given in *Biological Chemistry of the Elements,* cited earlier.

Conclusion

The emerging picture of the role of metals in biology is increasingly one in which it appears that all the metals in each of the main subgroups of the periodic table possess unique properties that are fit for particular vital and essential biological roles. Without them life remotely as rich and complex as it exists on earth would be impossible. The transitional metals, for example, give every impression of having been tailored to form the electronic circuits of the cell and to manipulate in various ways the oxygen atom. Moreover, it increasingly appears that even individual metal atoms such as calcium, iron, copper, magnesium, and molybdenum may be uniquely fit for some of the biological roles they serve. Iron may be uniquely tailored for the sort of reversible binding to oxygen which occurs in hemoglobin, and magnesium for the absorption of light in chlorophyll. Trying to envisage life without metals is every bit as difficult as imagining human technology without them, for, as Robert Williams concludes: "There is no biology without metals.[21] . . . metal elements in some organisation are of the essence of life as much as this is true of amino acids and nucleotides."[22]

Chapter 10

The Fitness of the Cell

In which it is argued that the cell is uniquely and ideally fit to function as the basic unit of carbon-based life. Cells are capable of carrying out any instruction, adopting any shape, creating the vast diversity of multicellular organisms and ultimately the whole world of life. Evidence is examined which suggests that the cell membrane is uniquely and ideally fit for its role of bounding the cell's contents and conferring on the cells of higher organisms the ability to move and adhere selectively to one another. These critical properties are also dependent on the size of the average cell being approximately what it is and on the viscosity of cytoplasm being close to what it is. The membrane is also fit, in that its selective impermeability to charged particles confers additional electrical properties, which form the basis of nerve conduction. A variety of coincidences underlying the ability of cells to selectively adhere and move are discussed. The known properties of cells are remarkable enough, but there is still much to learn. The possibility that cells may possess powerful computing abilities and may even be able to behave intelligently is considered.

The protozoan Stentor.

This protozoan organism consists of a single cell about one-fifth of a millimeter long. Small fragments of the cell, less than one-hundredth of its volume, are capable of regenerating a tiny but exact replica of the whole cell.

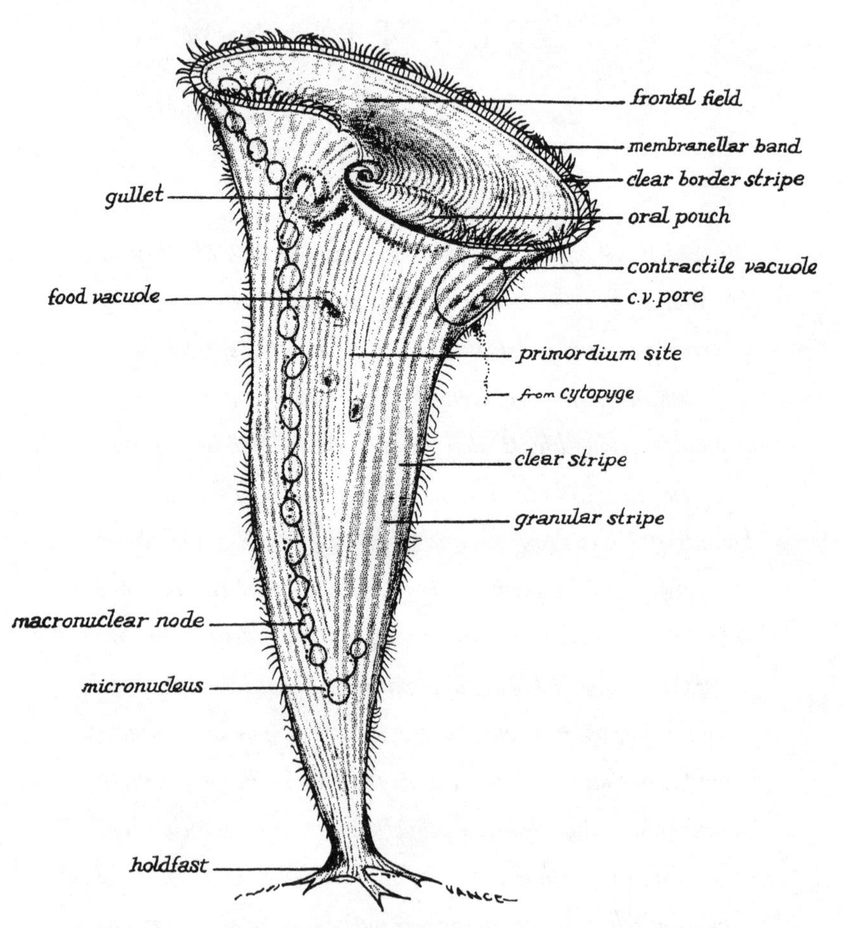

From V. Tartar, (1961) *The Biology of Stentor,* p. 8 (London: Pergamon Press). Reproduced by permission of Wanda Tartar.

Its talents are legion, but its size is minuscule. *E. Coli* is a cylindrical organism less than 1 nm in diameter and 2 nm long—20 would fit end to end in a single rod cell of the human retina. . . . Yet it is adept at counting molecules of specific sugars, amino acids, or dipeptides; at integration of similar or dissimilar sensory inputs over space and time; at comparing counts taken over the recent past; at triggering an all or nothing response; at swimming in a viscous medium . . . even pattern formation.

—H. C. Berg, *Cold Spring Harbor Symposium on Quantitative Biology,* 1990

Cells amaze. Even to a nonbiologist cells convey the impression of being very special, unique objects with extraordinary capabilities.

Considering their accomplishments, it is hard to believe that there could exist any other sort of organized material form, some other type of nano-erector device remotely equal to the cell. It is these remarkable specks of organized matter that have constructed every multicellular organism that ever existed on earth. It is cells that assemble the human brain, putting down a million connections a minute for nine months during gestation. It is cells that build blue whales, butterflies, birds, and grass. It is cells that built the dinosaurs and all past life on earth. Through the activities of some of the simplest of their kind, over the past 4 billion years they gradually terraformed the earth, generating oxygen via photosynthesis and thereby releasing its energizing powers for all the higher life forms.

The cell betrays every evidence of being uniquely and wonderfully adapted for its assigned task. The diversity of cell form is every bit as bewildering as the diversity of carbon compounds or proteins. Cells exhibit not only a vast diversity of form but also a diversity of functional and behavioral capacities of bewildering richness. They are miracles of nanotechnology. Some can move by the rowing action of cilia or by the propellerlike action of the bacterial flagellum. Others can creep and crawl. They can estimate the concentration of compounds in their immediate environment. They can change their form and chemical composition. They can put out pseudopodia and grasp small objects in their immediate vicinity. They possess internal clocks and can measure the passage of time. They can sense electrical and magnetic fields. They can synchronize their activities and can combine forces and crowd together to make a multicellular organism. Cells can communicate via chemical and electrical signals. They can encase themselves in various armorlike skins. They can replicate themselves with what seems to be surpassing ease and, as mentioned in chapter 7, even reconstruct themselves completely from tiny fractions of their mass.[1]

Cells can survive desiccation for hundreds of years, and so on and on. In short, they can *do anything, adopt almost any shape, obey any order,* and seem in every sense perfectly adapted to their assigned task of creating a biosphere replete with multicellular organisms like ourselves. The astounding nature of the nanotechnological miracle the cell represents is self-evident.

From the knowledge we now have of the molecular machinery that underlies some of their extraordinary abilities, it is clear that cells are immensely complex entities. On any count the average cell must utilize close to a million unique adaptive structures and processes—more than the number in a jumbo jet. In this the cell seems to represent the ultimate expression in material form of compacted adaptive complexity—the complexity of a jumbo jet packed into a speck of dust invisible to the human eye. It is hardly conceivable that anything more complex could be compacted into such a small volume. Moreover, it is a speck-sized jumbo jet which can duplicate itself quite effortlessly.

The fitness of the cell for its biological role in the assembly and functioning of multicellular life gives every indication, as with so many of life's constituents, of being unique. In the case of many of their key properties and abilities, it is very difficult to imagine how these properties and abilities could be actualized except in a material form with the precise characteristics of the living cell. In other words, if we were to design from first principles a tiny nanoerector about 30 microns in diameter with the capabilities of the cell—with the ability to measure the chemical concentration of substances in its surrounding medium; with the ability to measure time, to move, to feel its way around in a complex molecular environment, to change its form; with the ability to communicate with fellow nanoerectors using electrical and chemical messages and to act together in vast companies to create macroscopic structures—we would end up redesigning the cell.

Lipids

An important class of biomolecules that play a critical role in many aspects of the life of the cell are the lipids, which include the fats and the fatty acids. Lipids are found in all living things. They have many different functions. They are a major source of cellular energy. They function as electrical insulators and as detergents. They form the waxes which coat the feathers of birds. Some function in the gaseous state as pheromones, substances which attract other organisms of the same species.[2]

All types of lipids contain long hydrophobic chains of carbon and hydrogen atoms which are insoluble or only sparingly soluble in water. The structure of the fatty acid stearic acid is shown below.

The chemical structure of stearic acid.

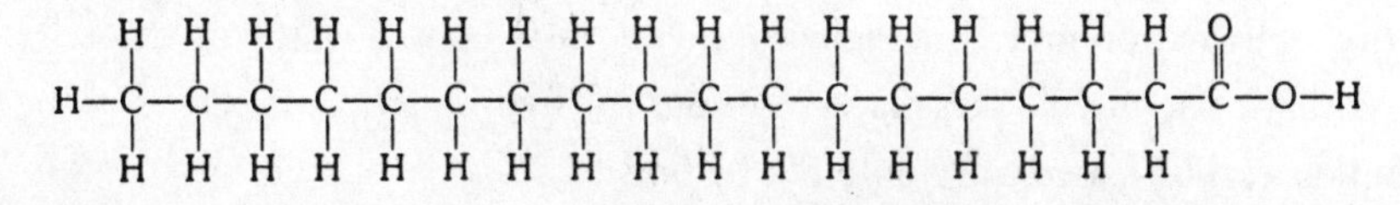

The fact that many types of lipids are insoluble in water is of great biological significance. Without insoluble components, the compartmentalization of the cell and the persistence of cellular structures would not be possible. Lipids are also the major component of the bounding bilayer membrane which surrounds every living cell. (See page 216.) If there were no carbon compounds insoluble in water, such as the lipids, organic chemistry would not be fit for life. Correspondingly, if water was truly a universal solvent, the alkahest of the alchemists, it would not be a fit medium for life because no compartments or stable structures would be possible and all the cells constituents would merely dissolve away.

The hydrocarbon chain length of most of the lipids which occur in the cell is generally between 16 and 18 carbon atoms long. This chain length is fit for a number of reasons. Chain lengths of more than 18 carbons long are too insoluble to be of biological utility—they cannot be mobilized at all in water—but less than 16, they are too soluble.[3] Fortuitously, lipids containing chains of this length are also fluid or near fluid over the temperature range in which most metabolic processes occur in living things. If lipids of these chain lengths had been solid at ambient temperatures, no structure constructed out of them would possess the necessary plasticity to function in the cell. Moreover, because they are also more viscous than water when fluid, they act to buffer the organs of higher organisms against shearing forces.[4] The lightness of lipids compared to water also has significant biological consequences. It gives aquatic life buoyancy. The subcutaneous lipid of warm-blooded animals also acts as a heat insulator, preventing heat loss. The extreme example of this is the cetacea with a layer of blubber up to 15 inches thick, which allows them to thrive in the polar seas.[5] (See Appendix, section 2, for more details on the fitness of lipids.)

In addition to providing the cell with stable structures, boundaries, and compartments, the nonpolar hydrophobic nature of lipids is also of great utility because lipid aggregates provide the cell with tiny nonaqueous microenvironments. Such hydrophobic microenvironments are vital to the life of the cell, because many of the synthetic and enzymic processes upon which the life of the cell depends can only occur in a microenvironment where water has been excluded. So their insolubility plays two roles in the cell, creating the stable insoluble boundaries between compartments and the vital hydrophobic microenvironments in which so much of the cell's synthetic chemistry takes place. Without the hydrophobic properties of the lipids, carbon-based life would not be possible.

The Cell Membrane

One of the most important structures in the cell, which is largely composed of lipids, is the cell membrane. It is difficult to see how a cell could survive without some sort of bounding membrane which was relatively impermeable to the cell's constituents, especially to small metabolites such as sugars and amino acids, to prevent its contents from diffusing away into the surrounding fluid. Such a membrane would also have to be relatively plastic and able to maintain a continuous barrier between the cell and its environment in the face of the ever-changing shape of the cell. As one leading biologist points out, it is essential that the cell membrane should behave like a "two-dimensional liquid" and be able to flow in all directions over the surface of the cytoplasm to maintain a continuous barrier between the cell and its surroundings in the face of "of the ever changing protrusive activities of the cell surface."[6]

The lipid bilayer satisfies these criteria admirably. Being hydrophobic, it satisfies the criterion that it must be impermeable to the majority of the cell's constituents, such as sugars, amino acids, and other organic acids which are soluble in water. Moreover, the lipid bilayer also has precisely that fluid character needed to preserve a continuous barrier surrounding the ever-turbulent and motile mass of cytoplasm. It is in fact highly fluid and has a level of viscosity like that of olive oil.

The structure of the lipid bilayer is shown below. Note the lipids making up the membrane are phospholipids, which are lipids containing a charged phosphate group at one end. The phosphate end is hydrophilic—water loving—while the fatty end is hydrophobic—water hating. When the structure assembles itself, the hydrophilic phosphate groups orient themselves facing the water, while the hydrophobic hydrocarbon chains orient themselves away from the water.

Another factor which contributes to the fitness of this remarkable structure is that it forms *automatically* around the outer surface of the cell, like the spreading of a monomolecular film of immiscible substance (typically a lipid) on the surface of water. As cell biologist John Trinkaus comments:

> Because water is itself a strongly polar molecule, the polar phosphate of the membrane lipids will inevitably be attracted to the surfaces of the membrane, both external and cytoplasmic. And just as inevitably their nonpolar fatty acid parts will tend to be squeezed into a nonpolar phase in the interior of the membrane. . . . The beauty of it is that everything arranges itself. . . . Simply

The structure of the cell membrane.

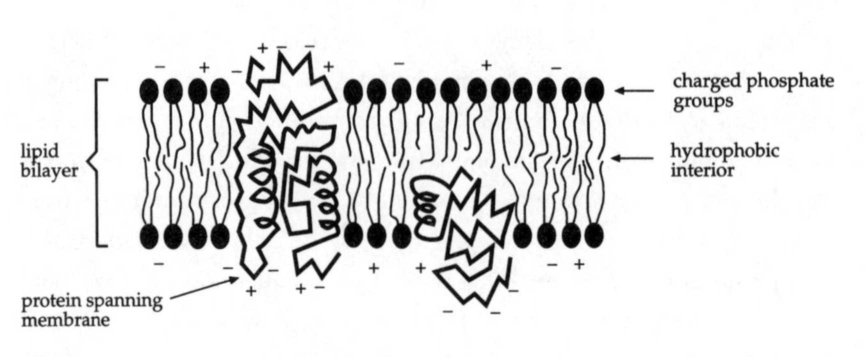

Redrawn from S. J. Singer and G. L. Nicholson (1972) "The fluid mosaic model of the structure of cell membranes," *Science* 175:720–731, by permission of Professor S. J. Singer.

> because of their intrinsic chemical nature phospholipids *naturally* and *spontaneously* self-assemble to form a bilayer in a watery solution. . . . It is, as it were, "the nature of the beast" for them to do so.[7]

There is not the slightest element of contingency in the fact that a lipid bilayer forms the boundary layer surrounding the cell. No other material is known which could substitute for this particular structure. The properties of impermeability, fluidity, and spontaneous assembly would seem to be essential characteristics in any membrane system surrounding the cell. Yet this unique combination of characteristics is only found in the lipid bilayer—another case where a key biological function is carried out by an adaptation which appears to be both *unique and ideal* for its assigned role. The existence of the cell is in effect absolutely dependent on the lipid bilayer possessing the precise suite of biochemical and biophysical characteristics that it does.

The lipid bilayer has another property inherent in its hydrophobic nature—that of an electrical insulator.[8] Because it is impermeable to charged particles such as the ions of sodium (Na^+) and potassium (K^+) and hence capable of restricting their movement across the membrane to specially designed ion gates, the cell is able to generate an electrical potential between the inside and the outside of the cell, the so-called membrane potential, by pumping charged particles through the ion gates in the membrane. If the lipid bilayer was not an insulator, cells would not be able to maintain the membrane potential, and many biological phenomena such as the transmission of nerve impulses, which depend on the membrane potential, would not be possible. In addition to its insulating characteristics, the lipid bilayer is fit for the generation of membrane potential in another way—it provides an ideal environment in which the membrane proteins which pump the ions in and out of the cell can reside and function.[9]

That one of the properties of this remarkable structure, its insulating character and the membrane potential it automatically generates, should provide precisely the electrical characteristics required for the transmission of electrical impulses between cells and ultimately for the construction of the nervous system is surely a fact and coincidence of very great significance. No less than its insulating, electrical properties and its ability to self-assemble, the nervous system itself is also, in a very real sense, in "the nature of the lipid bilayer beast."

The electrical properties of cells depend on many other factors in addition to the insulating properties of lipid membranes. The propagation, for example, of the nervous impulse depends on the rapid transmission of a wave of depolarization along the nerve fiber. The speed of depolarization is itself due to the speed of diffusion of sodium and potassium cations through special gates in the membrane of the nerve cell. This process is again greatly enhanced by the low viscosity of water and by the unique properties of the cations themselves.

Cell Adhesion

The ability of cells to selectively adhere to one another is one of their most important characteristics. According to one authority, "the adhesions that cells make with one another lie at the very basis of multicellularity. The form and functioning of all creatures that consist of more than one cell depend on their cells adhering firmly to one another and to the extracellular materials that intervene."[10]

The surface of a typical cell is not smooth but rather rugged, and many cells make initial contact with each other via microprotrusions on their surface. These are often .1 micron across the tip and have an area in the range of one-hundredth of a square micron. It is by using these microprotrusions, like a cat its whiskers or a man his hands, that a cell explores its microenvironment and is able to "taste" and "feel" the surface of the other cells in its immediate environment.

The actual process of adhering to another cell occurs via special adhesion molecules which are positioned on the microprotrusion as is shown in the diagram below. Pairs of adhesion molecules bind to each other via complementary matching surfaces using the same principle of lock-and-key matching recognition used by proteins to recognize their substrates. The bonds between two adhesion molecules are referred to as affinity bonds. Each affinity bond consists of a number of weak, or noncovalent, bonds. The strength of each affinity bond between two cells is made up of the sum of the various weak chemical bonds which bind the two adhesion molecules together. The diagram below illustrates two cells linked via an affinity bond at the end of a microprotrusion.

An affinity bond linking two cells.

Note one nanometer (nm) is one-millionth of a millimeter. A typical body cell, such as a lymphocyte, has a cell surface area of 200 square microns (a micron is one-thousandth of a millimeter). The number of adhesion molecules per cell is about 100,000, that is, 500 per square micron. Most cells contain a great many different types of adhesion molecules.

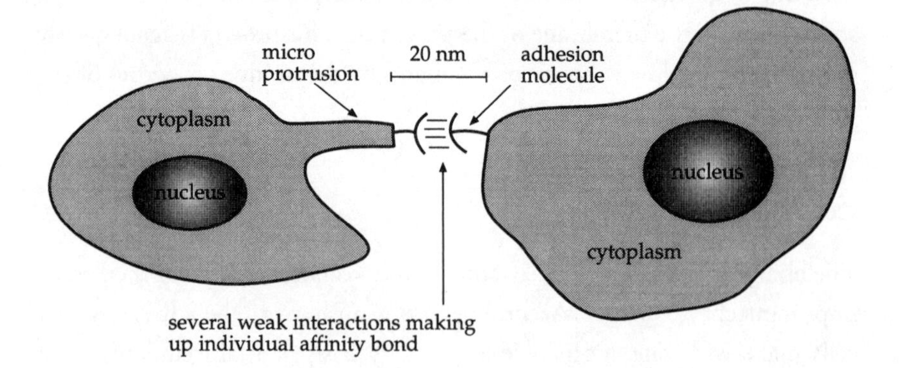

The fact that cells make initial contact via microprotrusions, often called filopods, is not accidental. The outer surface of cells is negatively charged, so cells tend to repel one another electrostatically. However, these repulsive forces become smaller as the contact area decreases in size down to the protrusion surface area of about .01 square micron, and the energy required to bring the tiny area of the tip of a microprotrusion into intimate contact with an adjacent cell membrane in order to make affinity bonds is no longer prohibitive.

Cell adhesion via microprotrusions also plays an important role in cell navigation. One of the most important mechanisms by which cells find their way as they migrate through the developing embryo is by successively adhering to a series of target cells or structures that lead them through the embryo to their assigned place. This inevitably involves their pushing and thrusting their way past other cells. To search for their next target they put out filopods in many directions until the correct contact is made. A cell unable to send out projections of this sort would be unable to feel any directional clue that was not in its immediate vicinity. Such a restriction would make direction finding by cells well nigh impossible, analogous to trying to find one's way through a dark room without using one's hands.

Consideration of the size of individual adhesion molecules (many are approximately 10 nanometers across) and the small area of a filopod tip suggests that packaging constraints will limit the number of adhesion molecules that can be fitted onto one tip.[11] Moreover, even when a filopod probe is very close to another cell surface, only a proportion of the adhesion molecules on the filopod tip will be able to make affinity bonds with the target cell surface. To make the affinity bond, the two complementary matching surfaces of the adhesion molecules must come into almost perfect alignment at distances of less than a nanometer. This is a very stringent requirement and suggests that initial binding can only invoke a small fraction of the available adhesion molecules on the tip of the filopod. Unless individual affinity bonds were relatively strong, sufficient to bind the protrusion to its target cell securely in the face of the various shearing forces usually met with in biological systems, no protrusion would ever bind to a target cell. The phenomenon of selective cell adhesion would be severely constrained and the use of tiny protrusions for pathfinding by embryonic cells would in all probability be impossible.

Recent studies have measured the strength of individual affinity bonds. Remarkably, these turn out to be on average equivalent to the force needed to lift 40 nanograms (there are 1 billion nanograms in one gram), which is enough to tie a filopod to a specific contact point. These studies have shown that only between one and ten affinity bonds are sufficient to hold two cells together against most of the common forces met with in biological systems. As few as thirteen affinity bonds will hold a lymphocyte to another cell in a fluid with the viscosity of water flowing at 1 centimeter per second. An individual filopod by which, as we have seen, many cells make contact with their environment and other cells, may be stabilized by as little as two affinity bonds. Individual affinity bonds, in short, are of sufficient strength to tether a filopod to its target cell.[12]

The contact of a particular filopod to its target is stabilized by the rapid migration and accumulation of additional specific adhesion molecules to the filopod tip. It is interesting to note that the two-dimensional mobility of proteins embedded in the membrane is only possible because of the unique fluidity of the cell membrane, a characteristic touched on above.

At any one time, most cells in developing embryos must be able to "feel" and make contact with a considerable diversity of cells and extracellular features in their immediate environment. It is hard to imagine how cells could do this any way except by using filopods.

Note also that the strength of the affinity bonds upon which the whole filopod adhesion system is based is ultimately determined by the strength of the weak chemical bonds. We have already seen that proteins would not be stable, nor enzymes bind their substrates, if these interactions were severalfold weaker. Here we have another instance in which an important biological process is also critically dependent on the absolute strength of these same weak interactions. Any weaker, and specific cell-cell binding via adhesion molecules on the surface of cells would be impossible. Any stronger, and, once formed, it would be very difficult for cells to detach themselves from one another.

The phenomenon of selective adhesion via filopods is only possible because, first, the cytoplasm is highly deformable and of low viscosity, and lends itself to being drawn into a long fingerlike extension; second, the repulsive negative electrostatic forces between the small area on the tip of a filopod and the target cell is relatively low; third, more than one specific adhesion molecule can be placed strategically on the end of a filopod tip, and fourth, the relative strength of affinity bonds is sufficient to bind cells against the various shearing forces tending to pull them apart.

Crawling

Yet another vital characteristic of cells is their ability to crawl, which is no less important and critical than their adhesive properties. Selective adhesion would be of little utility if cells could not move toward particular targets. Indeed, higher forms of life would not exist if cells could not crawl, and we would certainly not be here to ponder the phenomenon.[13] In crawling, a cell puts out fan-shaped extensions called lamellae, the leading edge of which make transient attachments with the underlying surface, and as they glide forward, they pull along the cell body passively behind them. The process is somewhat similar to the crawling of a snail, in which the snail's foot is analogous to the lamellae and the shell analogous to the cell body.

The process involves continual restructuring of the cell's substance and continual changes to its shape and form.[14] These necessitate that the cytoplasm be readily deformable—in other words, must have the quality of a relatively viscous colloidal material so that the interior of the cell can be drawn into the advancing protrusion. Yet the cell's interior must also contain stable structural elements making a mechanically rigid scaffold so that traction forces can be exerted between the adhesion points on the undersurface of the lamellae and the mechanically rigid scaffold.

It is clear that the ability to crawl requires the satisfaction of exacting criteria. The cytoplasm must be of the appropriate viscosity so that it can be deformed and drawn into the leading edge. If too viscous, the cell's contents would be immobilized. The cell must be able to reversibly adhere to the substratum, and, as we have seen, the phenomenon of adhesion depends critically on the strength of weak chemical bonds. The cell must be capable of generating sufficient traction forces to pull the mechanically rigid scaffold and the cytoplasmic contents enmeshed within it toward the leading edge. These same traction forces must also be sufficiently powerful to pull apart the affinity bonds as the cell moves away from a region to which it was previously tied. Discussion of just how finely balanced these various forces must be is a subject of many recent papers in cell biology.[15] If the value of the viscosity of water which forms the matrix of the cell, the energy levels of weak chemical bonds which determine the strength of affinity bonds, and the power of the traction forces of the cell's myosin motors were all slightly different from what they are, crawling would in all probability be impossible. And as crawling itself plays such an essential role in all developmental processes it is difficult to see how higher organisms could be assembled if these values had been even slightly different from what they are.

Interestingly, both crawling and adhesion are also dependent on cells being approximately the size they are. If diffusion had limited cells to a size range of ten to a hundred times smaller than they are, then crawling would be impossible, as the required molecular machinery and its regulatory mechanisms could hardly be packaged into a volume a thousand to a million times smaller than the average animal cell. As the surface area of such a cell would be a hundred to a thousand times less, the total number and diversity of cell surface adhesion molecules would also be drastically reduced. So too would be the utility of selective cell adhesion as a mechanism for organizing the arrangement of cells during development. Such tiny cells would hardly be able to put out complex arrays of protrusions to feel their way through a developing embryo. Indeed, the surface of such tiny cells might have to be entirely devoted to devices involved in transporting materials across the bounding cell membrane and leave little room for adhesion molecules. Lacking the ability of selective cell adhesion, even if such small cells could crawl, it is doubtful if this ability could be utilized to generate the complex patterns of cell movement and association which underlies much of the morphogenesis of higher organisms.

Again, one is struck with the central importance the low viscosity of water plays in so many aspects of biology. The relatively rapid diffusion rates of small molecules in water permits the relatively large size of animal cells: between 10 and 50 microns across, containing more than a trillion atoms, and large enough to contain the necessary molecular machinery for crawling. It also ensures that the cytoplasmic mass of such cells has a low viscosity and can be mobilized relatively easily by the molecular motors in the cell. Cells are also large enough to deploy a large number and diverse set of adhesion molecules on their surface so that their crawling can be specifically directed by selective cell adhesiveness during embryogenesis. If the various interactions and metabolic transformations on which the life and replicative abilities of the cell depends had only been possible at higher molecular concentrations of these key constituents, the cell's interior would have been too viscous to move. But if the cell's interior had been less viscous, then dragging its contents forward on a mesh of mechanically rigid components would have been problematical, as the cytoplasm would keep flowing back through the mesh.

What is particularly striking is that many of the unique abilities of cells, such as their ability to selectively adhere to objects in their environment and to crawl, are critically dependent on the unique global properties of the cell's cytoplasmic mass and on the cell membrane having almost precisely the properties it does.

The plastic, metastable character of cytoplasm, which is so fit for crawling and selective adhesion, has not been created by natural selection. On the contrary, it is an inherent property of an aqueous solution of the constituents of the cell. Start with those key constituents of life—DNA, protein, sugars, lipids, etc., which are uniquely and ideally adapted for their various roles in the living process, including self-replication—dissolve them in water at the precise concentrations necessary for cell function and self-replication, and as if by magic, a colloidal substance is formed—cytoplasm—possessing just those qualities required to permit cells to crawl and to selectively adhere to each other and hence to assemble higher organisms and ultimately to make a world of life. It is surely highly suggestive of design that a soup of these basic vital ingredients at precisely the concentration required to carry out the miracle of self-replication surrounded by the lipid bilayer should have, coincidentally, precisely that suite of biophysical properties of viscosity, density, excitability, etc., ideally and uniquely suited for the cell to carry out its designated task of building a biosphere of multicellular life.

It is hard to escape the conclusion that the ability of cells to selectively adhere and to crawl, twin abilities upon which the assembly of multicellular organisms during development is critically dependent, could only find instantiation in an entity of the size and with the global biophysical and biochemical properties of the average animal cell.

Osmosis

Another physical phenomenon which has a critical bearing on the design of the cell, and particularly on the design of the cell membrane and hence the ability of cells to crawl and adhere, is the phenomenon of osmosis and its consequence, osmotic pressure. Osmosis is an inevitable consequence of the process of diffusion. It occurs wherever two solutions, one dilute and one concentrated, are separated by a membrane that is permeable to water but not to the solutes. In such a situation water moves from the dilute to the concentrated solution—in other words, the solution containing the most dissolved particles. The influx of water can only be prevented by applying hydrostatic pressure to the concentrated solution. This reverses the influx and forces the water back across the semipermeable membrane into the dilute solution.

Cells, in the human body and in nearly all other complex organisms, tend to contain a greater concentration of dissolved particles than does the extracellular fluid which surrounds the cell. There are two causes for this. The first is the so-called Gibbs-Donnan effect,[16] which arises because the surface of the proteins in the cell contains many positively and negatively charged groups and these attract a large number of counterions, including small organic molecules. The net effect is an increase in the concentration of particles inside the cell compared with that in the extracellular fluid. The second is that the cell necessarily contains a large number of small organic compounds which result from the cell's own metabolic activities. As the cell membrane is permeable to water (but not to proteins and many of the ions and small organic compounds in the cytoplasm), and as water moves by osmosis from a dilute to a more concentrated solution, every cell in the body of man and all other animals would suffer a persistent influx of water which, if the cell took no action to reverse, would ultimately have disastrous effects. The pressure inside the cell would increase, causing swelling of the cell until eventually the fragile cell membrane would rupture.

There are only two strategies available to avoid this consequence. One is to surround the cell in a very strong, rigid wall. This is the strategy adopted by plant cells, which are encased in a firm cellulose lining that is strong enough to resist the pressure that develops inside the cell by the osmotically driven influx of water. Bacterial cells have also adopted a similar strategy.

The other strategy—the one universally used by animal cells—is to continually pump ions such as sodium out of the cell and thereby maintain "unnaturally" lower concentrations of small inorganic ions in the cytoplasm than in the extracellular fluid outside the cell, which in turn draws water out of the cell.[17] By this ingenious strategy animal cells can avoid rigid cell walls. The pliable, nonrigid, and relatively fragile cell membrane suffices to hold the cell's contents together. The sufficiency of the typical animal cell membrane is of great significance, for as we have seen, many of the critical properties of cells such as crawling and adhesion depend on the pliability and deformability of the membrane.

This continual pumping strategy requires enormous amounts of energy, up to one-third of the total energy supply of many cells (two-thirds in the case of nerve cells).[18] This is why poisons which interfere with the cell's ability to generate or utilize energy cause immediate swelling of the cell. There is yet another element of fortuity in the fact that this massive pumping activity is commensurate with the energy-generating capacity of cells. Had the Gibbs-Donnan effect been just very slightly more intense, the basic design of the cell as a concentrated soup of proteins bounded by a pliable semipermeable membrane would not have been feasible. Cells would have had to devote all their energies merely to avoid the inrush of water, cell swelling, and death.

It is also fortunate that the absolute pressures generated by osmosis are not any greater than they are. Surprisingly, even quite dilute solutions can exert enormous osmotic pressures. A solution of 7 grams of salt dissolved in 100 milliliters of water develops an osmotic pressure of about 60 atmospheres, equivalent to the pressure 2,000 feet below the sea.[19] To prevent the water moving from the distilled water into the salt solution, a hydrostatic pressure of 60 atmospheres would have to be exerted on the salt solution—this would be equivalent to a column of water 2,000 feet high. Clearly, if the absolute value of the pressures generated by osmosis had been much higher, then no feasible cell could be designed to survive contact with very dilute aqueous solutions. The cell walls in the roots of plants would have had to be many times stronger and thicker to resist the increased pressures. And even in the cells of animals where the solute concentration in the intracellular and extracellular fluids are almost exactly balanced, minor fluctuations in their relative concentrations could well prove disastrous. If osmotic forces had been, say, ten times greater, the minor but relatively sudden dilution of the body fluids which occurs on drinking water might create osmotic imbalances of catastrophic consequence. Only if the cell walls of animal cells were far more rigid (like plant cells) would they have been capable of existence in a world where the pressures generated by osmosis had been ten times greater. But then the cell membrane would not possess those many critical characteristics upon which the world-building abilities of cells depend.

Energy Balance

Cells require energy not only to defend themselves against osmotic pressures by continuously pumping ions out across the cell membrane. They are also faced with the uphill task of continuously replacing all their molecular constituents, and this also requires energy. The half-life of many proteins in the cell varies from a few minutes to several days. Even those proteins which have relatively long half-lives, such as the hemoglobins, the proteins forming the contractile apparatus in the muscles and collagen, the major component of tendons, all turn over eventually and have to be replaced.

One reason for the relatively short half-life of many proteins is the highly reactive oxygen radicals which are ever present in the cell. These cause deleterious chemical changes that render them nonfunctional. If unchallenged, the accumulation of abnormal nonfunctional proteins would eventually clog up the cell's machinery and bring the life of the cell to a halt. Not only must the cell continuously replace these damaged proteins, but an additional energy burden is imposed by the need for their selective degradation. In fact, cells contain a whole set of enzymes whose only purpose is to selectively remove damaged proteins.

In an adult human about 15 percent of the energy expended is devoted to protein synthesis alone.[20] Clearly, if the stability of proteins in the face of oxidative and other types of chemical degradation was even slightly less, then the energy burden imposed on cells would be insurmountable and the cell system would not be possible in any recognizable form.

Are Cells Intelligent?

Crawling and selective adhesion, etc., are just some of the remarkable properties packed into these tiny particles of matter. Cells possess many other properties in addition to these. Even the smallest, simplest cells dazzle us with their abilities. The abilities of the tiny bacterium *E. coli,* which is far smaller than an animal body cell, are simply amazing. It can, for example, estimate the concentration of many of the common and important molecules it comes in contact with, and from the result of such calculations change the direction in which it is swimming toward the richest source of nutrient.[21]

But some of the behavioral repertoires of unicellular organisms such as amoebas many million times the size of a bacterial cell are even more amazing and truly astonishing, for a colloidal mass of protoplasm, although far bigger than a bacterium, is still only the size of a speck of dust. The dramas played out in a drop of water are almost as rich and diverse as those played out on the plains of Africa. In 1910, one famous scholar gave a very graphic description of an amoeba hunting prey. It is worth citing at length:

> Amoeba frequently prey upon one another. Sometimes the prey is contracted and does not move; then there is no difficulty in ingesting it. . . . But the victim does not always conduct itself so passively as in this case, and sometimes finally escapes from its pursuer. This may be illustrated by a case observed by the present writer. . . . I had attempted to cut an amoeba in two with a tip of a fine glass rod. The posterior third of the animal, in the form of a wrinkled ball, remained attached to the rest of the body by a slender cord. The Amoebae may

One amoeba hunting another.

From H. S. Jennings (1905), *The Behavior of Lower Organisms,* figure 21.

be called Amoeba *a* while the ball will be designated *b* [see fig. 10.5, above]. A larger amoeba (*c*) approached, moving at right angles to the path of the first specimen. Its path accidentally brought it in contact with the ball *b,* which was dragging past its front. Amoeba *c* thereupon turned, followed Amoeba *a,* and began to engulf the ball. A cavity was formed in the anterior part of Amoeba *a* . . . [sufficient to engulf the ball]. Amoeba *a* now turned into a new path; Amoeba *c* followed. . . . After the pursuit had lasted for some time, the ball *b* had become completely engulfed by Amoeba *c.* The cord connecting the ball with Amoeba *a* broke and the latter went on its way. . . . Now the anterior opening of the cavity in Amoeba *c* became partly closed, leaving only a slender canal. . . . There was no adhesion between the protoplasm of the ball and Amoeba *c.* . . . Now the large Amoeba *c* . . . began to move in a different direction . . . carrying with it its meal. But the meal—ball *b* now began to show signs of life, sent out pseudopodia, and became very active; we shall henceforth refer to the ball as Amoeba *b.* It began to creep out through the still open canal, sending forth its pseudopodia to the outside. Thereupon Amoeba *c* sent forth its pseudopodia in the same direction, and after creeping in that direction several times its own length, again engulfs *b.* The latter again partially escaped, and was again engulfed completely. Amoeba *c* now started again in the opposite direction, whereupon Amoeba *b,* by a few rapid movements, escaped from the posterior end of Amoeba *c,* and was free—being completely separated

> from *c*. Thereupon Amoeba *c* reversed its course again, overtook *b*, engulfed it completely again. . . . Amoeba *b* then contracted into a ball and remained quiet for some time. Apparently the drama was over. Amoeba *c* went on its way for about five minutes without any sign of life in Amoeba *b*. In the movements of *c*, *b* became gradually transferred to its posterior end, until there was only a thin layer of protoplasm between *c* and the outer water. Now *b* began to move again, sent pseudopodia through the thin wall to the outside, and then passed bodily out into the water. This time Amoeba *c* did not reverse direction and attempt to recapture Amoeba *b*. The two Amoebae moved in opposite directions and became completely separated.[22]

The amoeba, although the size of a small speck of dust, exhibits behavioral strategies which seem objectively indistinguishable from those of animals far higher up the scale. If an amoeba were the size of a cat, we would probably impute to it the same level of intelligence as we do to a mammal. Just how do such minute organisms integrate all the information necessary to make such apparently calculated intelligent decisions? While the processes underlying the selective stickiness of cells, their crawling ability, and even the molecular counting ability of *E. coli* referred to above have been worked out at least to some degree, accounting for the amoeba's behavior, the way it integrates all the information necessary to pursue its prey, its decision to change direction, its persistence in the pursuit when its prey escapes, the sudden breakout of the smaller amoeba from its imprisonment in the interior of its captor at the moment when the wall of protoplasm was at its thinnest—all this remains to be fully explained in molecular terms.

The behavior exhibited by both the pursued and the pursuer must involve an extraordinary level of sophisticated information processing. It is not at all clear how this is done; despite all the advances in computer technology, building completely autonomous robots to mimic the behavior of an amoeba is quite beyond our capacities at present. The behavior of the amoeba in pursuing its prey and our inability to give a coherent account of this in molecular terms shows that there is still a vast amount of complexity in the cell still remaining to be uncovered, highlighting again how much we still have to learn.

The possibility that there could be radical new levels of complexity in the cell is raised by some recent ideas proposed by Stuart Hameroff of the University of Arizona. According to Hameroff, the traditional estimates of the brain's computing power based on a neural net assumption that each "neuron-neuron synaptic connection is a fundamental binary switch led to classical estimates of the brain's information processing capacity of $4 \cdot 10^{12}$ bits per second (40 billion neurons changing their state 100 times per second)."[23] However, Hameroff suggests that there might be another computing system within each cell based on the "microtubules" which make up the cytoskeleton of the neurons. The microtubules are tiny tubes made up of protein subunits that form an important component of the cytoarchitecture in virtually all the cells of higher organisms. Such a subcellular system would vastly add to the power of the brain's computing capacity, for it would mean that within each neuron there was a minicomputer with computing power equivalent to that of an IBM desktop computer![24]

Hameroff's ideas cannot be ruled out of court, although there is no direct evidence for his claims at present. But as he notes, "single-celled organisms like amoeba and paramecium perform complex tasks without benefit of synapse, brain or nervous system."[25] And as he points out, such subcellular minicomputers could regulate and organize processes like transport, cell growth, cellular movement, as well as the guidance and movement of single-celled organisms. Perhaps the amoeba is far more intelligent than our current knowledge of the cell suggests. Maybe its hunting strategies are aided by a subcellular computer of immense power. Even in the case of insects, it has always been something of a puzzle just how their complex behavior could be packed inside brains which, by vertebrate standards, are extremely small. The abilities of insects puzzled Darwin: "It is certain that there may be extraordinary mental activity within an extremely small absolute mass of nervous matter; thus the wonderful diversified instincts, mental powers, and affections of ants are notorious, yet their cerebral ganglia are not so large as the quarter of a small pin's head. Under this point of view, the brain of an ant is one of the most marvellous atoms of matter in the world, perhaps more so than the brain of man."[26] Although considerable progress has been made since Darwin's day in explaining insect behavior in terms of a few stereotyped routines, the possibility that subcellular computing devices may play some role in their behavior cannot be entirely ruled out.

The concept of subcellular computing devices using the microtubules was discussed by the British physicist Roger Penrose in his latest book, *Shadows of the Mind.*[27] If there is any truth in Hameroff's suggestion, then the human brain itself may be vastly more complex than currently supposed.[28]

There may be other types of subcellular computational devices in addition to the microtubules. Many protein networks in the cell may serve this function.[29] In short, there is much that remains to be discovered. Cells may possess many additional abilities, vastly more complex and sophisticated than any we know of at present. But as it is, from the scientific knowledge we have already acquired, there is no doubt the cell represents an exceptional and unique material form. And one which, as in so many other instances, seems ideally and uniquely fit for its biological role in creating the world of multicellular life. Is it conceivable that there could exist some other tiny material form which could compact, into so small a volume, so many extraordinary abilities as a living cell?

Conclusion

We are now in a position to consider playing again the game we borrowed from Robert Clark and started in chapter 6, when we imagined ourselves as Plato's *Demiurge* creating a world of life from scratch. As we saw starting the game with the carbon atom, two primary requirements would be a medium in which our carbon-based life could function and a vehicle to distribute the carbon atom to all parts of our imaginary biosphere. When we searched through the vast number of known fluids and the vast number of carbon compounds, we found only two—water, and the gas carbon dioxide—that would do. We found that they would not only do, but were amazingly fit for their respective functions, and in many ways. We found oxidation to be the obvious choice for energy generation and it too was amazingly fit for carbon-based life in many ways. As the game continued, for every new constituent we required, there was a ready-made solution that seemed ideally and uniquely prefabricated, as if by design, for the biological end it serves.

Having now reviewed, in the past few chapters, the mutual adaptations of DNA and proteins and the utility of the various metals for carbon-based life, we can continue our game toward the creation of a complex self-replicating system and again everything continues to fall into place with ridiculous ease. We would need molecular machines to manipulate atoms and we would find that proteins are ideal for this role. We would then find that they are also wonderfully fit to function in conjunction with the nucleic acids which we discover are themselves ideal as information bearers. We would want to control and manipulate oxygen, and we would find yet again ideal oxygen manipulators in the transitional metals. Finally, we would want to wrap up our constituents into a small packet or cell, and we would find to our astonishment that there exists a very simple means of bounding the cells—a lipid bilayer which has a suite of properties that are, again, just what we need. Even more remarkable we find that when we have packed all our constituents into our "cell," the viscosity of the resulting material and the size and properties of the membrane are all just about right for crawling and selective adhesion, properties we need to build multicellular organisms and a vast complex world of life. We would find an ideal source of energy in the radiant light of main sequence stars, and ideal habitats for our carbon-based life on the trillions of earthlike rocky planets which abound throughout the cosmos. At every step in the game we would find the same ready-made solution for each particular biological end we sought. And this would be repeated in case after case, leading down through a long, seemingly endless chain of coincidences from the carbon atom to the cell and eventually to a world of life very similar to that which exists on earth.

In short, the cell system as revealed by molecular biology has turned out to be a unique and peerless whole in which every component is uniquely fashioned by the laws of nature for its designated role, a three-dimensional jigsaw in which all the pieces fit together as perfectly and harmoniously as the cogs in a watch. As we have seen, every constituent—water, the carbon atom, dioxygen, carbon dioxide, bicarbonate, the lipids, the lipid bilayer membrane, the double helix, the proteins, the genetic code, the iron atom, the molybdenum atom, calcium, magnesium, sodium and potassium ions, and so on—appears unique and ideally adapted for its respective biological role. There are many other constituents, not reviewed here, which also give the appearance of being specially fit for their biological roles. These include the sugars and the storage form of sugar, the compound glycogen.[30] The phosphate radical which occurs in the energy-carrying molecule ATP and is also used as the joining radical between adjacent bases in the DNA is also uniquely fit for its many biological roles. And again, as in so many instances, there is no alternative—no other compound will substitute for the phosphate radical in the many functions it performs in the cell.[31] (See Appendix, section 3.)

Future work may well reveal that many other of the cell's constituents are also uniquely fit for the biological roles they play. This was hinted at by George Wald when he raised the possibility that chlorophylls, the heme pigments, the carotenoids, and vitamin A (involved in photoreception) may all possess properties that fit them uniquely to perform their specific functions, so much so that all carbon-based life forms anywhere in the cosmos would of necessity be forced to use these very same compounds in attempting to achieve the same ends for which they are now utilized on earth.[32] The fact that vitamin A is used in every known visual system throughout the animal kingdom is highly suggestive.[33] Is it possible that without vitamin A there might be no vision? The universal occurrence of the compound acetylcholine in organisms as diverse as plants, protozoans, and mammals, and its use as a neural transmitter in organisms belonging to virtually every animal phylum, struck Carl Pantin as suggestive that it might possess properties for neurotransmission that are not available in any other molecule.[34]

Even above the level of the individual molecule, many of the structural materials used by living things, materials such as bone, skin, tendon, calcareous shells, chitin, and wood, which are what a structural engineer would class as composites, are also remarkable for their apparently ideal biomechanical characteristics.[35]

A final and very remarkable aspect of the fitness of the constituents of life is that most of the key organic building blocks—sugars, amino acids, nucleotides, etc.—can be manufactured in a relatively small number of chemical steps from a small number of readily available simple molecules. It is a remarkable fact that the great majority of the atoms used in their synthesis are derived from only three very simple molecules that are available freely and in great abundance on the surface of the earth: water, carbon dioxide, and nitrogen. Not only are the key components of life wonderfully fit for their biological roles, they are all only a very small chemical distance away from such universally available starting materials. Indeed, there are not many steps from hydrogen itself—the starting point of atom creation in the stars—to the ingredients of life.

But not only is this remarkable set of key building blocks readily synthesized from available materials, they can all be readily interconverted via a small number of chemical steps. It is fortuitous, indeed, that so many of the key molecules of life, which possess so many unique chemical and physical properties, all exist within easy chemical reach of each other. The astonishing chemical proximity of all life's constituents is surely a fact of very great and crucial significance. The fitness of the individual ingredients, such as lipids, proteins, and DNA, although remarkable enough, is insufficient in itself; it is only because all the components of life can be derived easily from simple starting materials and interconverted readily that the miracle of the cell and self-replication is possible.

Contrast this with artificial systems, even fantastically simple ones quite incapable of replication, such as a motor car or a computer or a typewriter. In the case of such artificial machines, each individual component, such as a metal rod, a silicon chip, or a plastic disc, can only be manufactured by long circuitous routes involving complex industrial processes that may involve temperatures of 1,000°C and all manner of diverse chemical processes.

The emerging picture is obviously consistent with the teleological view of nature. That each constituent utilized by the cell for a particular biological role, each cog in the watch, turns out to be the only and at the same time the ideal candidate for its role is particularly suggestive of design. That the whole, the end to which all this teleological wizardry leads—the living cell—should be also ideally suited for the task of constructing the world of multicellular life reinforces the conclusion of purposeful design. The prefabrication of parts to a unique end is the very hallmark of design. Moreover, there is simply no way that such prefabrication could be the result of natural selection. Design in the very components which make an organism possible cannot be, as Carl Pantin pointed out some time ago, *the result of natural selection.*[36] The many vital mutual adaptations in the constituents of life were given by physics long before any living thing existed and long before natural selection could have begun to operate.

In the current molecular biological picture of life, we have found a "watch" more complicated and more harmonious than any conceived by William Paley, exhibiting in its design precisely what Richard Bentley was looking for, a "*usefulness conspicuous not in one or a few only, but in a long train and series of things.*" (My emphasis.)

Chapter 11

Homo Sapiens: Fire Maker

In which it is argued that our species may be uniquely fit to explore and understand the cosmos and that the laws of nature appear also to be uniquely fit for large organic forms of our size and dimension. The evidence is not conclusive, but highly suggestive. Our species exhibits a set of adaptations which are collectively unique among carbon-based life forms on earth. These include high intelligence, linguistic ability, the hand, high-acuity vision, the upright stance, sociability. Moreover, the design and dimensions of the human body are fit for the handling of fire—a crucial ability, because it was only through the conquest of fire that humans discovered metals, developed technology and science, and ultimately came to comprehend the laws of nature and grasp the overall structure of the cosmos. Many coincidences appear to underlie our fitness for handling fire and our fitness for understanding the cosmos. For example, the earth's size and atmosphere are fit both for beings of our size and dimension and also for fire. The strength of muscles is commensurate with mobility in a being of our size on a planet the size of the earth. The laws of nature conform to mathematical patterns which the human mind seems curiously adapted to grasp. In conclusion, the cosmos appears to be fit for our being and our understanding.

The first muscle man.

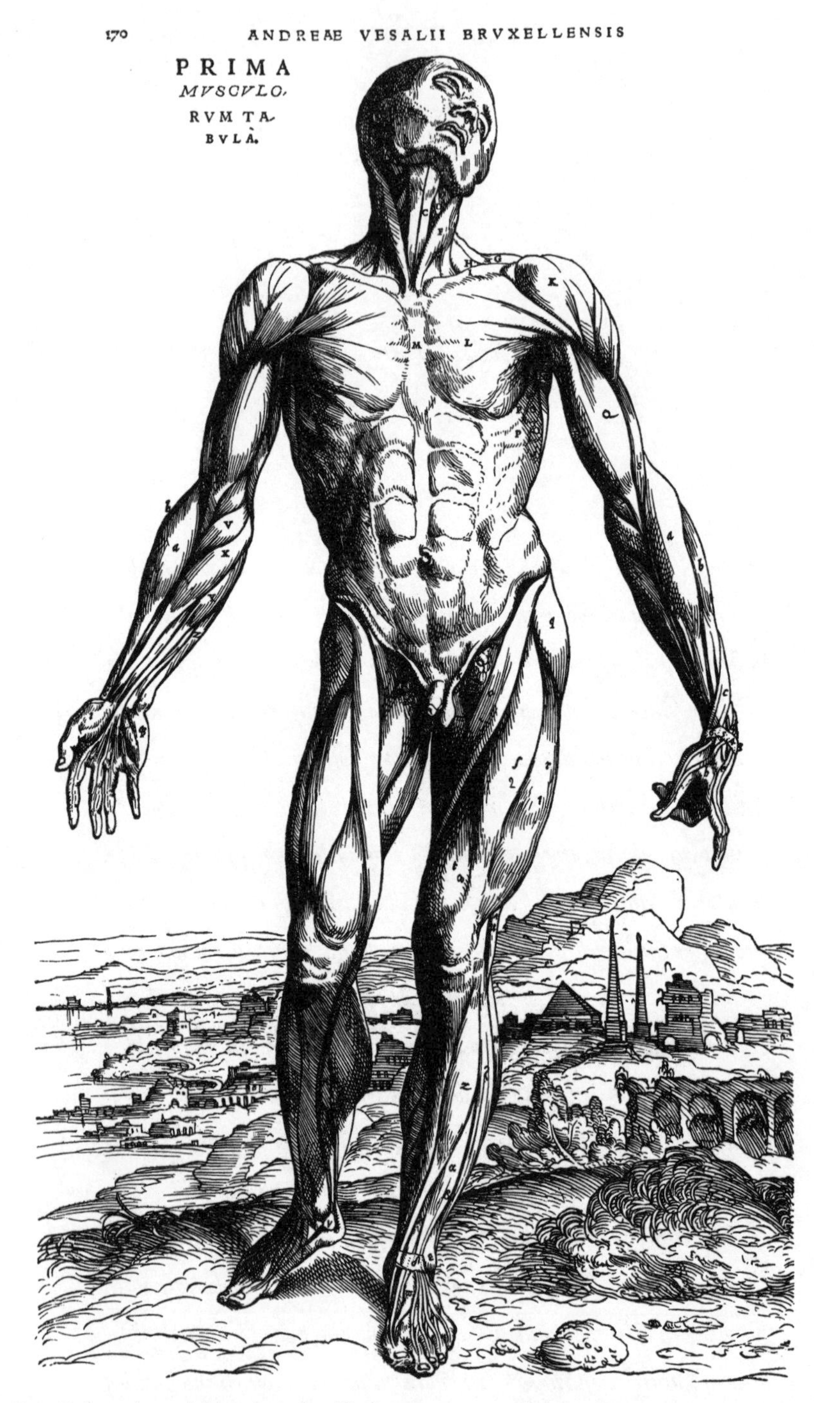

From *De humani corporis fabrica* by Andreas Vesalius.

How noble in reason, how infinite in faculties, in form and moving, how express and admirable in action, how like an angel in apprehension, how like a God! the beauty of the world; the paragon of animals.

—William Shakespeare, *Hamlet*

"For what purpose," asks Cicero, "was the great fabric of the universe constructed? Was it merely for the purpose of perpetrating the various species of trees and herbs which are not endued even with sensation?—the supposition is absurd. Or was it for the exclusive use of the inferior animals? . . . which, although endued with sensation, possess neither speech nor intelligence. For whom then was the world produced?—doutbless for those beings who are alone endued with reason."

—John Kidd, Bridgewater Treatise, *The Physical Condition of Man,* 1852

The ancient Greeks, who had an answer to most things, believed that Prometheus brought down fire from heaven—and got himself into much trouble with Zeus for doing so. "From bright fire," says Aeschylus in *Prometheus Vinctus,* "they will learn many arts."

—A. J. Wilson, *The Living Rock,* 1994

Of all the many varied life forms on earth, only our own species, *Homo sapiens,* is capable of any genuine understanding of the world. By any standards our success in comprehending and manipulating nature has been astounding. In the space of only four centuries since the scientific revolution, we have measured the diameter of galaxies, we have probed into the heart of the atom, we have peered back to the very beginning of time, and in the past few decades we have even contemplated traveling to the stars.

Our intellectual endowment is certainly remarkable, but are we unique, as the anthropocentric thesis predicts? Could such genius and abilities be instantiated in some other material form? Could some other thinking being radically different in design to *Homo sapiens* have been equally successful at unraveling the secrets of nature? Could there even be, as modern science fiction implies, a veritable phantasm of other beings utterly alien and exotic but just as "noble in reason and infinite in faculties" as ourselves?

From the evidence presented in the previous chapters, such a phantasm of alien beings—designed along entirely different principles and instantiated in an exotic chemistry—looks increasingly implausible. For as we have seen, it would appear that there are few if any alternative ways of putting together the atoms of the world into a complex self-replicating system as sophisticated as the living cell. If we start from the carbon atom, our route is highly constrained. Having chosen carbon, we must next choose water, then proteins, DNA, oxygen, and so on until we arrive eventually at the design of the living cell as manifest in all living things on earth.

But even if life based on the carbon atom is the only form allowed by physics, it is obvious from the variety of life on earth that the possible number of complex carbon-based multicellular life forms is immense and that our own species, *Homo sapiens,* is but one within a universe of possibilities. Could it be that within this plenitude the only type of organism capable of manipulating and exploring and eventually understanding the world is an upright bipedal primate of biology and design very close to that of *Homo sapiens*? I believe the evidence strongly suggests that the answer is yes.

Key Adaptations

Six adaptations have been widely cited as being crucial to the unique success of our species: (1) high intelligence, (2) linguistic communication, (3) highly developed visual ability, (4) possession of a superb manipulative tool—the hand, (5) our upright stance, and (6) our being a highly social species. In addition to these six adaptations, our technological success has depended on a crucial ability—the ability to handle and control fire, which led in turn to the development of metallurgy and ultimately, through the use of metals, to scientific and technological knowledge. Although the evidence currently available is insufficient to prove the case, it is hard to imagine these adaptations plus the additional ability to handle and manipulate fire (which poses its own rather specific design constraints on an organism) being possessed by any organism markedly different to a modern human.

As far as our cognitive capacities are concerned, it is true that other species—dolphins, parrots, seals, and apes—possess intelligence, but none, as far as we can tell, comes close to the intelligence of man. At present, the basis of our unique cognitive capacity is quite mysterious. (See discussion below.) It may be rooted in some curious wiring feature of the primate brain or may be in some way connected with the evolutionary path which led to modern man. Whatever it is about the architecture of the human brain that confers such a high level of intelligence, and whatever evolutionary processes led to such a prodigious development, as far as life on earth is concerned, our intelligence far surpasses that of any other known form of life.

Language

Language is another unique distinguishing characteristic of man. No other species possesses a communication system remotely as competent for the transmission of new information or abstract concepts as human language.[1]

Of course, in themselves neither intelligence nor the capacity for language are sufficient to provide an organism with the ability to understand the world. The body must also be fit for the task. The brain and its capacity for abstract symbolic manipulation must be conjoined with body and organ systems through which the brain can interact with the outside world. It is only because our brain can sense and experience the world and translate our thoughts into actions that we are able to explore, manipulate, and ultimately understand the world. A computer, no matter how "intelligent," is unable to communicate with the outside world via a "body," and is thus incapable of such exploration.

Further, capacity for speech would be of little utility without the appropriate equipment to produce it. Human speech depends not only on our special cognitive abilities but also on our possessing the appropriate organs to generate complex sound patterns. In fact, modern man's speech-producing apparatus is quite different from the comparable systems of living nonhuman primates. Nonhuman primates have supralaryngeal vocal tracts in which the larynx exits directly into the oral cavity. In the adult human the larynx exits directly into the pharynx. This confers on man the capacity to generate a far richer phonetic repertoire than that available to a chimpanzee.[2] Many vowels and consonants used in human language could not be reproduced by a chimp. A chimp with a human brain could formulate sophisticated thoughts but would lack the ability to communicate verbally as efficiently as a human.[3] Our vocal apparatus is in all probability at least as important as our bipedalism as a prerequisite to our becoming fully human.

Language also presumes a gregarious social animal. Man in common with all the other primates is highly social. No solitary species would develop a language. Sociality in general was probably an essential element in man's biological and intellectual evolution.

Vision

Our visual ability is hardly less significant than our ability to communicate by speech. Possessing intelligence and speech but lacking good eyesight, it is difficult to imagine how humans could have acquired extensive knowledge of the world. Aristotle, in this famous section from the beginning of his *Metaphysics,* acknowledges the importance of vision to our ability to comprehend the world:

> All men by reason desire to know. An indication of this is the delight we take in the senses . . . and above all in the sense of sight. . . . The reason is that this, most of all the senses, makes us know and brings many differences to light.[4]

We saw in chapter 3 that the resolving power of the human eye is close to the optimum for a camera type of eye using biological cells as photodetectors. Its visual acuity cannot be improved to any significant degree by making changes in its absolute size or the relative size of its components. For this reason nearly all the eyes of those higher organisms which possess high-acuity vision are approximately the same design and dimension, roughly between 2 and 6 centimeters in diameter. As we saw, compared with most other biological structures, the size of the high-acuity vertebrate camera eye is quite large. Although it is only about one-thousandth the length of a California redwood (the largest living structure), it is about 1 million times longer than a bacterial cell. The vast majority of organisms are far smaller than the human eye. Even some birds and mammals weigh far less than the eye. Neither an ant nor even a mouse could support an organ the size of the human eye. Neither can see as clearly as a man and neither could be creatures of genuine understanding. To see clearly, *Homo sapiens* must be a relatively large organism on the scale of all biological forms.

The Hand

In addition to our brain, our linguistic ability, and our highly developed visual ability, we possess another wonderful adaptation, the ideal manipulative tool—the human hand. No other animal possesses an organ so superbly adapted for intelligent exploration and manipulation of its physical surroundings and environment. Only the great apes, our cousins, come close. Yet the hand of the chimp and gorilla, although possessing an opposable thumb, is far less adapted to fine motor movement and control.[5] Although some chimps are remarkably dexterous,[6] when one sees them attempt even simple manual tasks, they appear clumsy and inept compared to humans. Even a chimp with the intelligence of a human would have considerable difficulty carrying out many of the manipulative tasks that we take for granted, like peeling an apple, tying a knot, or using a typewriter.

One of the earliest and still one of the most fascinating discussions of the adaptive marvel that is the human hand was given by the first-century physician Galen: "To man the only animal that partakes in the Divine intelligence, the Creator has given in lieu of every other natural weapon or organ of defence, that instrument the hand: applicable to every art and occasion."[7] And he continues: "Let us then scrutinise this member of our body; and enquire not simply whether it be in itself useful for all the purposes of life, and adapted to an animal endowed with the highest intelligence; but whether its entire structure be not such, that it could not be improved by any conceivable alteration."[8] The adaptive perfection of the hand was a popular topic among nineteenth-century natural theologians.

In the context of explaining man's biological preeminence on earth, the crucial question is not whether the human hand represents the absolute pinnacle of manipulative capability, but whether any other species possesses an organ approaching its capabilities. The answer simply must be that *no other species possesses a manipulative organ remotely approaching the universal utility of the human hand.* Even in the field of robotics, nothing has been built which even remotely equals the all-around manipulative capacity of the hand.

The hand not only provided man with the ability to manipulate and explore his environment but also with the ability to construct all manner of diverse tools and instruments, the use of which has been crucial to the acquisition of technological and scientific knowledge. It is impossible to envisage man progressing beyond the most primitive technology without the hand.

The hand, like any other organ, does not function in isolation. In fact, its utility is dependent to a large extent on that other crucial and unique adaptation of man, our upright stance and bipedal gait. Without these, the human hand would not be free to execute its manipulative explorations. All the great apes are basically quadrupeds, defined as knuckle-walkers by Owen in the nineteenth century.[9] Only among man's immediate antecedents (known only from fossil remains) is a habitual bipedal posture and gait achieved.

In addition to the above five adaptations, there are other aspects of our biology which have enabled us to be truly *Homo sapiens,* most notably that we are a social animal, a condition of great significance. Being social was not only almost certainly essential to the evolutionary development of language and other key aspects of our intellect, but only a social species could have ever developed an advanced technology through which to further the exploration of nature.

Fire and the Dimensions of the Human Body

In addition to the above adaptations, our ability to handle and manipulate fire is also critically dependent on the basic design and dimensions of the human body being close to what they are.

Our ability to handle fire is no trivial ability because it was only through the use of fire that technological advance was possible. Through fire came metallurgy and metal tools and eventually chemical knowledge. Because metals are the only natural conductors of electricity, the discovery of electromagnetism and electricity, even the development of computers, are all in the last analysis the result of our ancient conquest of fire. That fire is a phenomenon of great significance was perceived from the earliest of times. In many cultures it was invested with mystical and magical significance. Prometheus was condemned because he stole fire from the gods, and in ancient Persia it was worshiped as the manifestation of the Deity.

That fire is itself a remarkable phenomenon has already been noted. That the chemical reaction between carbon and oxygen is manageable at all is the result of the relative chemical inertness of the carbon atom and dioxygen at ambient temperatures. This chemical inertness is not only fit for oxidative metabolism, which provides energy for living systems; its attenuating influence also makes carbon combustion of utility to humans. It is only because of the slowness of the combustion of wood that fire can be handled by a large terrestrial organism on a planet like the earth.

Because the smallest sustainable fire is about 50 centimeters across, only an organism of approximately our dimensions and design—about 1.5 to 2 meters in height with mobile arms about 1 meter long ending in manipulative tools—can handle fire. An organism the size of an ant would be far too small because the heat would kill it long before it was as close as several body lengths from the flames. Even an organism the size of a small dog would have considerable difficulties in manipulating a fire. So we must be at least the size we are to use fire, to utilize metal tools, to have a sophisticated technology, to have a knowledge of chemistry and electricity and explore the world. It would appear that man, defined by Aristotle in the first line of his *Metaphysics* as a creature that "desires understanding," can only accomplish an understanding and exploration of the world, which Aristotle saw as his destiny, in a body of approximately the dimensions of a modern human.

Would an upright bipedal primate much larger than a modern human be feasible? Probably not. The design of a bipedal primate of, say, twice our height and several times our weight would be problematical to say the least. As it is, our upright stance puts severe strain on our lower back, especially on the intervertebral discs. Such a gigantic primate would almost certainly require thicker legs, suffer severe spinal problems, and be less nimble than modern man, and certainly no more capable of building a fire.

Being the size we are is also essential in another way. It is very unlikely that a brain the size of a bee's, which contains less than a million nerve cells, would be large enough to support intelligence remotely like that of man. Although, as Sir Julian Huxley concedes in his book *The Uniqueness of Man,* we have no way of knowing how big a brain built on biological principles out of nerve cells interconnected into a vast synaptic network needs to be to support intelligence, there is every reason to suppose that it must be quite large, as he notes: "The intelligence of a rat would be impossible without brain cells enough to outweigh the whole body of a bee, while the human level of intellect would be impossible without a brain composed of several hundred million cells and therefore reckoned in ounces, outweighing the majority of animals."[10] The fact that all other species exhibiting a degree of intelligence, including porpoises, parrots, apes, seals, etc., have brains close to our own in size, supports strongly the notion that a prerequisite for intelligence is a large brain.

The handling of fire would also be very difficult in an organism without a highly developed sense of vision. And again, only a relatively large organism can possess a high-acuity eye. It turns out, then, that to utilize fire we need to be approximately the size we are for several reasons: to be able to physically manipulate the actual fire itself, to have a brain sufficiently large to support the intelligence required to control that physical manipulation, and to have an eye to see the fire. As well, we need to have manipulative organs somewhat close to the design of arms and hands in modern humans.

Fire and the Size of the Earth

There are some intriguing coincidences related to our biological design and our ability to utilize fire. A carbon-based organism of our size and design possessing an upright bipedal posture is only feasible on a planet of approximately the size and mass of the earth. It is the size of the earth (or more specifically, its total mass) which determines the strength of its gravitational field. This in turn limits the maximum size of large terrestrial organisms like ourselves. If the earth had only twice the diameter, its gravitational field would be eight times stronger and a large upright bipedal creature like ourselves would not be feasible. In a very important sense, then, the earth's size is fit for the design of a bipedal animal of the dimensions of a man and therefore fit for our ability to handle fire.

But this is not all. As we saw in chapter 4, if a planet is to possess a stable hydrosphere and atmosphere fit for life, it must of necessity also possess a mass and consequently a gravitational field very close to that of the earth and undergo the same geophysical evolution. As we saw, its gravity must be strong enough to retain the heavier elements but weak enough to permit the initial loss of the lighter volatile elements, such as hydrogen and helium. Consequently, earth's mass is also fit for the evolution of an atmosphere similar to today's, in density and composition, containing oxygen and therefore capable of sustaining fire.

So the mass of the earth is not only fit for an atmosphere capable of sustaining a complex biosphere and supporting fire, it is also fit for an organism of the weight, size, and dimension capable of utilizing it. The use of fire is of course dependent on additional environmental factors—on the availability of wood, for example, and relatively dry conditions. Unless these additional factors were also favorable, then despite all the physical and mental adaptations which makes us *Homo sapiens* and despite the fitness of the earth as an abode for carbon-based life, neither fire, metallurgy, chemistry, nor any scientific progress would have been possible.

Muscles and Movement

The manipulation of fire necessitates movement. In the case of a large organism the size of a human, this in turn necessitates special structures—muscles—capable of exerting mechanical forces.

Moreover, if muscles are to be fit to give mechanical power and motility to an organism the size of a human, then they must be capable of generating considerable mechanical force, of controlling the generation of this force, and of exerting it repeatedly over short periods of time. It is now known that the muscles of the human body and indeed the muscles of all organisms have the same basic design, consisting of densely packed arrays of contractile elements, and are of approximately the same strength—i.e., they exert the same force per unit volume.[11]

Recent research into the molecular structure of the contractile machinery has shown that each basic working component in the muscle cell is an individual protein molecule consisting of a long tail and a short head rather like an elongated tadpole, and known as a myosin motor. Movement comes about as a result of a sequence of three conformational changes. First, the myosin head attaches itself to another long fibrillar molecule known as actin, indicated as point *a* in diagram 1 below. Second, as shown in 2 below, the head bends suddenly—the power stroke—and this bending causes the myosin molecule and the actin to move in opposite directions. Third, as shown in 3, the head unbends and attaches itself to the actin at point *b*. The sequence is repeated again, and gradually, via a series of small steps, the two molecules slide past each other.

Recent work has also shown that each myosin head moves about 8 nanometers with each power stroke and that the heads are stacked in the

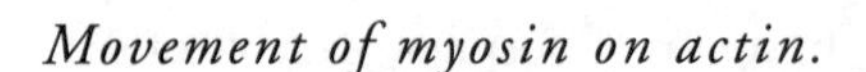

Movement of myosin on actin.

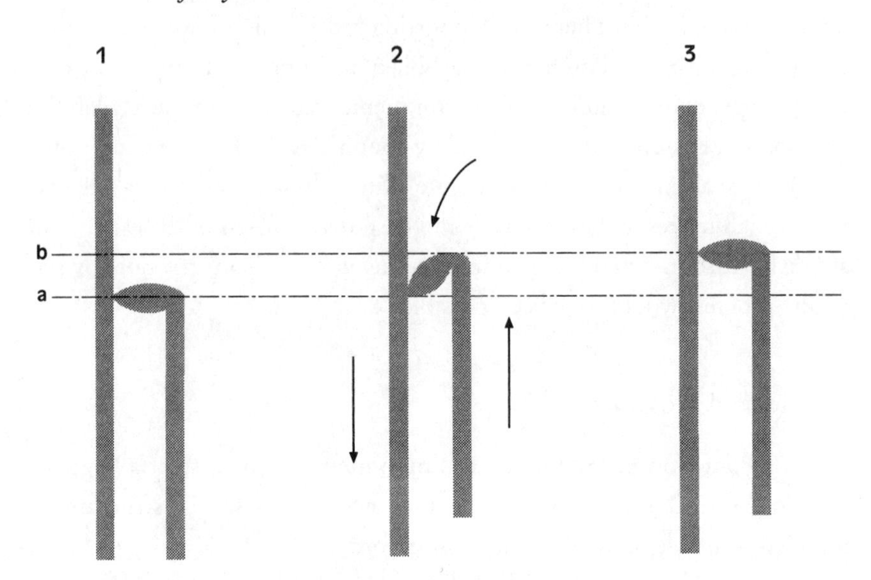

muscle fibrils in a helical conformation about 14 nanometers apart.[12] From consideration of the geometrical constraints consequent on the size and movement of the myosin heads, it is unlikely that further improvement in muscle power could be achieved by increasing the density of packing of the myosin motors.

One hypothetical strategy for increasing the power of muscles might be to envisage increasing the force of the individual power strokes that each myosin head makes as it bends and pushes on the actin fiber. However, recent measurements of the force of an individual power stroke show that this is about 3 piconewtons, and this is already several times greater than the strength of individual weak bonds.[13] Because it is the weak bonds that hold all the cell's constituents together, including the myosin motor and the actin fiber on which it pushes, then it is impossible to increase the force of the power stroke to any significant degree or each stroke would cause damage not only to the myosin motor itself but also to the actin fiber and other delicate components of the contractile machinery.

The movement of muscles is not the only type of biological movement that is based on a sliding fiber design involving a power stroke delivered by a moving head. The same sliding filament system is now known to underlie all biological movement, including the intracellular transport of mitochondria and other organelles in the cytoplasm. (The only biological movement not empowered by the same type of molecular motor is the rotary flagellum system of the bacteria.) The power strokes of the various other types of molecular motors have also been measured, and these are very close to the 3-piconewton level of the myosin motors in mammalian muscle.[14]

Overall, the evidence suggests strongly that, for fundamental reasons, the maximum power stroke of any sort of molecular motor cannot be much greater than it is. And since the packing of the myosin motors in muscle tissue is virtually crystalline and just about as tight as possible, then muscles cannot be designed, on biological principles, to generate any greater degree of power.

If either the tightness of packaging or the power of the motors had to be less for some reason, then organisms of our size and weight would not be feasible, because the muscles would be unable to generate the necessary mechanical forces to lift the body off the ground and hardly any movement of any sort would be possible. As it is, the human body is between 40 and 50 percent muscle, and as every medical student comes to learn when first dissecting the human body at medical school, our limbs are almost entirely composed of muscles. It would be simply impossible to redesign the human body to function with muscles ten times less powerful per unit volume. Indeed, no feasible large terrestrial organisms built on biological principles could be designed to move with muscles ten times less powerful. Even muscles only two or three times less powerful per unit volume would create considerable design problems.

In this context it is interesting to note that the strength of the grip of the human fingers is generated by extrinsic muscles in the forearm and not by the small muscles in the hand itself. Given the existing contractile power of muscle, this placement of the grip muscles in the forearm is not in the least bit gratuitous but of absolute necessity because the muscle bulk necessary to provide the required strength of grip cannot be accommodated in the hand. The fact that it is necessary, even with the strength of muscles as they are, to place the muscle-generating grip in the forearm indicates the tremendous difficulties that would be encountered in attempting to redesign the human frame to handle fire and to inhabit a planet the size of the earth if muscles were even slightly less powerful. It is astonishing that the design of the musculature of the arm of man and even the placement of specific muscle groups can be rationalized to a very large degree from consideration of the force delivered by one individual molecular motor. And it is clear that our existence is critically dependent on that force being almost precisely what it is. Any less, and we would be unable to move; any more, and each power stroke would tear the contractile machinery apart.

It is evident, then, that to make a fire both the power of our muscles and our dimensions must be very close to what they are. A miniature human, one-half or one-third our size, would have considerable difficulty in manipulating logs of more than one kilo in weight. Such a being would be restricted to making fires using small twigs, and whether the heat and sustainability of such fires would have sufficed for the discovery of metals and for the development of metallurgy is open to question. Metallurgy necessitates high temperatures of several hundred degrees.[15] The melting point of gold is 1,064°C, of silver 960°C, of copper 1,083°C, and that of iron 1,535°C. Only tin has a low melting point of 231°C, but tin is far too malleable to be of any great utility in itself for toolmaking.

The strength of muscles is not just relevant to the movement of our limbs and the maintenance of an upright posture. It is muscles which drive the circulation and provide the heart with its ability to pump the blood. It is muscles which move the chest during respiration. If muscles were less powerful, then despite the low viscosity of water, the circulatory system would be impossible. The work of breathing, which necessitates the movement of air through the respiratory passages, is only possible because the density and the viscosity of air is very low. As it is, during periods of maximal ventilation the work of breathing takes up 30 percent of the oxygen consumption of the body. If the power stroke of the myosin motors were somewhat less than it is, it is easy to see that the work of breathing would soon become prohibitive. It is doubtful if any type of large, complex air-breathing organisms would be possible. The circulatory and respiratory systems are only feasible because the viscosity and density of water and air, and the power of the myosin motors, are all very close to what they are.

And this is not all. Not only must the power stroke of each myosin motor exert about the force it does; the energy requirements to drive an individual stroke must also be close to what they are. We have seen that the delivery of oxygen to the human body requires complex adaptations. It is virtually impossible to envisage any sort of radical redesign of either the circulatory or respiratory systems in complex organisms that would double or treble the delivery of oxygen to muscle tissues. As it is, during strenuous activity much of the volume of active muscles is made up of blood capillaries. If the power stroke of molecular motors were only half or a third as efficient—i.e., if it required two or three times more ATP or metabolic energy—then motile complex forms of life would in all probability be impossible.

The Speed of Nerve Conduction

Muscles, no matter how powerful, would be of little biological value unless their movements could be carefully controlled. In the human body the control of muscular movement is carried out by the nervous system.

Catching a ball, rowing a canoe against a current, dodging a wave breaking on the beach, blinking an eye to prevent small objects from impacting on the cornea, handling and manipulating the moving embers in a fire—indeed, manipulating any mobile structures and elements in the environment—all such activities require fast movement and this in turn necessitates rapid reflexes. In an organism the size of a human, rapid movement and reflexes are possible only because the speed of nerve conduction is very rapid. If the maximum possible speed of nerve conduction had been one hundred or one thousand times slower than it is, life as we experience it would be unimaginable and even the simplest of tasks would be full of enormous danger. Constructing and manipulating a fire would be hazardous indeed!

In fact, the speed of nerve conduction imposes an absolute limit on the maximum size that an animal can attain. No animal can be 200 meters long and at the same time nimble. Even at the fastest conduction speeds, in a 200-meter-long organism, a nerve impulse will take four seconds to travel from the brain to its extremities and back.

Among organisms, nerve conduction speeds vary over more than three orders of magnitude, from 10 centimeters per second in simple invertebrates to 120 meters per second along the high-speed myelinated axons in the nervous system of mammals.[16] Consideration of the basic characteristics of nerve impulse propagation suggest that the speed of conduction in the myelinated axons of mammals is close to the maximum possible compatible with the electrical properties and general design of cells. Obviously, fine control and coordination of muscular activity in organisms of our size is only possible because such relatively rapid speeds of nerve impulse conduction are in fact attainable. An organism the size of man could never handle fire or undertake any sophisticated exploration of the world if the maximum speed of nerve conduction was a hundred times less. Indeed, such a creature would probably be unable to think in any way imaginable to us. If nerve conduction speeds were this much lower, only very tiny organisms the size of a protozoan could possess rapid reflexes.

The speed of nerve conduction is itself determined by a number of biophysical constants, such as the speed of diffusion of sodium and potassium ions across the lipid bilayer membrane. And of course, the existence of the membrane potential is itself inherent in the insulating character of the lipid bilayer that surrounds all animal cells and, as we have seen (in the previous chapter), is the only structure known which is fit to serve as the bounding membrane of the cell.

The Size of Nerve Axons

There is another aspect of the design of the nerve cell which is critical, and that is the diameter of an individual nerve cell axon, or fiber, which carries the nerve signal from the spinal cord to the muscle cell. In man and most mammals the diameter of the high-speed myelinated nerve axons is about 20 microns. Classic physiological studies some decades back showed that the diameter of axons is linearly correlated with conduction velocity up to about 20 microns and that this diameter is optimal for high-velocity conduction.[17] In fact, this is quite large for a nerve axon because of the layers of myelin—a fatty material—which are wrapped around the axon. But it is still small enough to make it feasible to pack thousands of individual nerve cell axons into a small nerve bundle a few millimeters across—the approximate size of the nerve tracts in the human arm. If individual nerve axons could not be designed to function with diameters as small as 20 microns, the number of nerve axons which could be conveniently carried into the arm would be decreased. If high-speed nerve axons had to be, for some reason, 1 millimeter in diameter instead of 20 microns, then carrying the many thousands of axons necessary for the fine motor control of the limbs would necessitate nerve tracts several centimeters in diameter—larger in fact than the limbs themselves. Clearly, the fact that the diameter of the nerve axon can be as small as 20 microns is another critical factor in the design of *Homo sapiens* as important as the necessity for conduction speeds of 120 meters a second!

Without nerve conduction speeds of 120 meters a second, without nerve axons approximately 20 microns in diameter, without molecular motors capable of generating forces of 3 to 5 piconewtons, the whole design of the human musculoskeletal system and our capacity to perform coordinated movements would be impossible. We would not be able to move quickly, if at all, let alone have the strength and capability to construct a fire.

Of course the functioning of the muscular system is also dependent on other adaptations, including the unique characteristics of bone that provide the skeletal framework upon which the muscles work and the unique strength of collagen fibers that form the high-tensile-strength tendons, which transmit the force of muscular contraction throughout the body. An intriguing aspect of the strength of collagen already referred to in chapter 8 is its dependence on the apparently esoteric fact that the amino acid glycine has a very small side group—in fact the simplest possible—consisting of one hydrogen atom that allows the three amino acid helices which form the collagen fiber to twist tightly round each other into a super strong helical cable. Without the very small side group of glycine, the unique super packing of the helices and great tensile strength of collagen could not be achieved, and there is no obvious alternative design to achieve the equivalent strength in a protein fiber. It is an intriguing thought that the functioning of the entire muscular system may be ultimately dependent on the existence of an amino acid with the precise characteristics of glycine.

The Relative Size of Organs

As we have seen, the muscular system necessarily (given the strength of the weak bonds) occupies about 50 percent of the mass of the body. Providing sufficient energy for this muscle mass necessitates in turn a certain mass or volume of tissue devoted to the delivery of adequate amounts of oxygen to the muscles.

In man the proportion of the body that is devoted to the respiratory and circulatory systems is about 20 to 25 percent of the body's volume. Their function is just about as efficient as possible, given the constraints imposed by the solubility of oxygen, the viscosity of water, airway resistance (determined by the density of air), and so forth. It is hard to envisage an organism in which the lungs occupied more than 50 percent of the total volume of the body. Such an organism would be in effect a gigantic air sac! As it is, in many mammals the volume of the respiratory system is already about 50 millimeters for every kilogram of body weight.[18] And in birds, despite the smaller volume of the lungs, the respiratory system as a whole (lungs plus air sacs) occupies an even larger fraction of the total volume of the body.[19] The efficiency of oxygen delivery to muscles could hardly be improved much by increasing the throughput of blood. As we have seen, in highly active muscles much of the volume of the muscles is taken up by the blood capillaries. Increasing the blood volume would only reduce the muscles to a mass of blood capillaries.

Similarly, we have seen that the power generated by muscle per unit of muscle volume can hardly be improved. If efficient oxygen delivery to the muscles had necessitated an oxygen delivery system ten times the weight of the muscle mass, then no large organism would ever have moved. The muscles would be incapable of moving the sheer weight of the oxygen delivery system. Clearly, if a ratio between the mass of muscle and the mass of oxygen delivery system of about 2 to 1 could not have been at least approximately achieved, for whatever reason, then the design not only of man but of all mammals and higher vertebrates would not be feasible.

It is not only the ratio of muscle mass to the mass of the oxygen delivery system which must be very close to what it is. The nervous system, for example, could hardly occupy a volume several times that of muscle. Unless an efficient nervous system could be designed to take up only a fraction of the volume and mass of the muscle, the muscle-nerve ratio would not be fit for life.

It follows that the coherent functioning of the human body and indeed that of all advanced vertebrates is critically dependent on the minimum mass required for the efficient function of each organ system—the circulatory system, the urinary system, the nervous system, the muscular system, and the respiratory system—being very close to what it is. Moreover, the functional efficiency and design of each organ system is largely determined in turn by the laws of nature, by the rate of diffusion of oxygen, by the capacity of transitional metals to reversibly associate with oxygen, by the strength of the weak bonds, by the viscosity of water. If these constants were very slightly different, large complex organisms of design and biology similar to ourselves would be impossible.

Inertia

A further very interesting twist to the deepening teleology in this chain of fortuity is the influence the phenomenon of inertia has on the size and dimensions of man. Inertia is the name we give to the property of things to resist a change in their velocity. An undisturbed body remains at rest and requires the exertion of a force to impart motion to it. A moving car requires force to slow it down or to make it change direction. Like gravity, inertial forces are related to mass. It requires more force to make a large object move or change the direction of its motion than for a small object. The wind may set a feather in motion but not a boulder.

If inertia had been less, then the wind could well have set a boulder in motion. In such a world we would be subjected to a continual bombardment by all types of objects in our environment. However, if inertia had been much greater, then unless the strength of muscles was much greater, we would have profound difficulty even in starting to move our finger. And once in motion, control of its direction and speed would be next to impossible. It is clear that the inertia of matter must be very close to what it is for an animal of our size to function in an environment similar to the earth's. Extraordinary as it may seem, physicists have proposed that the inertial forces experienced by objects on the earth are generated by the total combined gravitational attraction of all matter in the cosmos, including the most distant stars and galaxies. Because most of the matter in the universe is far from the earth, this means that the greatest contribution to the inertia of objects on earth is made by the most distant galaxies. As Dennis Sciama comments in his *Unity of the Universe:*

> The idea that distant matter can sometimes have far more influence than nearby matter may be an unfamiliar one. To make it more concrete, we may give a numerical estimate of the influence of nearby objects in determining the inertia of bodies on the earth: of this inertia, the whole of the Milky Way only contributes one ten-millionth, the sun one hundred-millionth, and the earth itself one thousand-millionth. . . . In fact, 80 per cent of the inertia of local matter arises from the influence of galaxies too distant to be detected by the 200-inch telescope.[20]

In a very real sense, then, the existence of beings of our size and mass with the ability to stand, to move, and to light a fire is only possible because of the influence of the most distant galaxies, whose collective mass determines the precise strength of the inertial forces on earth. If this view is correct, then it means that our existence is critically dependent on both the *mass of the earth* and the *total mass of the universe* being very close to what they are. There is a distinct echo in these curious coincidences of the old medieval doctrine of man the microcosm, which held that the dimensions of the human body reflect in some profound sense the dimensions of the macrocosm.

Alternative to *Homo Sapiens*

It is sometimes claimed that the unique capabilities of *Homo sapiens* could be actualized in a bipedal reptile. This possibility cannot be entirely discounted. However, no known reptile is remotely as intelligent as even a primitive mammal. Moreover, lacking the neocortex of the mammalian brain, the design of the reptilian brain may not permit the evolutionary development of the flexible and adaptable behavior so characteristic of the mammals, especially groups like the primates, cetaceans, and carnivores. Similar constraints may also apply to the evolution of high intelligence in birds. One advantage mammals may possess over reptiles is their highly sensitive skin. This is probably superior to the scaly skin of a reptile for a manipulating device which must be exquisitely sensitive to the texture and form of the objects in its grasp.

Neurological constraints may also mitigate against the evolution of a bipedal marsupial with the cognitive capacities of man. No known marsupial approaches the intelligence of a seal or porpoise. It would appear that the placental mammals have the most advanced brains on earth.

It would seem that no aquatic species could be as fit as *Homo sapiens* to explore and ultimately comprehend the world. The limbs in aquatic vertebrates are adapted for swimming and not suited for grasping and sophisticated manipulative tasks. However, the major problem in envisaging how an intelligent aquatic species could advance and develop a technology is the impossibility of utilizing fire in an underwater environment. And without fire, it is difficult to see how an aquatic organism, no matter how intelligent, could develop a technology or acquire chemical knowledge.

The necessity to derive an alternative humanoid being via a natural evolutionary process imposes additional constraints on the range of possible alternatives. Even if we could specify the design of an intelligent humanoid radically different to man, we might well find that we could not derive it via a plausible evolutionary process. If a computer could be built that was more intelligent than man, this would still not threaten our uniqueness in at least one important sense. The "intelligent computer" would not be a natural form and would not be derivable by natural processes from the materials available on a primitive planetary surface. Even if an intelligent silicon android could be designed, could it be derived by a natural evolutionary process? I suspect the problems would in all probability be insurmountable.

To get from atoms, via molecules, to the cell, and from the cell, via primitive multicellular life forms, to the first mammal, and finally from a primitive mammal to an upright bipedal organism with good eyes, grasping hands, and possessing intelligence and language will surely necessitate a relatively unique evolutionary path. An arboreal stage may be an evolutionary necessity somewhere on the path to *Homo sapiens,* to ensure the development of a grasping hand, high-acuity vision, and perfect coordination between the two. A subsequent terrestrial stage may also have been an evolutionary necessity to ensure the evolution of bipedality.

In a thoughtful analysis of man's evolutionary history and the acquisition of our unique biological adaptations, Sir Julian Huxley concluded:

> Writers have indulged their speculative fancy by imagining other organisms endowed with speech and conceptual thought—talking rats, rational ants, philosophic dogs and the like. But closer analysis shows that these fantasies are impossible. A brain capable of conceptual thought could only have been developed in a human body.[21]

Moreover, as Huxley points out, the evolutionary generation of *Homo sapiens* has come about via a unique path:

> The essential character of man is . . . conceptual thought. And conceptual thought could only have arisen in a multicellular animal, an animal with bilateral symmetry, head and blood system, a vertebrate against a mollusc or an arthropod, a land vertebrate among vertebrates, a mammal among land vertebrates. Finally it could have arisen only in a mammalian line which was gregarious, which produced one young at birth instead of several, and which had recently become terrestrial after a long period of arboreal life.
>
> There is only one group of animals which fulfils these conditions—a terrestrial offshoot of the higher Primates. Thus not merely has conceptual thought been evolved only in man: it could not have been evolved except in man. There is only one path of unlimited progress through the evolutionary maze. The course of human evolution is as unique as its result. It is unique not in the trivial sense of being a different course from that of any other organism, but in the profounder sense of being the only path that could have achieved the essential characters of man. Conceptual thought on this planet is inevitably associated with a particular type of Primate body and Primate brain.[22]

The Human Brain

If the anthropocentric thesis is correct, then the human brain should be the most powerful possible thinking machine—biological or artificial—that can be built out of the atoms of our world. It should be peerless.

As far as biological brains are concerned, unfortunately we still know an insufficient amount about the relationship between intelligence and the structure to be able to judge with any certainty whether or not the human brain represents the most advanced biological brain possible. However, what little knowledge we have seems consistent with the presumption that it is. To begin with, the human brain is one of the biggest in the animal kingdom in terms of the number of neurons it contains and in terms of its gross volume. The brain of a fly is about 1 milligram,[23] small mammals have brains which weigh only a few grams, while the human brain weighs about 1.4 kilograms—more than 1 million times more than that of a fly. Only certain whales, porpoises, and the elephant have larger brains than man.[24] The largest cetacean brain, about 9 kilos (that of the sperm whale) is about six times the size of the human brain, that of an elephant nearly four times larger.[25]

However, brain size alone seems to bear little direct relationship to intelligence. In man, for example, there is no obvious correlation between brain size and intellectual ability.[26] And although the brain of a dolphin may be larger than that of man or any other primate species, its neurons are far simpler and the cortical layer in the dolphin is also only about half the thickness than it is in man.[27] (Moreover, the overall design of the cetacean brain appears to have retained many primitive features.) It seems likely that intelligence is related not only to the sheer size of the brain or number of neurons but to many more subtle factors, including the thickness and convolutions of the cerebral cortex, the complexity of the individual neurons, the density of synaptic connections, and the development of those parts of the cerebral cortex associated with higher integrative functions, such as the frontal lobes. By these criteria the primate brain is the most complex in the animal kingdom.

And among the primates, the human brain is by far the most developed. The human brain is three times larger than that of a chimpanzee and contains absolutely more neurons than any other primate brain and a vastly increased frontal region.[28] Moreover, the neurons in the cortex of humans and other primates are far more complex than those in the cortex of a rat.[29]

Compared with the rat, each neuron in the human brain makes between ten and one hundred times more synaptic connections.[30] Altogether, the human brain contains about 10 million more synapses than the brain of a rat. The compaction of synaptic connections in the human brain is in fact staggering. Each cubic millimeter of the human cortex contains, in addition to 100,000 cells, some 4 kilometers of axonal wiring, 500 meters of dendrites, and close to 1 billion synapses.[31]

If intelligence is related to the total number of nerve cells, the total number of connections between them as well as the density of the connections, then as a recent *New Scientist* article puts it, "on this basis the human brain is the most complex in the animal kingdom." Moreover, as the article continues, "no radical improvement in synaptic density may be possible because of the need to maintain the fine balance between the size and number of neurons and the blood vessels which nourish them. To produce a significant rise in processing power, the axons would have to be wider than they are now to speed up the rate at which they pass signals. This in turn would demand equivalent increases in the amount of insulation along the axons and a better blood supply, which would take up extra space in the brain cavity, leaving less room for axons." As the article puts it, "Humans are about as smart as they are going to get."[32]

The evidence is certainly consistent with the possibility that the human brain does indeed represent the most advanced information-processing device that can be built according to biological principles, that we are indeed "as smart as biological systems can get." It may be that the size of what may be the smartest biological brain, capable of the miracle of understanding the world, is perfectly commensurate with the design and dimensions of the human frame which, as we have seen, is itself wonderfully fit for the exploration and physical manipulation of the world.

Artificial Intelligence

The possibility that the human brain may be the most sophisticated biological brain within the entire realm of carbon-based life naturally raises the far more radical possibility that the human brain may represent the only material form, out of all possible assemblages of atoms, in which genuine cognition and high intelligence can be instantiated, a possibility that would follow from the anthropocentric thesis.

Again, as our knowledge is so preliminary, no final judgment on this issue is possible, but from what we know of the structure and functioning of biological brains, it is not hard to be persuaded that no other alien assemblage of atoms in any other exotic realm of chemistry would come close. To begin with, each one of the basic building blocks of biological brains is a living cell—a veritable molecular microcosm in itself, consisting of a cell body plus a major axon and a vast dendritic tree ending in 100,000 branches. The neuron and its dendritic tree is not a mere frozen network of silicon threads but a living, ever-changing network, learning, reacting, responding, and integrating a vast number of different electronic and chemical signals. In the words of a recent *Nature* reviewer, "The latest work on information processing and storage at the single-cell level reveals previously unimagined complexity and dynamism." We are left with "a feeling of awe for the amazing complexity found in nature. Loops within loops across many temporal and spatial scales."[33]

The enormous problems now being encountered as artificial intelligence (AI) researchers try to create genuinely intelligent computers is consistent with the extreme view—that only in the precise molecular architecture of the human brain can intelligence be instantiated in our world. Although the mechanistic faith in the possibility of AI still runs strong among many researchers in the field, there are also many detractors, including John Searle[34] and Roger Penrose.[35] And there is no doubt that to date, as Penrose argues, no one has manufactured anything which exhibits intelligence remotely resembling that of man.

Final confirmation that the human brain is indeed the most sophisticated thinking machine possible can only come from future advances in the neurosciences and in the field of AI early in the next century. Potential advances that might support the notion could include, for example, evidence that brains built out of biological neurons possess unique properties which cannot be exactly mimicked in nonbiological brains and that self-conscious reflection and genuine cognition is only possible in brains with these particular unique properties. Again, increased knowledge of the various biological design constraints alluded to in the *New Scientist* article cited above could confirm that no further improvements in the information-processing capacity of the human brain is possible. Future research might also reveal that our cognitive abilities are critically dependent on unique neuronal wiring patterns which can only be actualized in the primate brain.

Homo Mathematicus

Our success as a biological species has depended on many factors: on our being smart, on our being terrestrial, on our possessing a body of a dimension and design appropriate to handle fire and explore the environment, on the fitness of the earth's atmosphere to support fire and technological advance. However, there is another intriguing aspect to our success—the mutual fitness of the human mind and particularly its propensity for and love of mathematics and abstract thought and the deep structure of reality, which can be so beautifully represented in mathematical forms. In other words, the logic of our mind and the logic of the cosmos would appear to correspond in a profound way. And it is only because of this unique correspondence that it is possible for us to comprehend the world.

If the laws of nature could not be formulated in simple mathematical terms, it is unlikely that science would have advanced so quickly. It might, in fact, never have advanced at all. The physicist Eugene Wigner, who was much struck by the correspondence between mathematics and the physical world, spoke for many mathematicians and scientists when he remarked:

> It is hard to avoid the impression that a miracle is at work here. . . . The miracle of the appropriateness of the language of mathematics for the formulation of the laws of physics is a wonderful gift which we neither understand nor deserve.[36]

Of course, the fact that nature's laws can be described in mathematical terms is only helpful to minds already fine-tuned for mathematical abstraction. If humans had not had the love and capacity for mathematics and abstract thought, then again no scientific advance would have been possible.

And there are other aspects of the structure of reality which give the impression of having been tailored to facilitate our understanding of nature and ultimately the scientific enterprise itself. On this point Paul Davies comments:

> It is easy to imagine a world in which phenomena occurring at one location in the universe or on one scale of size or energy, were intimately entangled with all the rest in a way that would forbid resolution into simple sets of laws. Or, to use the crossword analogy, instead of dealing with a connected mesh of separately identifiable words; we would have a single extremely complicated word

answer. Our knowledge of the universe would then be an "all or nothing" affair.[37]

That the structure of the world appears to be curiously fit for human comprehension also struck Aristotle. Jonathan Lear comments that for Aristotle "the inquiry into nature revealed the world as meant to be known; the inquiry into man's soul revealed him as a being who is meant to be a knower. *Man and the world are, as it were, made for each other.*[38] (My emphasis.) The stupendous success of science since 1600 is testimony enough to the remarkable fitness of our mind to comprehend the world.

Imperfections

I have argued that the cosmos is fit for only one form of thinking animal—our own species, *Homo sapiens*—and that our biology uniquely equips us with the capacity to comprehend the world. But it does not necessarily follow that the biological design of *Homo sapiens* is ideal or perfect. Such a deduction is unwarranted and in fact absurd. The human body is a wonderfully crafted machine, but its design is not perfect in any absolutist sense.

Our design is constrained due to our evolutionary origin. We suffer spinal problems because the spinal column was not designed originally for an upright stance. Childbirth is painful and difficult in humans because the relative size of the human infant's head at birth is far larger than in the case of any other primate. Because air and food both pass through the same passage; the pharynx, inevitably there is always the possibility of choking. The recurrent laryngeal nerve loops around the aorta and back up to the larynx instead of taking a more direct route.

Moreover, the human body is a material object subject to the laws of nature like everything in the cosmos. And as Plato pointed out, all material objects are imperfect to some degree. Even a geometric form like a triangle can only be imperfectly represented in the world of matter. All material objects from a sand castle to a galaxy are subject to turnover and decay. Nothing lasts forever. And this is true also of the human machine, which ages, runs down, and finally fails at death. The genetic system is a wonderful contrivance, but like all such complex systems, artificial or natural, errors inevitably occur and genetic disease is the result. Similarly, the same tectonic activity that ensures the recycling of the elements also causes earthquakes and volcanoes which are at times massively destructive to life. The cosmos may be uniquely fit for life on earth, but this does not mean that it is so crafted to ensure that every individual living organism will exist in a nirvanalike state of absolute contentment and plenty.

The fact that individuals suffer pain, that individuals die, may raise all manner of questions for a religious believer, but these have no relevance to the fitness of the cosmos for *Homo sapiens* as a biological species. Such imperfections are inevitable in a material world of flux and change. We cannot be material and immortal. The central mystery of human existence is not whether we are here by design but why the design, by its very material and transitory nature, inevitably entails suffering and death. Blake touched on the same enigma when he asked of the tiger, "Did He who make the lamb make thee?"

Conclusion

The evidence that the laws of nature are fit for only one unique thinking being capable of acquiring knowledge and ultimately comprehending the cosmos may not be compelling, but it is eerily suggestive. Is it really just a coincidence that what may well be the most advanced possible biological brain possesses sufficient insight and intelligence to comprehend the world; that a biological brain capable of such feats need not be so large that it would require a clumsy elephantine quadruped to house it, but is of a size fit for an organism of the design and dimensions close to *Homo sapiens*; that this physical design is itself so fit for the manipulation of fire, the key to technology and knowledge; that the muscles provide sufficient power to move the body and limbs; that the speed of nerve conduction is fast enough to permit rapid coordinated movement; that the diameter of the highest-velocity nerve fibers is small enough to efficiently wire the musculature of the body; that a planet of the size of the earth is fit both for a bipedal primate of our size and dimensions and also for an atmosphere capable of supporting fire? The chain of coincidences underlying our existence, our ability to make fire, to develop technology, and ultimately to comprehend the cosmos, is simply too long and the appearance of contrivance too striking.

There is diminishing room in the cosmos for the fantastic alternatives of science fiction. If we want to build out of the matter of the cosmos a creature of understanding, a being possessing the defining characteristics of our species—high intelligence, the capacity to manipulate and investigate the environment with a highly developed visual capacity, the capacity for language and abstract thought, to make fire, to use metals, to do science, and to have power over the natural world, and so on—we will be led via a long chain of fortuity or mutual fitness in the nature of things to an air-breathing vertebrate of about our size and dimensions, and eventually to a gregarious mammal with a highly developed visual sense and endowed with a hand—in other words, toward one unique end—*Homo sapiens.* Moreover, such a being can only come to understand the cosmos, to use fire, make metal tools, and develop a technology on a planet of about the size and characteristics of the earth, with an atmosphere containing between about 12 to 20 percent oxygen, sufficient to sustain a fire and to support oxidative metabolism in advanced, active air-breathing organisms.

What is so striking is that the cosmos appears to be not just supremely fit for our being and for our biological adaptations, but also for our understanding. Our watery planetary home, with its oxygen-containing environment, the abundance of trees and hence wood and hence fire, is wonderfully fit to assist us in the task of opening nature's door. Moreover, being on the surface of a planet rather than in its interior or in the ocean gives us the privilege to gaze into the sky to see the Milky Way. Because of the position of our solar system on the edge of the galactic rim, we can gaze farther into the night to distant galaxies and gain knowledge of the overall structure of the cosmos. Were we positioned in the center of a galaxy, we would never look on the beauty of a spiral galaxy nor would we have any idea of the structure of our universe. We might never have seen a supernova or understood the mysterious connection between the stars and our own existence.

Part 2

EVOLUTION

Chapter 12

The Tree of Life

In which the possibility is considered that the origin of carbon-based life and the whole subsequent process of organic evolution on earth has been the result of an immense built-in generative program. The concept of the tree of life as a natural form whose evolutionary development was directed and determined by natural law was familiar to many nineteenth-century biologists. It is pointed out that DNA is remarkably fit for directed evolution. The possibility that the direction of evolution may have been partly the result of an emergent spontaneously generated order as proposed by Stuart Kauffman is also considered. Also discussed is the growing consensus that the origin of carbon-based life is built into the laws of nature and that carbon-based life is therefore inevitable on any planetary surface where conditions permit it. It is hard to envisage how life could have originated unless there exist generative laws to guide a series of self-replicating systems from chemistry to the cell. The point is also made that, if life's origin is built in, then it is hard to dismiss the possibility that the whole tree of life may also be built in. Evidence consistent with the possibility of direction includes the phenomenon of parallel evolution, the uniformity of molecular evolution, the speed of molecular evolution, and the rapidity of the major morphological transitions.

The tree of life, drawn up by Ernst Haeckel in 1866.

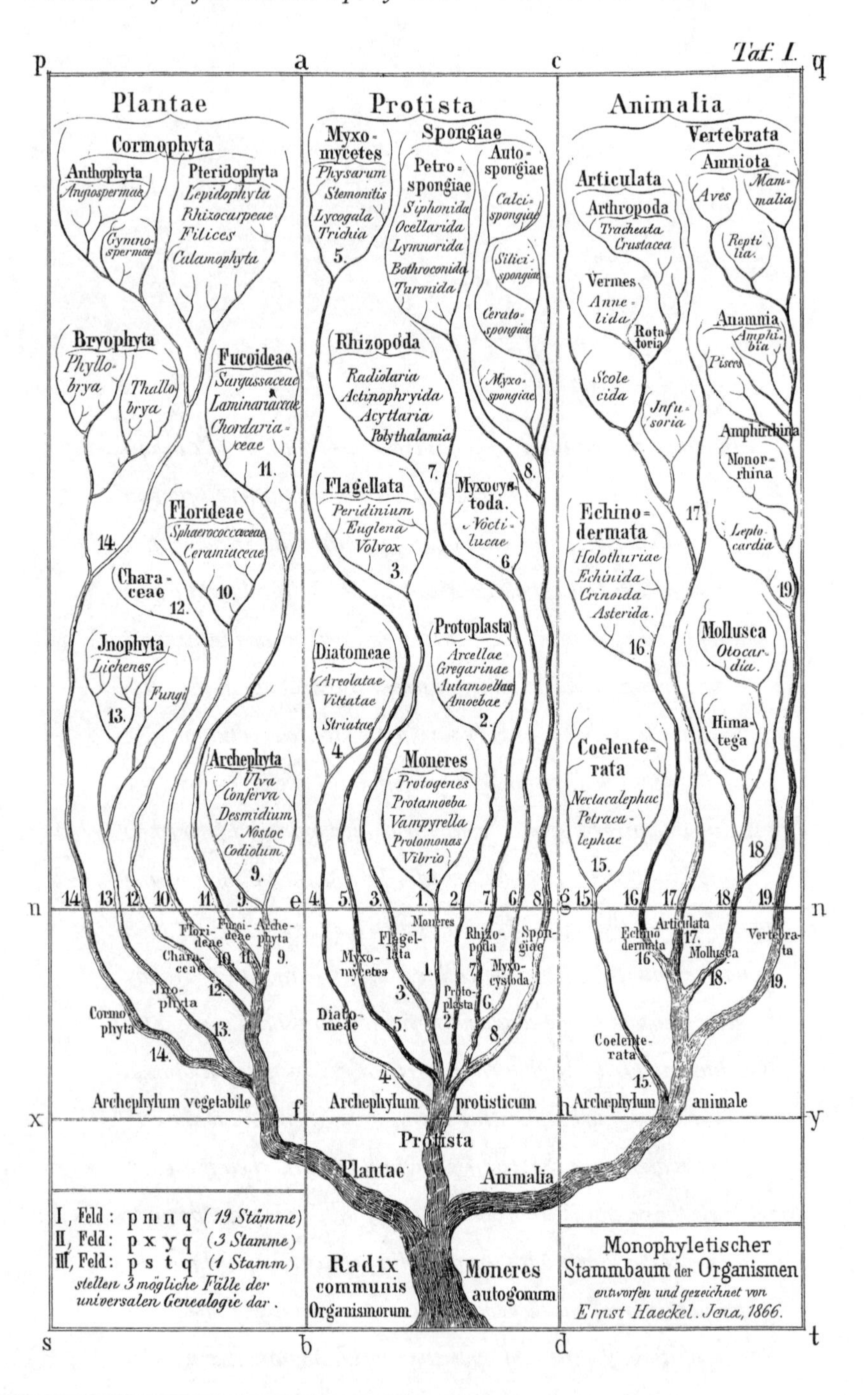

The affinities of all the beings of the same class have sometimes been represented by a great tree. I believe this simile largely speaks the truth. The green and budding twigs may represent existing species. . . . At each period of growth all the growing twigs have tried to branch out on all sides. . . . The limbs divided into great branches and these into lesser and lesser branches, were themselves once, when the tree was young, budding twigs. . . . From the first growth of the tree, many a limb and branch has decayed and dropped off; and those fallen branches of various sizes may represent those whole orders and families which have no living representatives, and which are known to us only in a fossil state. . . . As buds give rise by growth to fresh buds . . . so by generation I believe it has been with the great Tree of Life, which fills with its dead and broken branches the crust of the earth, and covers the surface with its ever-branching and beautiful ramifications.

—Charles Darwin, *On the Origin of Species,* 1859

Nearly everyone is familiar with the story of life on earth from its first primeval stirrings in the ancient Paleozoic ocean to its gradual flowering over the past 4 billion years into many exotic and diverse types of life. From the first simple photosynthetic autotrophs—blue-green algae, which as far as we can tell appeared shortly after the formation of the oceans and thrived in the original anoxic atmosphere—it took more than 3 billion years before sufficient oxygen had accumulated in the atmosphere to support more complex multicellular forms of life. Evolution had to wait until the early Cambrian era, a mere 600 million years ago, before multicellular life forms finally emerged in an explosion of diversity. In a very short period of time, perhaps no more than 25 million years, all the major groups of life were generated, an event we now call the Cambrian Explosion. In the early Paleolithic seas appeared the first representatives of the vertebrates, molluscs, annelid worms, arthropods, echinoderms, and so on. During this short period of time the tree of life underwent a great and never-to-be-repeated burst of creative growth, sprouting all its main branches that were subsequently to grow more sedately down through the geological eras to the present day. Since that great explosion, while the continents formed and re-formed over the past 500 million years, one branch of the tree, which was merely an inauspicious twig in the Cambrian seas, was to flower spectacularly, and to ultimately overshadow all other branches, generating a unique and diverse assemblage of large life forms both aquatic and terrestrial, including such familiar terrestrial forms as dinosaurs and mammals. And it was from one of the mammalian twigs of this great vertebrate branch that in the very latest stages of the drama the human species was born.

As the basic outlines of the story were gradually worked out in the first decades of the nineteenth century, it became increasingly apparent that all the past and present forms of life on earth could be arranged into a vast evolutionary tree. Consequently, it became the major goal of biology to provide an explanation of how this remarkable progression of life forms had come about. Eventually, Darwin's theory of natural selection would come to dominate. But before 1859, and even for a considerable period after the publication of the *Origin,* as Neal Gillespie in his *Charles Darwin and the Problem of Creation* points out, "most explanations invoked some sort of design."[1]

The *Vestiges* and Directed Evolution

An interesting early attempt to provide a comprehensive account of the evolution of life as a process directed by natural law was presented by Robert Chambers in his famous *Vestiges of the Natural History of Creation,* published in 1840 and one of the best-selling books of its day. Chambers proposed that the whole pattern of evolution had been written into the cosmic script from the beginning and that all the laws of nature had been specially arranged or programmed at the original creation to generate the tree of life.

Chambers's view of evolution is worth recounting in some detail because it represents one of the most important attempts in the early nineteenth century to formulate a completely naturalistic interpretation of evolution. Moreover, Chambers's aim to account for evolution by natural law and his view of the tree of life as a natural form correspond closely to the view being presented here. Reviewing critically the creationist position which was the orthodox view in the early nineteenth century, he comments:

> In what way was the creation of animated beings effected? The ordinary notion may, I think, be not unjustly described as this—that the Almighty author produced the progenitors of all existing species by some sort of personal or immediate exertion. . . . How can we suppose an immediate exertion of this creative power at one time to produce zoophytes, another to add a few marine molluscs, another to bring in one or two conchifers again to produce crustaceous fishes. . . . This would surely be to take a very mean view of the creative power—to, in short, anthropomorphise it, or reduce it to some such character as that borne by the ordinary proceedings of mankind. . . .
>
> Some other idea must then be come to with regard to *the mode* in which the Divine Author proceeded in the organic creation. . . . We have seen powerful evidence, that the construction of this globe and its associates, and inferentially of all the other globes of space, was the result not of any immediate or personal exertion on the part of the Deity, but of natural laws which are expressions of his will. . . . More than this, the fact of the cosmical arrangements being an effect of natural law, is a powerful argument for the organic arrangements being so likewise, for how can we suppose that the august Being who brought all these countless worlds into form by the simple establishment of a natural principle flowing from his mind, was to interfere personally and specially on every occasion when a new shell-fish or reptile was to be ushered into

> existence on *one* of these worlds? Surely the idea is too ridiculous to be for a moment entertained.[2]

In Chambers's view the "Divine attributes must appear not diminished or reduced in any way by supposing a creation by law, but infinitely exalted." Chambers continues:

> If the properties adopted by the elements at the moment of their creation adapted themselves beforehand to the infinity of complicated useful purposes which they have already answered, and may still further to answer, under many such dispensations of the material world, such an aboriginal constitution, so far from superseding an intelligent agent, would only exalt our conceptions of the consummate skill and power that could comprehend such an infinity of future systems, in the original groundwork of his creation.[3]

Chambers saw the origin of life as being analogous to crystallization:

> Crystallisation is confessedly a phenomenon of inorganic matter; yet the simplest rustic observer is struck by the resemblance which the examples of it left upon a window by the frost bear to vegetable forms. In some crystallisations the mimicry is beautiful and complete; for example, in the well known one called the *Arbor Dianae.* An amalgam of four parts of silver and two of mercury being dissolved in nitric acid, and water equal to thirty weights of the metals being added, a small piece of soft amalgam of silver suspended in the solution quickly gathers to itself the particles of the silver of the amalgam which form upon it a crystallisation precisely resembling a shrub. . . . Vegetable figures are also presented in some of the most ordinary appearances of the electric fluid. . . . The correspondence here is curious. A plant thus appears as a thing formed on the basis of a natural electrical operation—*the bush* realized.[4]

Chambers was led from his deterministic evolutionary model of the universe to the view that all reality, biological and physical, was in the end one immense interconnected Divine artifact.[5]

One of the most fascinating insights in Chambers's *Vestiges* is the author's analogy between embryology—the development of the individual from an egg cell—and phylogeny—the development of individual species from an original progenitor. This very obvious similarity was taken up by evolutionary biologists later in the nineteenth century and used by biologists such as

Ernst Haeckel as evidence for the fact of evolution, and as a basis for the famous and now discredited law—ontology recapitulates phylogeny.

However, Chambers uses the analogy for a different purpose, one of far deeper significance and one that has subsequently been ignored. Chambers is claiming in effect that the growth of the tree of life is analogous to the growth during development of an individual organism, that the tree of life is in fact a sort of super organism which has grown under the direction of laws of nature that are every bit as determining as those which govern the growth of the individual organism.

Chambers also believed that life existed on many other planets, and, being the product of the same life generating laws as on Earth, it would closely resemble life on Earth.[6]

Chambers's views of the origin of life have a remarkably modern feel, especially in light of the news from Mars. If life arose rapidly on both Earth and on Mars shortly after the planets were formed, as the NASA report implies, then the process can certainly be thought of in Chambers's terms as being analogous to crystallization. And if the origin of life, that is, the transition from chemistry to the cell (biochemical evolution) is programmed into the laws of nature, then why not also the transition from the cell to man (biological evolution) as well. The deduction seems eminently plausible.

The *Vestiges* was mercilessly attacked by the Victorian scientific establishment. As Gavin de Beer puts it in his introduction to a recent reprint of the first 1844 edition: "Adam Sedgwick is on record as saying 'from the bottom of my heart I loathe and detest the Vestiges' and he went on to devote 90 pages of the Edinburgh Review to vent his spleen on it. Huxley also pilloried *Vestiges* unmercifully as 'pretentious nonsense,' 'foolish fancies,' and said that the author knew no more science than can be picked up by reading some popular science journal."[7]

The fact remains, however, that although in de Beer's words: "Chambers's biology was poor, even by contemporary standards. . . [he] had constructed an argument of a forest of loose ends for the erection of a system. . . . It was an incredible amalgam of sound ideas and gratuitous blunders."[8] Nonetheless, read in the light of today's knowledge of cosmology and molecular biology, it has a remarkably modern feel. The *Vestiges* is one of the most remarkable cases of how, in Gavin de Beer's words, "the amateur like Robert Chambers can see a wood, although his detail of the trees may be wrong; the expert will only accept the wood after testing tree by tree."[9]

The basic idea of the *Vestiges* is so immensely attractive that it has simply refused to die. The idea that behind the apparently random ramifications of the evolutionary tree there is direction imposed by the order of nature has been reiterated over and over again since 1844. One of the most recent examples is in Arthur Koestler's *Ghost in the Machine:*

> Several eminent biologists have in recent years toyed with the idea, but without spelling out its profound implications. Thus von Bertalanffy wrote: 'While fully appreciating modern selection theory we nevertheless arrive at an essentially different view of evolution. It appears to be not a series of accidents, the course of which is determined only by the change of environments during earth history and the resulting struggle for existence, which leads to selection within a chaotic material of mutations . . . but is governed by definite laws, and we believe that the discovery of these laws constitutes one of the most important tasks of the future.' Waddington and Hardy have both re-discovered Goethe's notion of archetypical forms; Helen Spurway concluded from the evidence of homology that the organism has only 'a restricted mutational spectrum' which 'determines its possibilities of evolution.'. . .[10]
>
> [Such restrictions may mean that] given the conditions on our particular planet, the chemistry and temperature of its atmosphere, and the available energies and building materials, life from its inception in the first blob of living slime could only progress in a limited number of ways. . . . If this conclusion is correct, it sheds some additional light on man's status in this universe. It puts an end to the fantasies of science fiction regarding future forms of life on earth. . . . [It means]. . . that the evolution of life is a game played according to fixed rules which limit its possibilities but leave sufficient scope for limitless variations. The rules are inherent in the structure of living matter.[11]

Koestler, like Chambers, is viewed with suspicion by the orthodox academic community—another amateur who would presume to tell the professionals something about evolution. Gould pointedly refers to Koestler as a "nonscientist" in a recent review,[12] and Richard Dawkins refers to him pejoratively in the *Blind Watchmaker* as "another distinguished man of letters who could not abide what he saw as the implications of Darwinism."[13] Perhaps, however, even today amateurs may be able to see a "wood" where professionals only see the individual trees.

The idea of directed evolution was popular particularly among paleontologists in the late nineteenth and early twentieth centuries. Just how popular such ideas were is stressed by Bernhard Rensch in his *Evolution Above the Species Level.* Evidence for direction was seen in many phenomena, including parallelism and long-term evolutionary trends. The Russian zoologist and paleontologist Lev S. Berg, for example, believed that the variation of characters in an evolutionary lineage is confined within certain limits, that it follows a "definite course, like an electric current moving along a wire."[15] As evidence for direction in the course of evolution, Berg cited the evolution of plants from the simple mosses through the ferns to the conifers. During this evolutionary sequence a very clear long-term trend is manifested which involves the continual reduction of the sexual form of the plant (the gametophyte). This sexual form is most prominent in the mosses, where the gametophyte is a small self-supporting plant, while in the conifers it is reduced to a few cells.[16] Berg saw evidence for directed evolution in the many long-term evolutionary trends manifested in many vertebrate lineages. As examples, he cited the gradual ossification of the vertebral column, a reduction in the number of bones in the skull, and the transformation of a two-chambered heart into a three- and four-chambered organ associated with a corresponding increase in the complexity of the circulatory system.[17]

The Problem of Direction

A primary difficulty with all such notions of directed evolution, especially in the nineteenth century, was that the *known* laws of nature were not evidently biocentric in any obvious way. Although Chambers cites Whewell's Bridgewater Treatise as providing evidence of nature's biocentricity, the only significant evidence presented by Whewell was the thermal properties of water! Why, if the *known* laws were not obviously biocentric, should there be all manner of *unknown* laws which, according to Chambers in the *Vestiges,* were responsible for directing the course of evolution?

In the nineteenth century there was simply not the slightest justification for viewing the cosmos as biocentric in any way or life on earth as anything other than an inconsequential peripheral thing and certainly not in the least predestined in the order of things. The prevailing view of the cosmos as fundamentally nonbiocentric, governed by lifeless mechanical laws, is strikingly conveyed in this section from Hugh Miller's *Footprints of the Creator* published in 1849:

> Nature lay dead in a waste theatre of rock, vapour, and sea, in which the insensate laws, chemical, mechanical, and electric, carried on their blind, unintelligent processes: the creative fiat went forth; and, amid waters that straightaway teemed with life in its lower forms, vegetable and animal, the dynasty of the fish was introduced.[18]

The cosmos of nineteenth-century science, even seen through religious eyes, bore no evidence of a biocentric design. Rather than being "life-giving," the laws of nature are unintelligent, blind, mechanical. There is even a hint of the *sinister* in the descriptive terms used.

The worldview of nineteenth-century science, in which biology was an unnecessary contingent phenomenon divorced from physics, had no room or rationale for providential evolution or biogenic laws. Owen, Lyell, Mivart, Gray, and the many other leading biologists who opposed the Darwinian model and postulated a providential design were simply being true to the tradition of English natural theology. In rejecting the concept of a contingent biology, they were affirming the faith in the wholeness of the cosmos and the connectedness of all things, in an ultimate biocentricity for which they had no convincing evidence. Like Roger Cotes in his introduction to Newton's *Principia,* they were again affirming a faith for which neither nature nor science provided any justification. Their belief in directed evolution was, as Neal Gillespie points out in his *Charles Darwin and the Problem of Creation,* in effect an article of "faith held despite what nature indicated."[19] Even St. George Mivart, one of the most ferocious critics of Darwinism, conceded in his *Genesis of Species* (1871) "that one could not find in nature such evidence of design that no man could sanely deny."[20]

But perhaps an even more fundamental problem facing the proponents of directed evolution was the complete absence of any idea how the supposed "design or direction" could have been executed. Biological knowledge was simply too rudimentary. Neither Chambers nor any other advocate of directed evolution were able to provide any sort of detailed explanation of how precisely the direction of evolution might have come about.[21] If there were undiscovered laws of evolution, no one could imagine how they could be executed.

Compounding the problem was the additional challenge, to all evolutionary theories both directed and undirected, that at a gross morphological level the organic world appears to be markedly discontinuous. There are innumerable examples of complex organs and adaptations which are not led up to by any known or even, in some cases, conceivable series of feasible intermediates. In the case, for example, of the flight feather of a bird, the amniotic egg, the bacterial flagellum, the avian lung, no convincing explanation of how they could have evolved gradually has ever been provided.[22] The morphological discontinuities and especially organs or adaptations of extreme complexity, exhibiting what Michael Behe terms "irreducible complexity,"[23] have, ever since Darwin, provided ammunition for special creationists who have claimed that these "morphological gaps" could not have been closed gradually by natural evolutionary processes and that they represent *prima facie* evidence for Divine intervention in the course of nature.

The Fitness of DNA for Directed Evolution

Today the whole situation with regard to the possibility of directed evolution has been dramatically transformed. There is first the growing evidence, presented in the previous chapters, that the laws of nature, rather than being as they were viewed in the nineteenth century as insensate, mechanical, and lifeless, give every impression of having been preordained for life as it exists on earth. The concept of directed evolution is therefore no longer an anomaly in a nonbiocentric world. On the contrary, it is merely a logical deduction from a rapidly emerging new teleological worldview. Second, there is the revolutionary new DNA world revealed by modern genetics, the world of the genes, a world undreamt of in the nineteenth and early twentieth centuries. And it is a world which provides the basis for relatively detailed and plausible speculation as to how the whole pattern of evolution might have been written into the DNA script from the beginning.

We have already seen that every living organism is specified in a precisely determined way by a set of instructions encoded in the sequence of bases in its DNA. The fact that all living organisms are now known to be specified in the linear DNA sequence consisting of a string of several thousand genes has provided science with a new and utterly different representation of the organic world. These discoveries imply that not only is DNA remarkably fit for its hereditary role, it is also remarkably fit in a number of different ways for directed evolution.

One important consequence of the new molecular biological picture of nature has been the establishment of the principle that biological informa-

Genotypic and phenotypic space.

DNA sequence/genotypic space		**Morphological/phenotypic space**
DNA	→	Protein
Genes	→	Organism
Genotype	→	Phenotype
Linear DNA space	→	3-D morphological space

tion flows in only one direction (see above), from the genes or DNA to the organism, never in the reverse direction. This is known as the central dogma. This unidirectional flow of information from DNA to organism is clearly "fit" for directed evolution. In a world where the central dogma did not hold, where the genetic system was designed on different principles, where, for example, organisms had the ability to intelligently manipulate their DNA sequences at will, or where environmental factors could direct changes in DNA sequences, where information flowed from organism to DNA, it is very difficult to imagine how a long-term evolutionary program based on a programmed succession of changes in the DNA could have been feasible.

The Closeness of All Life in DNA Sequence Space

Over the past twenty years a vast number of DNA sequences drawn from many different species have been determined. The number of known sequences is growing exponentially as sequencing methods continually improve. By the end of the century much of the DNA sequence of the human genome will be determined. Already many large sections of the human genome have been sequenced and the DNA sequence of many regions of the genomes of other mammalian species, such as the chimpanzee, mouse, rat, and cow, have also been determined. In the case of certain microorganisms, the entire sequence has already been determined.

One of the most surprising discoveries which has arisen from DNA sequencing has been the remarkable finding that the genomes of all organisms are clustered very close together in a tiny region of DNA sequence space forming a tree of related sequences that can all be interconverted via a series of tiny incremental natural steps. So the sharp discontinuities, referred to above, between different organs and adaptations and different types of organisms, which have been the bedrock of antievolutionary arguments for the past century, have now greatly diminished at the DNA level.

Organisms which seem very different at a morphological level can be very close together at the DNA level. One of the most dramatic cases of this is that of the cichlid fish species in Africa's Lake Victoria. As Jared Diamond points out in *The Rise and Fall of the Third Chimpanzee:*

> Cichlids are popular aquarium species, of which about two hundred are confined to that one lake, where they evolved from a single ancestor within the last 200,000 years. Those two hundred species differ among themselves in their food habits as much as do tigers and cows. Some graze on algae, others catch other fish, and still others variously crush snails, feed on plankton, catch insects, nibble the scales off other fish, or specialise in grabbing fish embryos from brooding mother fish. Yet all those Lake Victoria cichlids differ from each other on the average by about 0.4% of their DNA studied.[24]

Another example is the very great similarity of the DNA sequences in the human and chimpanzee genomes. In fact, extensive comparisons of long sections of the DNA of man and chimpanzee show that the differences are extraordinarily trivial. Human and chimpanzee DNA sequences differ on average at only one base in a hundred. As far as we can tell, not only are the DNA sequences virtually identical, but every gene identified in the human genome has its counterpart in the chimpanzee genome. So all the morphological differences between man and chimpanzee, involving the form and relative shape of the limbs, the genital organs, sperm morphology, etc., *and all the mental differences* are generated from DNA sequences which are virtually identical. The distance between man and chimp which seems so significant and obvious at a gross morphological level is trivial in DNA sequence space. In fact, the differences between the DNA of man and chimp can be accounted for by simple well-known mutational processes which are occurring all the time in nature at present. In the case of primate DNA, for example, all the sequences in the hemoglobin gene cluster in man, chimp, gorilla, gibbon, etc., can be interconverted via single base change steps to form a perfect evolutionary tree relating the higher primates together in a system that looks as natural as could be imagined. There is not the slightest indication of any discontinuity. Indeed, human and chimpanzee DNA are closer together than the DNA sequences of many so-called sibling species of the fruit fly drosophila, that is, species which are almost indistinguishable in morphological characteristics.

Even more remarkable is the fact that the genetic programs underlying the development of most of the major metazoan phyla, such as the vertebrates, arthropods, and molluscs, which appear so far apart in terms of their fundamental body plan and morphology, are not so far apart in DNA sequence space as might be imagined. As Simon Conway Morris comments, "The story emerging from molecular biology is that what may look very different in anatomical terms can be founded on a basically identical genetic architecture."[25] In the case of organs as dissimilar as the heart in insects and vertebrates, for example, their development may involve common pathways.[26] In short, evolution is far easier to conceive of in DNA sequence space than in morphological or phenotypic space. By analogy, it is far simpler to move from mountain to mountain on a two-dimensional map than it is to move from mountain to mountain in actual three-dimensional space.

From the DNA perspective the whole evolutionary tree of life is in essence nothing more or less than a vast set of closely related DNA sequences clustered close together in the immensity of DNA sequence space, where each individual sequence is capable of specifying a viable life form, and where all sequences are interrelated and ultimately derivable via a series of steps from an original primeval sequence, which was the genome of the first life form on earth.

Escape from Selective Surveillance

DNA sequence space is fit for directed evolution in another important way. Unlike evolutionary change at a morphological level, where it is only possible to move from one adaptation to another through functional intermediates, in DNA sequence space it is possible to move at least hypothetically from one adaptation (position) to another in DNA space via functionless or meaningless intermediate sequences.

This is because a DNA sequence does not have to be functional to survive and be passed on through the generations. In fact, the greater part of all the DNA in nearly all the cells in higher organisms, although it is copied faithfully at each cell division, is never expressed. The genes, for example, specifying the development of the embryo, while present in all the cells of adult humans, are never expressed. These genes may be thought of as being "cryptic" in the adult. It is very easy to imagine how an evolving DNA sequence might be passed silently down through several generations before being expressed. Crucially, while in "evolutionary transit," such DNA sequences are not under any selective surveillance. The fact that DNA sequences can persist and be transmitted over many generations while suffering continuous mutational change which is not subject to selectionist surveillance means that new sequences and hence new evolutionary innovations can be generated, at least hypothetically, via functionless intermediates. Thus, new organs and structures that cannot be reached via a series of functional morphological intermediates can still be reached by change in DNA sequence space.

Overall, the new DNA sequence space is fit for directed evolution in a number of ways: (1) because of the closeness of all life forms at the DNA level and because all known sequences can be interconverted in small natural steps via well-known mutational processes, (2) because information flows only from the genotype to the phenotype, and (3) because functional DNA sequences can be derived via functionless intermediates, a new phenotype or organ system can be generated by saltation.

Directed Mutation and Development

That genes can direct biological change through time is evidenced in the process of development. The development of a complex organism is specified by a genetic program which contains a set of instructions that are deciphered by the organism as development proceeds. One set of instructions may provide information to make a red blood cell, while another may provide information for the laying down of a precise pattern of neuronal connections in the brain. The reading of each set of instructions involves the expression of different genes at different times and places in the embryo. What we have in effect is a fantastic process of biological change from an egg cell to an adult human, from a caterpillar to a butterfly, and in the case of some parasitic life cycles a whole succession of different larval forms, some of which are so dissimilar that for many years it was suspected that they were different species. These changes, directed by the genes, are in many cases quite as dramatic as those changes which occurred during the course of evolution. If genes can direct developmental changes, there is no reason why, at least in principle, they cannot also direct evolutionary change.

During development of an organism from egg to adult form, a variety of genetically programmed gene rearrangements are known to occur. These include, for example, the development of the immune system, where gene segments which are separate within the germ line are somatically rearranged to bring them into new functional relationships. Another example is the insertion of mobile sections of DNA known as transposons, which can change the expression of adjacent genes. In a well-known case in the bacterium *Salmonella typhimurium,* an inversion of the DNA sequence can switch the expression of one gene to another, a change which results in changes to the antigenic properties of the cell. Many other types of programmed genomic reorganization are utilized during development in different species.[27]

These rearrangements are strictly programmed and occur at precisely predetermined times in the development of the organism. There is no compelling reason why similar types of changes could not have been genetically programmed to occur during the far longer time course of evolution. Although most of us have been brought up to be skeptical of any sort of directed evolution, there is no doubt that the analogy here is striking and suggestive. Consider that, first, most genetic change underlying evolution, especially in higher organisms, has been largely a matter of the rearrangement of preexisting genes rather than the emergence of new genes; and second, we know that cells do measure the passage of time during development and count the number of cell divisions that have elapsed since a particular developmental event. So it is premature to reject out of hand the possibility that during the course of evolution specific preprogrammed genetic rearrangements have occurred at specific times.

Directed Sequential Change

Directed sequential change in DNA sequences could conceivably come about from the integration of a host of different genetic and mutational phenomena. The rate and pattern of evolutionary change in DNA sequences is known to be influenced by the sequence itself. Studies of the influence of the local sequence environment in primate DNA, for example, has revealed biases of varying magnitude in the rate of evolutionary substitution of the same base pair in different sequences. The DNA sequences in all organisms are subject to compositional constraints that affect both coding and noncoding sequences, and these largely determine the types of substitution that occur.[28] The proportion of the two base pairs, A-T and C-G, in DNA varies considerably between species and between different regions in the genome in any one species. These variations in sequence can be explained by directional mutational pressure and selective constraints which are inherent in the genome itself.[29]

During the late seventies and early eighties it became apparent that the genomes of nearly all organisms contain so-called gene families, which consist of multiple identical copies of the same gene. Surprisingly, these copies are often identical not only within the genome of one individual but in the genomes of all the individuals in the species. A variety of genetic mechanisms have been identified which act to maintain the sequential *identity between all the copies of the same gene in any one species.* In the early eighties Cambridge geneticist Gabriel Dover suggested that the integrated effect of these various internal mechanisms is potentially capable of causing synchronous genetic changes in all the members of a population. He termed the effect "molecular drive."[30]

It is relatively easy to envisage how such processes could be utilized by a grand instructional program to bring about cohesive directional mutational change during evolution.

Constraints in Sequence Space

Another conceivable mechanism by which evolution could have been directed along highly specific trajectories in the DNA space is if the DNA sequence space had been ingeniously contrived so that it contained preordained, highly restricted, or even unique functional paths. Consequently, an organism at any point on one of these preordained trajectories would have only a few alternative functional pathways available through DNA space.[31] If only certain routes through DNA space are possible, it is relatively easy to explain how directed evolution might have occurred by relatively conventional and quite natural mechanisms. In fact, because of their fundamental nature, living systems are ideally suited to search out the permissible trajectories. Given that living organisms are inevitably subject to mutational changes, and given also that all living things are inevitably also subject to changing environments, which in turn impose ever-changing selection pressures, then it is clear that living systems are in fact compelled to test the DNA space in their immediate neighborhood for trajectories which facilitate their survival.

In short, if indeed a tree of life consisting of a unique branching pattern of permissible or functional trajectories had been written into the DNA space, just so long as islands of function are within short mutational distances, the very nature of living things as self-replicating biochemical automata, subject to mutational change at each replicative cycle and subject to changing survival pressures as their environments gradually change, will inevitably lead to the successive discovery of these preexisting islands of biological function and to the gradual tracing out of the main branches of earth's great tree of life.

Given the fundamental nature of organisms, and given a specially prearranged DNA sequence space, the evolutionary process of tracing out the tree of life becomes a perfectly natural phenomenon; the inevitable unfolding of a preordained pattern, written into the laws of nature from the beginning.

To be viable such a tree would certainly necessitate a very special distribution of biological functions within DNA space. The functional trajectories through DNA space would have to be highly ordered, so that at the base of the tree, where the evolutionary search begins, are clustered functional DNA sequences specifying very simple unicellular plants (autotrophs) from which can be reached, in easy mutational steps, some way up the main trunk, slightly more complex organisms, including simple animals, and so on, as the mutational process reaches upward into new regions in the DNA space. Further, all the life forms recruited as the tree grows would have to be co-adapted and functionally interrelated. In other words, any *interconnected network* of functional islands is insufficient to make a viable tree. Permissible trajectories through DNA space must be preprogrammed so that as the branches of the tree are traced out, a succession of ecologically feasible systems forming a succession of integrated biospheres is generated.

Note in the context of constraints the distinction between directed and undirected evolution begins to blur. Everyone agrees there must be at least some sequential constraints on the types of changes permissible in DNA. Sequential changes are bound therefore to be "directed to some degree." The key issue is the question of the severity of the constraints. If they are very severe, then in effect they are bound to direct evolutionary change along restricted paths. For the teleologist such paths would be preordained in the nature of things; for the Darwinist they are merely contingent.

Other Sources of Direction

In addition to directed mutational processes acting within DNA sequence space, there are other possible sources of evolutionary direction. One source may be the emergent self-organizing phenomena discussed by Stuart Kauffman in his recent book *At Home in the Universe*[32] and in his previous book, *The Origins of Order*.[33] Such phenomena may be responsible for or related in some way to the deeply embedded morphological rules that British biologist Brian Goodwin believes may govern and constrain biological form.[34]

Kauffman's research involves the use of computers to simulate the behavior of complex systems and is basically directed at looking for those biogenic laws that eluded the biologists in the nineteenth century and which might determine the emergence of evolutionary innovations, the general direction of evolution, and the overall pattern of life on earth. According to Kauffman, "we are not precluded from the possibility that many features of organisms are profoundly robust" and that "deep and beautiful laws may govern the emergence of life and the population of the biosphere.

"We can never hope to predict the exact branches of the tree of life but we can uncover powerful laws that predict and explain their general shape. I hope for such laws. I even dare to hope that we can begin to sketch some of them now."[35] If Kauffman is right, his work may have provided the first definite glimpse of key emergent self-organizing processes which have driven and constrained evolutionary change along restricted paths.

Brian Goodwin also believes that the basic form of the tree of life was determined by natural law. He has recently proposed that "all the main morphological features of organisms—hearts, brains, guts, limbs, eyes, leaves, flowers, roots, trunks, branches, to mention only the obvious ones—are the emergent results of morphogenic principles."[36] Such features are therefore not contingent but would recur if the tape of life was rerun. Goodwin cites the example of the spiral patterns which are observed in many flowers and in the arrangement of leaves along a stem or branch. These spiral patterns belong in a great many cases to the mathematical series known as the Fibonacci series. This pattern, like so many other morphological patterns, is clearly not the result of natural selection but rather of mechanical and natural laws. As Goodwin comments, "plants generate this aspect of their form by doing what comes naturally—following robust morphogenetic pathways to generic forms."[37] By this logic we can conclude that wherever in the cosmos there are plants with spiral leaf or flower arrangements we may expect to see this conform to the Fibonacci series. Goodwin raises the possibility of "a logical or a rational taxonomy . . . a theory of biological forms whose equivalent in physics is the periodic table of the elements, constructed on the basis of a theory that tells us the dynamically stable patterns of electrons, protons and neutrons. Biology would begin to look a little more like physics in having a theory of *organisms* as dynamically robust entities that are natural kinds, not simply historical accidents that survived for a period of time. This would transform biology from a purely historical science to one with a logical, dynamic foundation."[38]

In effect, both Kauffman and Goodwin are seeking to identify "directive laws" which would of necessity exist in a cosmos designed for life as we see it on earth and programmed from the beginning to arrive at the life forms we observe. Their work certainly raises the possibility that there exist additional mechanisms by which the course of evolution might have been directed along prearranged paths, by mechanisms which would not have necessitated any sort of specific directed mutation in DNA sequence space. Of course, the two different means of evolutionary direction are not mutually exclusive.

Lamarckian-type Mechanisms

The notion of evolution as directed and the tree of life as a natural form inherent in the nature of the cosmos itself should not be taken to imply that every adaptation in every organism that ever lived is part of nature's grand design. It is hardly conceivable that all the many fascinating and sometimes bizzare adaptations of life, such as those found on isolated islands, were written into life's design from the moment of creation. In such cases it seems we must either assume that their evolution has been the result of natural selection or consider the heretical but tantalizing alternative possibility—that living things may possess some unidentified and mysterious emergent property which endows on them an inventive capacity enabling them to direct their own evolution in a completely autonomous fashion, at least to a limited degree. This was a view that was quite widely held in France in the early decades of this century.[39] Recent studies of adaptive mutations in bacteria (referred to below) which have led some researchers to suggest that microorganisms may be able to intelligently alter their DNA sequences in response to various adaptive challenges would, if confirmed, lend support to the notion that life has inherent inventive capacities.[40] Still, no matter how appealing such notions may be, there is however to date, no convincing evidence that organisms do possess such abilities and even if such a capacity does exist, it would have to be severely constrained. Because as mentioned above, any model of directed evolution leading to the manifestation of a unique tree of life will work only if genetic determinism is the rule. So that any capacity for autonomous, intelligent evolution that life may possess would have to be limited to minor microevolutionary adaptive changes.

The Question of the Spontaneity of Mutation

One of the major obstacles within the biological community in the way of any widespread acceptance of the idea of directed mutation is the very deeply held belief in the so-called spontaneity of mutation.[41] According to the authorities Dobzhansky, Ayala, Stebbins, and Valentine, writing in a standard text on evolution, "Mutations are accidental, undirected, random or chance events in still another sense very important for evolution; namely that they are unorientated with respect to adaptation."[42]

The idea of the spontaneity of mutation is taken as a proven fact by a great many biologists today. And this is the fundamental assumption upon which the whole Darwinian model of nature is based. If it could be shown that some mutations, even a small proportion, are occurring by direction or are adaptive in some sense, then quite literally the whole contingent biology collapses at once. What is very remarkable about this whole issue is that, as is typical of any "unquestioned article of faith," evidence for the doctrine of the spontaneity of mutation is hardly ever presented. Its truth is nearly always assumed. In nearly all the texts on genetics and evolution published over the past four decades, whenever the author attempts to justify the doctrine of the spontaneity of mutation, he refers back to a series of crucial experiments carried out in the late forties and early fifties on the bacterium *E. coli* that were associated with the names of Salvador Luria, Max Delbruck, and Joshua Lederberg.[43]

These experiments were based on the very simple observation that when bacterial cells are suddenly subjected to a particular selection pressure (for example, the addition to a culture of cells of an antibiotic which is lethal to wild-type cells) invariably a small proportion of cells survive because they contain a mutation that confers resistance to the antibiotic. Ingenious tests were carried out which proved conclusively that the mutations were present in the surviving cells *before the antibiotic was added* to the culture. It was concluded that the mutations were spontaneous events.

But the fact that some mutations in bacteria are spontaneous does not necessarily mean that *all* mutations in *all* organisms throughout the entire course of 4 billion years of evolution have *all been entirely spontaneous.* This very point was made by Max Delbruck himself, who carried out with Salvador Luria some of the crucial experiments proving the spontaneity of mutation. As he conceded at a Cold Spring Harbor Symposium over forty years ago, "One should keep in mind the possible occurrence of specifically adaptive mutations."[44] Recently, John Cairns, a leading microbiologist in the United States, commented in *Nature* on the subject of the spontaneity of mutation: "It seems to be a doctrine that has never been put to the test."[45]

During the course of the past 4 billion years of evolution, countless trillions of changes have occurred in the DNA sequences of living organisms. There is simply no experimental means of demonstrating that they were all spontaneous. And even if all mutations are entirely undirected in themselves, it is still possible to reconcile such a model of mutation with directed evolution by envisaging the existence of prearranged functional paths through DNA sequence space. For as we have seen above, such paths could conceivably have been traced out by an entirely random search, in the same way as an entirely blind trial-and-error search eventually leads to the unique exit from a maze.

The Trees of Life

The postulate of a unique tree does not necessarily imply that the full flowering of the tree would occur on every earthlike planet on which the tree initially took root. As with any real tree, the environment in which it grows will influence its final shape. On a planet entirely covered with water, those branches which lead to terrestrial forms would never appear. A planet like Mars, which has been subject to major climatic changes and contains at present less water than Earth, might only permit the growth of the base of the trunk so that the process of unfolding might only realize very simple unicellular plants and bacteria. Even on Earth different successions of climates and other vagaries of chance would influence growth of the tree. For example, had the meteor strike which is now believed to have wiped out the dinosaurs not occurred, then perhaps the dominance of the mammals that followed the extinction of the dinosaurs might never have occurred. Alternatively, had the dinosaurs died out 10 million years later, then perhaps the small mammals which had already waited through 100 million years during the Jurassic and Cretaceous eras with little change might still have given rise to the same mammalian world we have today. It is also quite possible that many of the branches of the tree never sprouted on earth because the specific succession of environmentally imposed selection pressures required to draw life down those particular trajectories never occurred at the propitious moment.

Nor does the postulate of a unique tree necessarily imply that its form, even in identical environments, would be exactly the same. Just because evolution is determined does not necessarily mean that it was rigidly determined in all details. Even if the major phyla and classes are natural forms and built into the order of the tree, there could still be room for almost infinite microevolutionary variation within each major group.

Marsupial and Other Doppelgängers

It is often claimed by supporters of undirected models of evolution that if the tape of life were played again the pattern that would emerge would be entirely different. On the postulate being advanced here, although the details would be different, the overall form and many of the major types and patterns that would emerge would be the same. Curiously, the evolution tape has been played again, at least in part. This has occurred on several occasions when the fauna on great continental land masses became isolated for millions of years. Two of the most celebrated examples of this phenomenon occurred in South America and Australia, which were isolated from the other continents for most of the past 60 million years.

The diversification of the marsupials in Australia is very instructive. Almost every type of placental mammal has its counterpart among the marsupials. There is a marsupial lion, cat, wolf, mole, anteater, jerboa, and flying squirrel. There was even a giant wombat equivalent to the placental rhino. Only the kangaroo is moderately unique, although it could be thought of as a giant jumping rat! The similarity in some cases is very striking. The skull of the marsupial wolf is amazingly similar to that of the placental wolf.

There are, though, some noticeable absences—there is no marsupial bat or elephant or whale. However, in the case of the whale, it is hard to see how the reproductive system of the marsupial could be adapted for marine life. And in the case of the elephant, perhaps given more time, even the evolution of elephant doppelgängers may have occurred. In the case of the bat, there is some evidence that bats reached Australia shortly after the continent was isolated so the niche available for bat evolution may have been occupied from the beginning of the marsupial radiation.

This remarkable set of doppelgängers is all the more striking considering that the starting point for their evolution was not the same as that of the placentals. Therefore, even from a very deterministic point of view, we should not expect to see an absolutely identical pattern.

The marsupials are not the only example of this phenomenon. Many of the mammals that developed in South America during its time of isolation are also doppelgängers of types that developed elsewhere during the same period. Another example is the parallel evolution that occurred in the various mammal-like reptile lineages over many millions of years. Another example is closer to home. Based on anatomical and behavioral evidence, modern humans and Neanderthals have always been considered to be quite close biologically. Both manufactured quite advanced tools, both used fire, both wore clothes and buried their dead. However, recent DNA comparisons suggest that the lines leading to modern humans and Neanderthals may have diverged 600,000 years ago from a primitive ancestral stock and that the evolution of sophisticated toolmaking techniques and even the evolution of religious belief might have occurred independently in the two different lineages.[46] None of these examples are exact replays of the tape, but they are at least suggestive that evolution may be constrained in very specific ways by as yet unknown mechanisms.

As mentioned above parallelism and long-term evolutionary trends have always struck many biologists, especially paleontologists, as difficult to account for in terms of undirected Darwinian models of evolution.

Molecular Hints of Direction

Although there is no direct evidence that mutational processes were directed during the course of evolution, there are two curious aspects of molecular evolution which strongly hint at the possibility. The first is the curious and surely suggestive fact that the rate of evolutionary substitution is almost equal to the mutation rate. The mutation rate is the number of changes which occur in the DNA sequence over one generation, calculated by counting the number of differences in the DNA sequences between parent and offspring. The evolutionary substitution rate is the average number of changes per generation which have occurred in the DNA since two species separated during evolution, calculated by counting the number of differences in the DNA sequence between two species and then dividing by the number of generations elapsed since their evolutionary divergence.

Comparisons of these two rates, the rate of mutation and the evolutionary substitution rate, have revealed the very surprising fact that the two rates are the same. This remarkable finding implies that *the differences between the DNA sequences of different species have been generated by mutation and that other factors such as natural selection could only have played a relatively minor role.* Which would imply that if all the DNA in the genome is informational, mutational processes pure and unfettered without any surveillance by natural selection must have created biological information.

The discovery of this relationship caused a considerable crisis when it first came to light in the late sixties and early seventies. The most popular explanation, subsequently adopted by most evolutionary biologists, was to suppose that the great majority of the DNA sequences in the genome of organisms was without any specific function and was therefore under no selectional constraints. This line of thought led to the "neutral theory of evolution" and was one of the main observations on which the "junk hypothesis" was based—the hypothesis that the vast majority of DNA is nonfunctional junk. The other observation which supported the junk hypothesis was the finding, during the 1970s, that only a small proportion of the DNA sequence in higher organisms coded for proteins. Most was intronic material, cut out of the initial RNA product of transcription, and discarded. Later the junk hypothesis concept was elevated into a whole new paradigm popularized by Richard Dawkins in *The Selfish Gene,* whereby much of the DNA was viewed as not only nonfunctional, but actually parasitic, perpetrating itself at the expense of the host organism in whose genome it "lived."

If it is true that a vast amount of the DNA in higher organisms is in fact junk, then this would indeed pose a very serious challenge to the idea of directed evolution or any teleological model of evolution. Junk DNA and directed evolution are in the end incompatible concepts. Only if the junk DNA contained information specifying for future evolutionary events, when it would not in a strict sense be junk in any case, could the finding be reconciled with a teleological model of evolution. Indeed, if it were true that the genomes of higher organisms contained vast quantities of junk, then the whole argument of this book would collapse. Teleology would be entirely discredited. On any teleological model of evolution, most, perhaps all, the DNA in the genomes of higher organisms should have some function.

While there is no doubt that at present no specific function can be attributed to most of the DNA in higher organisms, the idea that it is really junk is now under increasing attack. The first evidence that at least some of the noncoding DNA previously classed as junk does contain at least some genetic information is now emerging. Some of this evidence was referred to in a *Science* article entitled "Mining Treasures from Junk DNA" (February 4, 1994) and in a recent editorial in the same journal titled "Hints of a Language in Junk DNA" (November 25, 1994), describing the work of Eugene Stanley of Boston University, who used statistical techniques borrowed from linguistics and found evidence that much of the nonprotein-coding DNA has informational characteristics resembling those of a human language. Further evidence that at least some of the junk may be functional is the recent finding that many nonprotein coding sequences have been conserved over millions of years of evolution.

Another major discovery which arose from comparative studies of the gene sequences in different organisms was the so-called uniformity of molecular evolution. This phenomenon first came to light during the 1960s when it became possible to compare the sequences of genes from different organisms. By counting the number of differences between the sequences of a particular gene in a man and in a mouse, the percentage sequence difference could be determined.

By comparing sequences a curious pattern was observed. For example, in the case of the cytochromes, all the higher organism cytochromes (yeasts, plants, insects, mammals, birds, etc.) exhibit an almost equal degree of sequence divergence from the bacterial cytochrome in *Rhodospirillum.* This means that all their cytochrome genes have changed to about the same degree—in other words, have evolved at a uniform rate.[47]

The same phenomenon of uniform rates of evolution is observed in the case of nearly all the genes coding for proteins that have been examined. There are very few exceptions. The phenomenon raises all sorts of evolutionary questions. As Roger Lewin pointed out in a *Science* article entitled "Molecular Clocks Turn a Quarter Century," although the idea of a clock is "counterintuitive in the sense that anything in evolution might tick in a regular manner, the notion of and evidence for a molecular clock, nevertheless, has become even more pervasive than originally conceived." The development of the idea faced a considerable conceptual barrier, "that of associating the idea of any kind of regularity with the process of evolution." On the whole, however, as Lewin admits, "there is enough data to show that clocks can and do work, even if there appear to be many cogent reasons why they should not."[48]

At present, there is no consensus as to how this curious phenomenon can be explained. Some comprehension of the difficulties in attempting an explanation in selectionist terms can be gained by considering the evolution of the hemoglobins in the higher vertebrates. For example, the hemoglobins of man and salmon are equidistant from those of the hagfish, a primitive vertebrate. From this we may presume that the hemoglobins in the line leading to man and in the line leading to salmon have suffered the same number of substitutions since their common divergence. But since the two lines diverged, the line leading to man has undergone profound physiological and morphological changes, while the salmon remains fairly close in terms of its cardiovascular and respiratory system to its fish ancestor. In the case of the line leading to man, the heart has changed from a simple tubelike organ to a four-chambered efficient pump. The gills and bronchial arteries have been replaced by lungs and the pulmonary circulation. The red blood cells themselves have become completely different. From the large round red cells with a diameter of approximately 20 microns, possessing a nucleus, mitochondria, etc., they have changed into small platelike structures of 8 microns in diameter without nucleus or mitochondria and containing very much more hemoglobin per unit volume. While this dramatic series of changes was going on, the morphological and physiological organization of the cardiovascular and respiratory system on the line leading to the salmon must have remained fundamentally unchanged. Why, under selectionist explanations, has a protein functioning in the basically unchanging physiological environment of the salmon red cell accumulated precisely the same number of changes as a related protein in a line subject to such global adaptational changes?

Explanations of uniform rates of evolution in protein genes in terms of genetic drift of neutral mutations fare no better. The rate of genetic drift in a population is determined by the mutation rate. This is not controversial. Although mutation rates for many organisms are somewhat similar per generation time—10^{-6}/gene/generation—the problem is that generation times are vastly different, so that the rate of mutation per year in, say, yeast, may be 100,000 times greater than in a tree or a mammal such as man or elephant, organisms which have long generation times.

These twin discoveries—that the mutation rate equals the evolutionary substitution rate, and that the rate of change in many genes is regulated by a clock which seems to tick simultaneously in all the branches of the tree of life—may represent the first evidence, albeit indirect, that the mutational processes that are changing the DNA sequences of living things over time are indeed being directed by some as yet unknown mechanism, or more likely, mechanisms. Of course, these discoveries do not prove directed evolution, but it is far easier to imagine them as the outcome of some sort of direction than the outcome of purely random processes. There is no doubt that Asa Gray, Richard Owen, and especially Robert Chambers would have seen these discoveries as greatly supporting the idea that the course of evolution is determined in some way by natural law.

The Origin of Life

Before the tree of life can be manifest, before any sort of evolution can start, life must originate, and explaining how this happened is a major challenge for all models of evolution.

If the cosmos is uniquely fit for life of the carbon-based type that exists on Earth, and if the whole pattern of evolution was indeed written into the cosmic script, then it seems reasonable to suppose that the origin of life—the transition from chemistry to the cell—might be also written into the cosmic script. If it is true, as NASA claims, that carbon-based life—somewhat similar to that on the primitive Earth—arose on both the Earth and Mars shortly after their formation, then this would provide powerful support for the notion that the transition from chemistry to life is built into the cosmic design. As already alluded to in the "Note to the Reader" at the front of the book, many facts—such as the synthesis of carbon and the more complex atoms essential for life in stars throughout the cosmos, the fact that interstellar space contains vast quantities of organic carbon compounds, and the fact that the light of main sequence stars is ideal for photochemistry—makes eminent sense if the becoming of life is in some way programmed into the laws of nature.[49]

But even if it seems very likely that the becoming of life is built in, it has to be admitted that at present, despite an enormous effort, we still have no idea how this occurred, and the event remains as enigmatic as ever.

There have been only two significant developments in this whole area in the past forty years. The first was Stanley Miller's famous spark discharge experiment carried out in 1953. As described in a recent *Scientific American* article, Miller re-created the atmosphere of the primeval earth in a sealed glass apparatus:

> He filled it with a few liters of methane, ammonia and hydrogen and some water. A spark discharge device zapped the gases with simulated lightning, while a heating coil kept the water bubbling. Within a few days the water and glass were stained with a reddish goo. On analyzing the substance Miller found to his delight that it was rich in amino acids. These organic compounds link up to form proteins, the basic stuff of life. . . . Miller's results . . . seemed to provide stunning evidence that life could arise out of simple chemical reactions in the "primordial soup." Pundits speculated that scientists . . . would shortly conjure up living organisms in their laboratories and thereby demonstrate in detail how genesis unfolded. It didn't work out that way.[50]

In fact, the next stage of the process, from a soup of organic compounds to the current cell system, has remained enigmatic. At the heart of the problem lay a seeming paradox—proteins can do many things, but they cannot perform the function of storing and transmitting information for their own construction. On the other hand, DNA can store information, but cannot manufacture anything nor duplicate itself. So DNA needs proteins and proteins need DNA. A seemingly unbreakable cycle—the ultimate chicken-and-egg problem. As Monod put it in *Chance and Necessity*:

> The modern cell's translating machinery consists of at least fifty macromolecular components *which are themselves coded in the DNA: the code cannot be translated except by the products of translation.* It is the modern expression of *omne vivum ex ovo.* When and how did this circle become closed? It is exceedingly difficult to imagine.[51]

And Crick comments about the problem in *Life Itself*:

> An honest man, armed with all the knowledge available to us now, could only state that in some sense, the origin of life appears at the moment to be almost a miracle, so many are the conditions which would have to be satisfied to get it going.[52]

Just about the only real significant breakthrough after Miller's pioneering work was the discovery by Thomas Cech in the early eighties that RNA might have the ability to make copies of itself without the assistance of enzymes. This was very exciting.[53] As the *Scientific American* put it after this new finding: "Some investigators concluded that the first organisms consisted of RNA and that an early 'RNA world' had provided a bridge from simple chemistry to prototypes of the complex DNA-based cells found in modern organisms."[54] When the intermediate RNA world was first proposed back in the eighties, a major problem was the limited set of enzymic activities possessed by RNA molecules—these were insufficient to carry out the very many enzymic activities which would be needed to sustain an RNA world cell. But recent work in this area suggests that RNA molecules might in fact be able to carry out all the catalytic activities needed by any hypothetical RNA cell, thus removing one of the major obstacles to the RNA world hypothesis.[55] The scenario looks promising, although other studies have revealed serious drawbacks:[56] "Tests of the RNA-world hypothesis have shown that RNA is difficult to synthesize in the conditions that probably prevailed when life originated and that the molecule cannot easily generate copies of itself."[57]

And of course, even if the catalytic properties of RNA *were* exploited in some sort of intermediate self-replicating system which led eventually to the emergence of the modern cell, the actual pathway from the RNA world to modern DNA-based life has not been worked out even in outline.

Although many exotic hypotheses far more speculative than the RNA world have been proposed to close the gap between chemistry and life, none are convincing. There is the proposal that organic matter was first synthesized in interstellar space and that life was seeded on earth from space. Others suggest that life originated in the deep oceanic hydrothermal vents. Recently, a German lawyer, Günther Wächtershäuser, has developed a theory which proposes that life arose on the surface of iron pyrite, or fool's gold.[58] Most of these recent models do not merit serious attention, according to Miller. He refers to the organic-matter-from-space concept as "'a loser,' the vent hypothesis 'garbage' and the pyrite theory 'paper chemistry.'"[59] Nonetheless, completely novel and unexpected phenomena which may be highly relevant to the problem are continually coming to light, such as the very fascinating discovery recently reported in *Nature* of a self-replicating peptide.[60]

The problem has been compounded by the possibility that the early atmosphere in which the basic organic precursors of the cell were supposedly synthesized might have been far different from that assumed by Miller in his famous experiment and may not have supported the formation of the various organic compounds he reported. So the mystery may have deepened to include the origin of the basic building blocks themselves.

The mystery is further compounded by the fact that the time interval for life's emergence on earth is now known to be fairly short. In fact, it increasingly appears that life originated on earth shortly after the cessation of the meteor bombardment associated with the formation of the solar system, which ceased about 3.8 billion years ago. It seems that life appeared almost as soon as the planetary hydrosphere had cooled sufficiently to support it. The time available is certainly short—nothing like the supposed thousands of millions of years that was once assumed to be available.

Assuming that life arose as a result of natural processes, and nearly everyone working in the field accepts this assumption, then the very intractable nature of the problem raises the possibility that abiogenesis requires a completely new set of natural phenomena and processes, of which we have at present no idea. The fact that life emerged on the early earth as soon as conditions could support it points to the notion that life's origin was a natural and highly probable event which was inevitable given certain critical conditions. In this context Robert Chambers's notion that the process is analogous to "crystallization" seems remarkably apt, even though the *Vestiges* was written 150 years ago. When we finally hit on the mechanism of life's genesis, it will probably be as Miller confesses: "so damned simple that we'll all say, why didn't I think of that before?"[61]

One phenomenon, already touched on above, which could conceivably have played a role in the origin of life is the surprising tendency of complex dynamical systems to fall into highly ordered states. Spontaneous self-organization is a surprising phenomenon, as Kauffman points out: "Atoms and molecules are always doing their best to randomise themselves into maximum disorder. But on the other hand there are snowflakes, organised weather patterns, recurrent sunspot cycles on the sun . . . order and organisation seems ubiquitous in nature."[62] In 1965, Kauffman carried out a simple experiment in which he simulated on a computer a network of randomly connected interacting "genes." The result was quite counterintuitive. He found that even when the network of interactions was constrained by the simplest of rules, remarkably ordered patterns emerged in the interactions.[63]

As pointed out in a recent *New Scientist* article, "Order for Free," "Complete chaos is what most people would predict from such a system, known as a Boolean network, but instead order emerges. Kauffman calls it order for free."[64]

The notion that the transition to life was directed or facilitated by the laws of nature is perfectly consistent with the biocentric model of nature. Indeed, in a biocentric universe, where all the laws of nature have their ultimate meaning in the existence of life, it is hardly conceivable that the origin of life would have been left to chance. From a teleological perspective the origin of life *must* be viewed as something quite inevitable and built into the laws of nature from the beginning, just as were the properties of water and the mutual fitness of DNA and protein and all the other coincidences in the physical and chemical properties of life's constituents.

Curiously, many biologists are willing to accept the possibility that the origin of life might be built in but not the subsequent path of evolution. For example, Stephen Jay Gould, in a recent article entitled "War of the World Views" in the journal *Natural History*,[65] proposes "that the simplest kind of cellular life arises as *a predictable result* of organic chemistry and the physics of self-organizing systems but that *no predictable directions* exist for life's later development."[66] (My emphasis.) But surely it is far more likely that, if the chemical evolution of the first cell was built in, then the far less complex process—the biological evolution of life—will also turn out to be built in.

The Mode of Evolution

Evidence consistent with directed evolution is also emerging from recent studies of the mode of evolution. It has generally been accepted that the slower the rate of morphological change or the more slowly a new adaptation is acquired during the course of evolution, the more readily the process can be explained by simple nondirected Darwinian mechanisms. Conversely, the more discontinuous the mode of morphological or adaptive change, the more difficult it becomes to explain the process by undirected mechanisms and the more credible the concept of direction becomes. The relationship between saltation and direction is obvious if taken to the limit. The sudden emergence of an entirely new type of organism, or of a functionally perfect novel organ system, would be almost impossible to account for except within some kind of directed evolutionary or teleological framework. Gradualistic models of evolution have therefore always been favored by Darwinists and evolutionary theorists advocating undirected models of evolutionary change. For Darwin, *Natura non facit saltum* was virtually a creed, and for good reason.

Naturally then, when Niles Eldredge and Stephen Jay Gould proposed their theory of "punctuated equilibrium" in 1972,[67] it created some controversy, as it implied that the course of evolution consisted of long periods of stasis during which a species undergoes virtually no change and very short periods of explosive evolution when it suddenly gives rise to one or several new species that appear to burst into the fossil record per saltum.

Despite the considerable controversy surrounding this issue, the growing consensus and the best available evidence suggests that the punctuational mode is in fact the norm. The March 10, 1995, issue of *Science* presented new research on invertebrate paleontology which provided quite unambiguous evidence for the first time that at least in some lineages the evolutionary pattern was one of millions of years of stasis interrupted by periods of no more than 100,000 years of rapid and sudden change.

Accounting for a punctuated model in terms of classical Darwinian gradualism is not so straightforward. As R. A. Kerr, the author of the *Science* article, points out, "One mystery is what would maintain the equilibrium . . . keeping the new species from evolving in spite of environmental vagaries." He continues: "One possibility might be 'adaptive gridlock,' which arises because there are so many conflicting selection pressures pulling in different directions. . . . If a shellfish could reduce the weight of its shell, for example, it might have a better chance of escaping from some fast-moving predator. But that evolutionary route could become closed because a lighter, thinner shell, for example, would also decrease its resistance to other predators that bore into their victims. So the species remains unchanged for millions of years."[68] (The integrative complexity of biological systems and the sort of constraints this is bound to impose against bit-by-bit gradual change is raised again in chapter 14.) But if stability is the rule, and if selection tends to freeze organisms against change, then this raises the question of how selection ever transformed organisms so dramatically: "How do you get from funny little Mesozoic mammals to horses and whales? From *Archaeopteryx* to hummingbirds."[69]

The most celebrated of all evolutionary explosions is of course the sudden appearance of nearly all the major types of animal life in the early Cambrian seas. It has always been accepted that this explosion was compressed into a very short time span, geologically speaking, about 30 million years. However, recent work has revealed that the time span of the explosion may have been far shorter than this. A joint study carried out by Harvard and Russian geologists,[70] has drastically reduced this time span to perhaps as little as 5 million years. As Gould comments: "The entire Cambrian explosion, previously allowed 30–40 million years, must now fit into 5–10 million years and almost surely nearer the lower limit, . . . in other words, fast, much faster than we ever thought."[71] Explosive evolution is not only a phenomenon of animal evolution; the same pattern is seen in plants. Most of the modern flowering plants appeared in a few million years in the middle of the Cretaceous era.

Conclusion

The convincing grounds for interpreting the laws of nature as in some deep sense biocentric completely undermines the contingent *a priori* on which the Darwinian worldview was founded in the nineteenth century. It also provides a perfectly rational sanction for the concept of directed evolution. We have seen that water, the carbon atom, oxygen, the double helix, and many of the other constituents of life possess unique properties which seem so perfectly adapted to the biological ends they serve that the impression of design is irresistible. Many of these adaptations not only serve the end of microscopic life but also give every appearance of having been adjusted to serve the end of macroscopic terrestrial life forms such as ourselves. This raises the very natural but heretical idea, which has been explored in this chapter, that if the cosmos is fit for the being of higher life forms, then surely it is not inconceivable that an evolutionary mechanism for their actualization could also have been written into the order of things and that perhaps the entire process of biological evolution, from the origin of life to the emergence of man, was somehow directed from the beginning. I believe that our current knowledge of molecular genetics sanctions such possibilities.

Chapter 13

The Principle of Plenitude

In which it is argued that the diversity of life on earth approximates to the maximal diversity possible for carbon-based life. Such a plenitude is precisely what one might expect if the whole evolutionary process was itself built into the laws of nature. A variety of lines of evidence are considered. To begin with, there are those restricted cases where every possible variation on a particular biological theme is manifest. Examples cited are the early development of the animal embryo, the variety of bacteriophages, and the variety of image-forming eyes. Considering the complex constraints, embryological, physiological, and evolutionary, the diversity of life manifested on earth is remarkable. The possibility that life on earth approximates to the plenitude of all possible biological forms is perfectly in keeping with the teleological thesis that the cosmos is uniquely prefabricated for life as it exists on earth.

The plenitude of nature.

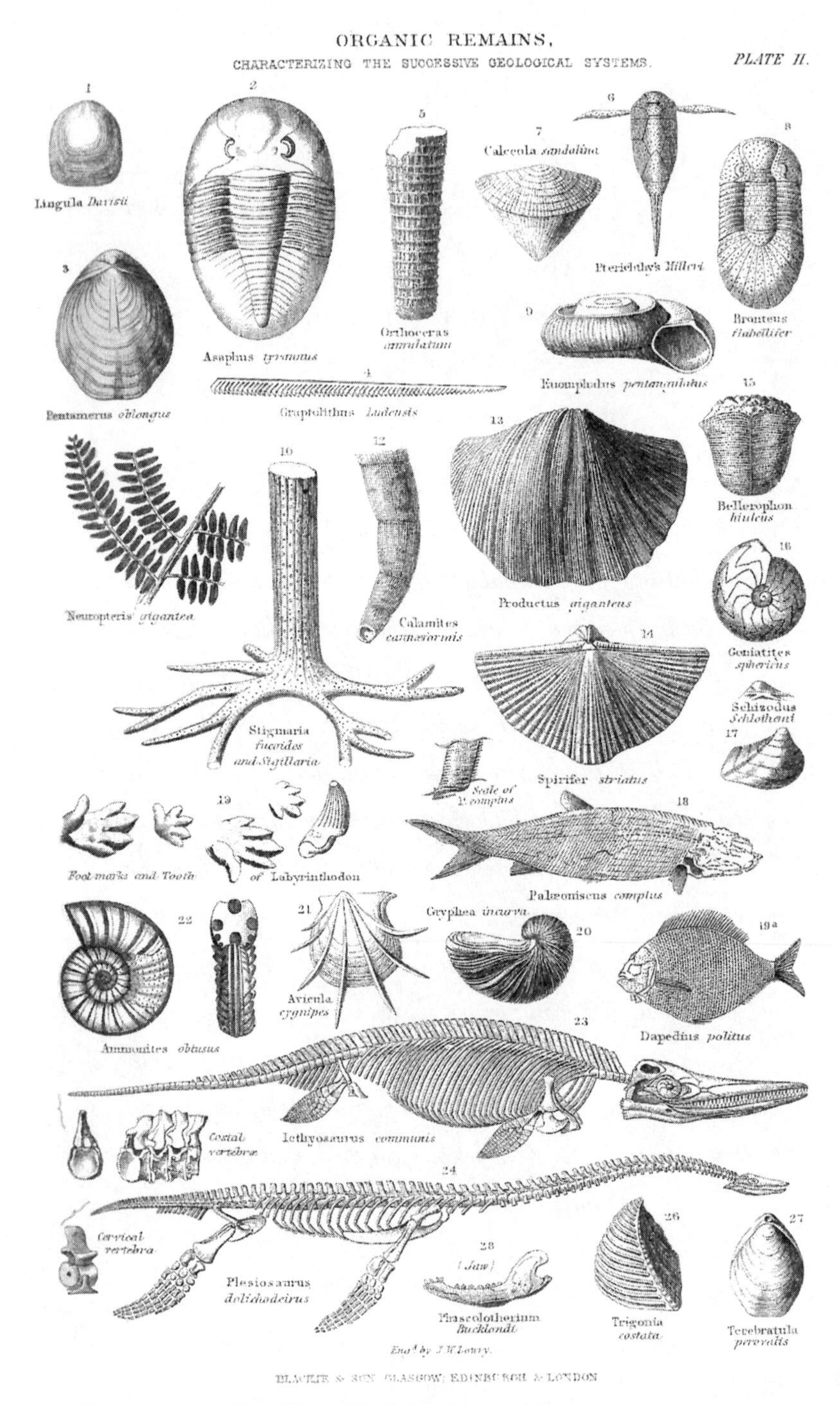

From Oliver Goldsmith (1876) *A History of the Earth and Animated Nature.*

Nature, inexhaustible in fecundity and omnipotence, has been settled in the innumerable combinations of organic forms and functions which compose the animal kingdom by physiological incompatibilities alone. It has realised all those combinations which are not incoherent and it is these incoherencies, these incompatibilities . . . which establish between the diverse groups of organisms those gaps which mark their necessary limits and which create the natural embranchments, classes, orders, and families.

—Georges Cuvier, *Leçons d'anatomie comparée,* 1835

No doubt we can imagine a greater variety of animals than do actually exist; such are the words of Archdeacon Paley. But what is the fact? Suppose we take the fabled animals of antiquity; not one of them could have existed. But we venture to say that every animal form, not actually existing in nature, but the invention of the artist or poet, would be discovered to have some defect in the balance of the exterior members, or were the exterior and moving parts duly balanced, some internal organ would be found unconformable or displaced.

—George Bell, Bridgewater Treatise, *The Hand,* 1832

It is impossible not to be struck by the enormous functional, structural, and behavioral diversity manifested by life on earth. Is it conceivable that there could be a world of life more varied, a phantasma more diverse than the one existing on our watery planet? From the tiniest bacterial cell to the immensity of the blue whale, from the great sequoias of California to the albatross of the southern ocean, with its wings inches above the waves, gliding perpetually on the eternal westerlies; from the giant squid of the ocean deeps, their bodies ever flickering with a sidereal bioluminescence to the exploding puffball fungi on an autumn evening; from the sinister cacophony of cries and screams as dusk falls across the Ngorongoro crater to the gentle piping of the Australian bellbird and the soft droning of a bumblebee; from the darting flight of the dragonfly to the plodding step of the elephant; from a tyrannosaur to an orb-weaving spider; earth exhibits a diversity of such exuberance that even the most bizarre imaginings of science fiction seem to have found realization in some exotic form of life. Our senses reel before the fantastic panoply of carbon-based life forms which clothes the earth. Is it possible that there could be a world as rich and as multifarious as the spectacular flowering of earth's great tree of life?

Before the rise of Darwinism it was widely believed that all possible living forms had actually been realized in nature. It was argued that an omnipotent Creator who had fashioned all the laws of nature to the end of life and man would surely have so organized these laws to make manifest in material form all possible biological types. This was the view of the leading French biologist of the early nineteenth century Georges Cuvier, who presented it with vigor and clarity in his classic work *Leçons d'anatomie comparée.* As Cuvier put it, nature had realized "all those combinations which are not incoherent."[1]

The idea that all possible organic forms have been actually manifested on earth is an old doctrine termed the Principle of Plenitude by the historian A. O. Lovejoy in his classic *The Great Chain of Being,* which long predates Cuvier and early-nineteenth-century biology. The doctrine was very influential in medieval and early Renaissance thought and remained influential in biology right up to the late eighteenth century. According to Lovejoy, the principle implied "that no genuine potentiality of being can remain unfulfilled, that the extent and abundance of creation must be as great as the possibility of existence and commensurate with the productive capacity of a perfect and inexhaustible source."[2]

The doctrine of plenitude would seem to be an inescapable corollary of the teleological position. For if a program for the evolution of life has been written into the cosmic script from the beginning, and if the earth is ideally fit for its evolutionary manifestation, then life on earth should indeed represent that plenitude that Cuvier envisaged and every possible type of carbon- and water-based life consistent with the laws of nature should have made its appearance on earth. The question, therefore, as to whether or not the pattern of life on earth approaches a theoretical plenitude of all possible forms has a critical bearing on the credibility of teleological claims.

The Diversity of Life

There is no doubt that the diversity of life is stunning. Nowhere is this more obvious than among microbial life. Microbes survive in every conceivable environment: from deep in the crustal rocks, to the frozen deserts of Antarctica, to the hydrothermal vents on the ocean bottom, to the hot springs in Yellowstone National Park. There could hardly exist an ecological niche on earth not filled by some exotic bacterial type. The biochemical diversity of microbial life is no less astonishing than the variety of environments they have exploited. For example, in the case of energy metabolism, bacteria seem to utilize almost every available reaction.[3]

Diversity is just as evident among the higher forms of life, from the protozoa up through the various multicellular groups. Consideration, for example, of the diversity of the unique and distinct bauplans of the major multicellular phyla suggests that at least a considerable proportion of all possible basic body plans have been actualized by life on earth. The basic arrangement of the body compartments in twelve of the more important phyla is shown in the figure below.[4]

Altogether there are currently around seventy known phyla, each of which has a unique and distinct body plan.[5] However, the existence of early Cambrian and Pre-Cambrian fossils of bizarre morphology and unknown affinity suggests that several additional types of body plan may have once existed in the early Paleozoic seas and that the total number of radically different basic animal body plans actualized on earth may be considerably larger than the seventy or so presently known.[6] However, from even a cursory consideration of the radically different arrangements of basic body compartments already realized in the seventy known phyla, it is possible to argue that there could not be a vastly greater number of radically distinct bauplans. The number of different ways in which the basic divisions of the animal body—an outer skin layer, body cavities, the gut, the gonads, the blood system—can be

Arrangements of body compartments in twelve animal phyla.

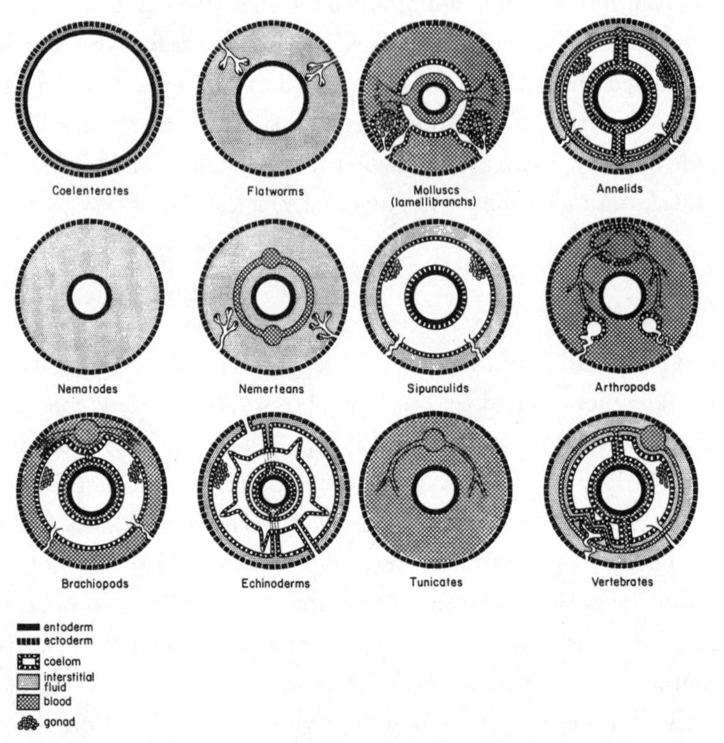

From Ernst Florey, *An Introduction to General and Comparative Animal Physiology;* © 1966 by Holt, Rinehart & Winston and renewed 1994 by Ernst Florey. Fig. 3.1. Reproduced by permission of the publisher.

arranged is limited. It would be surprising if the total number of possible different body plans for carbon-based life were many more than a few hundred.

Diversity within Phyla

Among the phylum Protozoa (the unicellular organisms) the diversity is so spectacular that it is indeed hard to believe that it does not represent a large proportion of all possible forms. As the German zoologist Bernhard Rensch commented in his *Evolution Above the Species:*

> In Protozoa, almost all vital processes show this undirected testing of nearly all imaginable possibilities. Thus, asexual reproduction may occur by simple fission, schizogony (simple separation of cells), or sporogony (spore formation); or there may be sexual reproduction by copulation, conjugation, pedogamy, automixis, or various types of metagenesis and heterogenesis.[7]

Even within small subgroups of the Protozoa such as the ciliates, the diversity of form and function is astonishing. There are about 5,700 ciliate species ranging in size from about 10 microns to 3 millimeters. As one authority put it, "about the same as between a blue whale and a rat."[8]

The diversity of insects is no less remarkable. They crawl, run, fly, swim, hop, and jump. There may be more than 3 million different species, ranging in size from tiny wasps and beetles which weigh only a few micrograms to the largest beetles, which are more than 10 million times heavier. Some dragonflies in the Cretaceous era had wingspans of 70 centimeters. As Rensch notes: "Some are carnivores, herbivores, omnivores, some are even coprophages (feces-eating), bloodsuckers, keratin and wax eaters, xylophages, nectar suckers and pollen eaters, and some that eat nothing during the span of their adult life. The differences are equally manifold as regards display or mating patterns, structures and appendages of genitalia in bumblebees, butterflies, and so forth."[9]

Similarly, within the Vertebrata, among the reptiles and mammals we have carnivores, herbivores, quadrupeds, bipeds, flying and aquatic forms. Among birds, a relatively specialized type, we have flying, terrestrial, and aquatic forms.

Possible Cases of Plenitude?

The question as to what proportion of all possible designs have in fact been actualized in the diversity of insects, vertebrates, or the basic bauplans, is clearly only a matter of subjective judgment. In most cases, it is impossible at present to determine what the theoretical limits might be—for example, what constructional or developmental constraints might apply.[10]

However, there are some cases where the constraints are relatively obvious and where we can be relatively certain that all possibilities have been realized.[11] Bernhard Rensch has suggested some instances, including hibernation strategies in butterflies and the incubation strategy of birds. Regarding the former, he comments:

> There are four possible stages of hibernation in butterflies of the temperate and cold zones: the insect can spend the winters in the stage of an egg, a caterpillar, a pupa, or an adult insect. All these possibilities are realised in nature.[12]

Rensch then goes on to show that in the case of the horns of antelopes the whole range of structural possibilities, apart from those that would be biologically discordant, are actually realized: "straight and smooth, straight with transverse ridges, slightly curved with a smooth surface, slightly curved with transverse grooves and ridges, twisted like a screw with a smooth surface, etc."[13] Likewise, as he shows, the same is true for the range of shells of Gastropoda: "These shells are shaped like a bowl, a cap, a tube, a flat spiral, a more or less tapering cone, a needle, a ball, and so forth."[14] He then considers the possible ways in which the stridulating organs of insects, which produce sound by rubbing body parts together, may be formed. It turns out that in beetles alone there are at least fifteen different ways in which a "pars stridens" can be moved against a "plectrum."[15]

A fascinating case cited by Rensch involves the early development of the egg. Here again, where the total number of possibilities is strictly limited, the evidence strongly suggests that every conceivable possibility has been realized:

> There are equal, unequal, discoid, and superficial cleavage; and there are various intermediate and modified forms (radial, leiotropic [left], or dexiotropic [right] spiral cleavage); and finally there may be a separation of the blastomeres (early embryonic cells) as in polyembryonic forms of development. Hence again we find all possible modes of development.[16]

Another example where all possibilities consistent with biological design may have been realized is seen in the ways by which organisms move. We have, in air, jet propulsion, gliding flight, flapping flight, ballooning; on land, articulated legs; in water, jet propulsion, swimming, and even by propeller in the case of bacteria.

Viruses appear to provide another instance in which there are reasonable grounds for believing that all possible biological forms may have been actualized. For example, all conceivable ways of storing genetic information—using single-stranded RNA, double-stranded RNA, single-stranded DNA, and double-stranded DNA—are found among various types of viruses. Again, of the two possible forms of virus capsule, the cylinder and the icosahedron, which are the only stable hollow structures that can be built up out of single subunits, both occur in nature.

Even among one small subset of viruses, the so-called bacteriophages, which replicate themselves inside bacterial cells, it would appear that all possible forms are actually found in nature. The bacteriophages are of peculiar relevance to the question of whether the actual forms of life represent the limits of the possible, because we are far closer to having a complete description of every aspect of their biology than in the case of any other comparable group of living things. Indeed, in the case of one class of bacteriophages, the so-called RNA phages, which are the simplest of all the viruses, their entire chemical structure has been determined, every gene has been identified, so have virtually all the genetic regulatory devices they utilize during the course of their replicative cycle. In addition, the three-dimensional structure of the viral capsid has been elucidated at a molecular level. This is one area of biology where genetic engineering holds out the fascinating prospect of attempting the creation of completely novel types of organisms quite different from those forms already known, thereby bringing the question as to whether the actual equals the possible into the realm of experimental science. But already, from what we know of the biology and the diversity of these fascinating types of nanoreplicators, it is very difficult to envisage the construction of a bacteriophage radically different from those already existing. Every possible variant seems already to be manifest.

All Possible Eyes?

Another case where it appears that every possibility has been realized is in the design of image-forming optical devices.[17] These include the familiar camera-type of eye found in vertebrates, molluscs, and various invertebrates; the reflecting eye of the scallop pecten and the crustacean *Gigantocypris,* which form an image by reflection from a concave mirror onto a retina situated at the focal point of the mirror; and the three different types of compound eye of the insects and arthropods. One type of compound eye found in diurnal insects is made up of a hexagonal array of tiny lenslets, each of which has its own photoreceptor cell that receives light only from its own lenslet. A second type (the superposition type) is found in nocturnal insects, again made up of a hexagonal array of tiny lenslets which bend the light rays so that light is focused by refraction through many lenslets to one point in the photoreceptor layer. A third type is also a superposition eye, but in this case the light is focused by reflection from a series of tiny square mirror-lined units onto the photoreceptor layer (see the description of the lobster eye in chapter 15). Finally, there is even what appears to be a scanning eye, utilized by a small marine crustacean which scans an image formed by a simple lens by rapidly moving a single photoreceptor back and forth across the image. In addition to the realization of what appears to be all possible image-forming devices, there are of course a near-infinite variety of nonimage-forming simple eyes, from subcellular photosensitive pigment spots in Protozoa to the simple photoreceptor eyes of many invertebrates, such as spiders.

It is also of interest that living things utilize or detect the entire range of electromagnetic radiation reaching the earth's surface from the ultraviolet to the infrared. The human eye, for example, can detect electromagnetic radiation from 0.38 microns to about 0.8 microns. Certain insects detect light in the ultraviolet around 0.3 microns,[18] and the chlorophylls absorb light in the blue (0.35–0.45 microns) and red regions (0.55–0.65 microns); vitamin D synthesis requires ultraviolet radiation near 0.3 microns, and so on.

Although organisms do not utilize infrared radiation (0.8 microns to 2.0 microns) to drive chemical reactions, nearly all organisms, including man, can detect it as warmth on the surface of the body. The warmth of the sun's infrared radiation on the skin is sensed by special cells called thermoreceptors. The precise mechanism by which they detect infrared radiation is not understood. However, they are remarkably sensitive. The thermoreceptors in the heat-sensitive pit organ of the rattlesnake, for example, can detect a change in temperature of as little as 0.003^{0}C.[19] The thermoreceptors are stimulated by all wavelengths of infrared radiation reaching the surface of the earth, i.e., from wavelengths of 0.8 microns to about 2.0 microns.

It would seem, then, that not only are all possible image-forming devices actualized in some branch of the tree of life but also that the full spectrum of radiant energy reaching the earth's surface is detected or utilized in some way by a variety of biological systems. Moreover, as we saw in chapter 3, radiant energy in the visible and the infrared is the only type of electromagnetic radiation of utility to biological systems.

In addition to their ability to see and detect heat, living things can detect sound over a large range of frequencies, in both air and water. They can detect vibrations transmitted through the ground; they can detect vanishingly small concentrations of a vast variety of chemicals in the air and water; they can detect gravitational fields, magnetic fields, and electric fields.

The Size of Organisms

One area where it is possible to be slightly more objective in judging how close actual life forms approach the limits of the possible is in considering the question of size. The difference in size between the smallest and the largest organism—between the smallest bacteria (the mycoplasma) and the blue whale—is 10^{21}. A blue whale may weigh more than 100,000,000 grams, a mycoplasma cell less than one-tenth of a picogram or 10^{-13} of a gram.[20]

As far as the mycoplasma is concerned, we can safely assume that it is very close to the lower limit of size for an autonomously self-replicating cell. The biochemist Harold Morowitz has speculated as to what might be the absolute minimum requirement for a completely self-replicating cell deriving all essential organic percursors—amino acids, sugars, etc.—from its environment but autonomous in every other way in terms of our current understanding of biochemistry.[21] Such a cell would necessarily be bound by a cell membrane and the simplest feasible would probably be the typical bilayered lipid membrane utilized by all existing cells on earth today. The synthesis of the fats of the cell membrane would require perhaps a minimum of five proteins. Energy would be required, and this might require a further eight proteins for a very simple form of energy metabolism. Altogether, probably a minimum of another hundred proteins would be required for DNA replication and protein synthesis. The size of such a cell, containing perhaps four mRNA molecules, a full complement of enzymes, a DNA molecule about 100,000 nucleotides long and bounded by a cell membrane, would be about one-tenth of a micron in diameter. Morowitz comments: "This is the smallest hypothetical cell that we can envisage within the context of current biochemical thinking. It is almost certainly a lower limit, since we have allowed no control functions, no vitamin metabolism and extremely limited intermediary metabolism."[22]

The physiologist Knut Schmidt-Nielsen considered the question of whether or not the blue whale weighing 100 million grams, the giant redwoods (1 billion grams) and the smallest existing organism, the mycoplasma, also known as PPLO, are close to the actual limits on what is possible and concluded: "There are cogent reasons to believe that the smallest and the largest organisms represent approximate limits to the possible size of animals under the conditions that prevail on our planet."[23]

There are certainly other cases in which there are very good grounds for believing that a particular species possesses the smallest possible body size compatible with its basic biological design. Rensch comments that, as the lower limit of body size is reached,

> organisms usually reduce special structures and special organs and only the indispensable mechanisms are maintained. A good example of this kind is provided by the minute marine snail *Caecum glabrum* (Opisthobranchia), 1 mm long and 2 mm wide. . . . In this species the cells are of about the same size as in larger relatives. . . . Consequently, the intestinal gland of this minute snail consists of two tubules, and the gonads are represented by a single folded tube (while in other snails these organs are made up of quite numerous tubules forming a solid network of glandular elements). Hence, it is quite evident that a further reduction of these organs is impossible. Here is another objective case where the limits of the possible are actualised in a particular form. Moreover in the case of virtually all the main subgroups of insects, vertebrates, molluscs and many other types of organism there is quite convincing evidence that the *actual range of size is close to the possible.*[24] [My emphasis.]

Oxygen Delivery

There are also firm grounds for believing that in the case of certain basic structural and physiological systems, such as oxygen delivery systems, skeletal systems, and excretory systems, all design possibilities have been exhaustively exploited. Consider, for example, the delivery of oxygen to the tissues of a terrestrial air-breathing organism. From first principles there are only two possible ways of achieving this: by the use of a circulatory system to carry dissolved oxygen via small tubes throughout the body, or by the use of a tracheal system of tubes to carry the oxygen gas directly to the tissues. Both these systems are actually exploited by the two major types of macroscopic terrestrial life on earth, the vertebrates utilizing the "circulatory solution" and the arthropods the "tracheal solution."

The vertebrate solution necessitates a lung in which the circulating fluid can be aerated and an efficient pump or heart to force the fluid through the small tubes, or capillaries, and into the tissues where the oxygen is consumed. Further, as we have seen in chapter 6, because of the relatively low solubility of oxygen, it cannot be carried in sufficient quantities in simple solution; consequently, a special oxygen-carrying molecule, hemoglobin, is required, and this in turn (for a variety of reasons already touched on in chapter 2) requires the hemoglobin molecules to be densely packed together in a special cell—the red cell. Thus, we are led from the necessity of delivering oxygen in sufficient quantities to support the needs of combustion in the tissues of a large terrestrial organism to the vertebrate respiratory and cardiovascular systems and even to a molecule with the functions of hemoglobin. As we saw in chapter 9, such a molecule would almost certainly have to be a protein and probably exploit the unique characteristics of the iron atom.

There is nothing in the slightest "accidental" about the fact that it is the larger vertebrates that use the circulatory system, while the tracheal system is utilized by the much smaller arthropods. No system of small air tubes, however modified, permeating the muscles of, say, an elephant or a whale, would be capable of efficient oxygen delivery. Even in the case of large, active insects, the air much be pumped through the tracheal network, because the diffusion of oxygen, although much faster in air than in water, is too slow to supply the needs of the organism. Air sacs which communicate with the tracheal system and which are periodically compressed by the movements of the insect's body are utilized to pump air through the tracheal system. To strengthen the trachea and prevent the tubes collapsing under various pressures exerted on them by the movement of adjacent muscles in the insect's body, many incorporate internal spiral ribs. The largest insects are in fact close to the maximum size possible for an organism obtaining oxygen via a tracheal system.[25]

For small active organisms such as flying insects, the tracheal system has great advantages. In fact, below a certain body size, as Schmidt-Nielsen shows, there are definite deficiencies in a circulatory system when compared to a tracheal system for the delivery of oxygen to metabolically active tissues in small organisms. In a fascinating section in which he discusses the various ways in which physiological constraints impose limits on the size of animals, he notes that the smallest birds and mammals are approximately the same size, between 2 and 3 grams, and asks whether this is the lowest possible weight for these two classes of organism.[26] He continues: "Is there some reason that birds and mammals are not smaller? To produce heat, we need both fuel and oxygen. Hummingbirds get most of their energy from nectar, exactly as do large moths; so fuel supply is an improbable constraint."[27] As he explains, the reason is that below a certain size a circulatory system is increasingly ineffective for oxygen delivery compared with the tracheal system.[28] On the other hand, the tracheal system is remarkably efficient for a small organism. The diffusion coefficient for oxygen in air is some 10,000 times higher than that in water and this ensures an adequate supply of oxygen without the need of blood.[29]

The fact that large insects are able to raise their body temperatures and metabolic rates during flight to that of warm-blooded organisms is testimony to the high functional capacity of the tracheal system.[30] The speed, power, and efficiency of the flight of dragonflies and some of the larger moths is every bit the equal of hummingbirds and small bats.

In passing, it is interesting to note that these two mechanisms of delivering oxygen to the tissues of terrestrial organisms are both critically dependent on the values of a number of key physical constants, including the diffusion rate of oxygen in air, the solubility of oxygen in water, the diffusion rate of oxygen in water, the viscosity of water, the density and viscosity of air, etc. It is fortunate that their values are close to what they are, because terrestrial life would have been impossible or enormously constrained if these values had been even slightly different. As it is, these constants limit the design of respiratory systems to two different designs—the vertebrate and the insect.

A final point in this context is of interest. In the case of the vertebrate lung, there are again two fundamentally different possible types. There is the familiar "bellows type" of lung in which air is drawn in and out via the same passage during respiration. This type occurs in most terrestrial vertebrates, including man. The other, radically different possible design, "the continuous throughput type," is realized in the avian lung, where air is inhaled through one passage drawn unidirectionally and continuously through a system of capillary tubes and then exhaled via another passage. Again, as with the two possible oxygen delivery systems, both possible types of lung have been actualized in nature.

The two basic and radically different types of skeletal system, the endoskeleton and the exoskeleton, have also been actualized in nature in the two major types of higher organism: the vertebrates and the arthropods. Vertebrates have an endoskeleton composed of several internal articulating rods, composed of bone, which is fundamentally a mineralized type of connective tissue. Insects have an exoskeleton composed of a very hard plastic-like polysaccharide, chitin. Again, this difference is not accidental. One can easily conceive of the difficulty of constructing an elephant-sized organism weighing five tons with an exoskeleton or the difficulties of attaching muscles to an articulating endoskeleton in an organism weighing 25 micrograms. The relative merits of the exoskeleton for small organisms and the endoskeleton in large animals are well known and have been discussed in a number of recent publications.[31]

There are surprisingly few simple materials available to living things which are mechanically rigid and hard, with some tensile strength, and therefore suitable for the construction of skeletal parts. The very hard polysaccharides, cellulose and chitin, are used by all plants and a great many invertebrates as the basic building material of their rigid components. Mollusc shells and vertebrate bones contain calcium salts—mainly carbonates in the molluscs, and phosphate in the vertebrates.

It is difficult to see how the vertebrate endoskeleton could be composed out of chitin like the insect cuticle. Chitin is very unreactive, being virtually insoluble in even strong acids and only digestible by enzymes under physicochemical conditions difficult to satisfy in the inside of an animal's body.[32] In this context it may be fortunate that the solubility of calcium phosphate is fit for the formation of the mineralized component in bone in an organism with a pH of 7 because the other readily available skeletal materials chitin and cellulose would be not be suitable for the formation of endoskeletal structures in a vertebrate.

Movement and Legs

Animals, both large and small, must move. Vertebrates have four legs and insects six. Can this difference also be explained, perhaps, by the difference in size? The answer is yes. An insect, having a very small body, is subject to instability caused by gusts of air. Six legs means that even in a fast gait an insect need only have three legs off the ground at any one instance. Calculation shows that this source of instability is significant and an insect with only four legs would be severely disadvantaged. Even the positioning of an insect's legs, well apart on either side of the body, is also related to the necessity for stability to prevent a small organism being destabilized in even a small gust of air.[33]

Virtually every conceivable way of moving on legs that is compatible with fundamental design principles is exploited by some type of terrestrial animal. Humans walk on two legs. Kangaroos hop on two legs. Most mammals walk, run, or amble on four legs. Insects run and walk on six legs, spiders on eight. Wood lice, centipedes, millipedes, and so on use more than ten. Snakes move by sliding without legs.

There would seem to be two basic biological designs for terrestrial macroscopic life, a large vertebrate type, which includes a circulatory system, lungs, heart, endoskeleton, and four legs, and a small insect type, with a tracheal system, exoskeleton, and six legs. These two designs are actualized in terrestrial life on earth, the insect type in organisms from about 25 micrograms to 100 grams and the vertebrate type in organisms from about 2 grams to 10,000 kilograms.

From this very brief consideration of the likely physiological criteria which would have to be satisfied by any type of viable macroscopic terrestrial life, constructed from organic carbon compounds, deriving energy from oxidation, and based in a matrix of water (which as we have seen is probably the only biochemical design capable of replication), it seems likely that only very few alternative basic designs for air-breathing terrestrial life exist. Again, the actual approaches the possible.

The Gaps in the *Scala Naturae*

Those many cases where gaps in the order of nature appear of necessity rather than by accident tend to strengthen the conclusion to plenitude.

Take, for example, the absence of intermediates between the unicellular Protozoa and the various primitive metazoan or multicellular groups (represented by the phylum Porifera, the sponges; the well-known group the Coelenterata, the jellyfishes; the less well-known group the Ctenophora, more commonly known as the comb jellies or sea gooseberries; the nematodes; flatworms; etc.). Collectively, the species comprising these primitive multicellular phyla are very diverse in morphology and behavior. Moreover, they are far from simple in structure and design. They are highly complex organisms made up of most of the basic cell types—muscle, nerve, epidermal, gland, etc.—found throughout the animal kingdom. The Protozoa are also a fantastically diverse group. But between these two groups there is an enormous gap filled only with a few types of simple colonial Protozoa.

That this gap is likely to be necessary is suggested by the difficulty of imagining realistic intermediate types of organism made up, say, of three, four, or five cells leading up to genuinely multicellular life. Precisely what form such organisms would take and what adaptive role their constituent cells might play is exceedingly difficult to imagine. It is not only my own imagination that is lacking here. After a century of intense speculation the evolutionary origin of the Metazoa is still problematical, primarily because no convincing series of functional intermediates between the unicellular and the multicellular level of biological organization has been envisaged. Simple life forms, it seems, can be composed of one cell or many cells but not readily of five or six cells.[34]

In the case of other apparent "gaps," it often turns out that the reason is obvious. There are no marsupial whales or seals. As already mentioned in the previous chapter, this is evidently because the marsupial reproductive system is difficult to adapt to an aquatic lifestyle. As we also saw in the previous chapter, there are no marsupial bats. This may well be because the niche was filled by placentals from the earliest beginnings of marsupial evolution. There are no large mobile terrestrial molluscs; no large terrestrial arthropods; no fish or amphibia capable of powered flight. To modify a frog for powered flight we would need to give it the cardiovascular system of a mammal or bird. To convert a mollusc into a mobile terrestrial form, we would have to give it an endoskeleton, rid it of its shell, clothe it in an impermeable skin—in other words, convert it into a vertebrate.

A great many of the seventy or so major phyla have never generated large complex forms. In many cases this is of necessity. The flatworms, for example, could hardly evolve into anything the size of a mouse. Their basic design prohibits such a prodigious development. Flatworms are flat for good reason—they have no circulatory system. As Huxley points out, "the flatness of the larger flatworms is partly due to the need for having every cell near enough to the surface to be able to get oxygen by diffusion. The elaborate branching of their intestines and all their other internal organs is needed to ensure that no cell shall be more than a microscopic distance away from a source of digested food."[35]

Constraints

In judging to what extent the phantasma of life on earth approaches a complete plenitude of carbon-based forms of life, it is necessary to note that the plenitude we are considering is a plenitude of possible forms, that is, *fully functional living systems that could possibly exist and survive in some conceivable ecosystem on earth.* Consider the fact, for example, that no large organism possesses rotating parts or exploits the wheel as a means of motion.

Wheels are only half as costly as legs in terms of energy consumption, but such a device is not compatible with biological design principles.[36] The reason wheels have not been utilized is because animals must maintain physical connections between their parts. As Stephen Jay Gould asks, "If wheels are so successful an invention, why do animals walk, fly, swim, leap, slither, but never roll?"[37] The problem is that "wheels require that two parts be in juxtaposition without physical connection . . . this cannot be accomplished in creatures familiar to us because connection between parts is an integral property of living systems. Substances and impulses must be able to move from one segment to another."[38]

The incompatibility of wheels with biology is in any case no defect in the scheme of things because of course wheels are only of utility on unnaturally flattened surfaces and are virtually useless on the uneven terrain which covers most of the earth. Advanced robots invariably use articulated legs, which gives them mobility over a far greater range of terrains.

Of course, there are a great many fundamental constraints that impose limitations and restrictions on the range of possible functional biological designs. An elephant with the legs of an antelope is impossible for obvious biomechanical reasons—the legs could never support its weight. Nor can an organism the weight of an elephant, or even a man, fly, because the energy necessary to provide the required degree of lift to make powered flight feasible is simply prohibitive. Even the largest existing flying birds, such as an albatross or a swan, have difficulty getting airborne from rest. The albatross generally takes off into the wind from a hillside or cliff, while a swan often has to run across the surface of a considerable stretch of water. Obviously, a vast number of imaginary organisms are simply impossible for a variety of self-evident biomechanical and other types of design constraints.

Because all the parts and organ systems of an organism are functionally interrelated, constraints on one organ system invariably impose constraints on many of the other organ systems. The need for functional integration and coherence if an organism is to be actualized is bound to restrict the functionally possible to a fantastically small subset of all conceivable organisms. As Schmidt-Nielsen emphasizes:

> The total number of interconnections in the living animal is overwhelming. We can just think of the many steps in supplying oxygen to match the metabolic rate. Structures and functions are all interconnected: breathing, lung size

and area, diffusional pathways, blood flow, heart, haemoglobin function, capillaries, mitochondria, enzyme concentrations, and so on, in a chain of seemingly unending interdependent variables.[39]

In addition to constraints imposed by the necessity of physiological integration, if an organism is to be actualized, it must also be ecologically feasible. A lion or a tyrannosaur without herbivores for prey is impossible. Similarly, herds of herbivores require plant life in abundance.

Evolutionary Constraints

Evolution, whether it is directed or not, can only proceed through functional intermediates, and this is bound to impose additional constraints on what life forms are possible. The origin of life, for example, is difficult to envisage, primarily because of the difficulty of imagining a credible functional sequence of increasingly more complex replicating systems leading from chemistry to the cell. (As was argued in the previous chapter, this suggests perhaps that there may be only one unique route.) A chemist, on the other hand, if he wished to create a living cell, *de novo,* would be free to choose a number of different strategies to synthesize the constituents of the cell and then artificially assemble them into a living whole (there is no reason why this could not be theoretically achieved). Being unconstrained by the necessity to move via a continuous functional series of intermediates, a chemist could at least in theory achieve the same end by far simpler means.

The evolution of the seventy different basic metazoan multicellular body designs from a unicellular ancestor which occurred at the beginning of metazoan evolution is problematical primarily because of the difficulty of envisaging viable intermediates leading from the single-celled ancestral form to a multicellular organism and then the subsequent diversification into the seventy different basic types of life. That all evolutionary transitions must occur through fully functional forms is obviously immensely constraining and suggests that perhaps not all possible fully functional life forms can be generated on a single unique evolutionary tree.

In addition to the evolutionary constraint, i.e., the need for functional continuity, there is another set of constraints which must act to restrict the range of the possible—those associated with the process of development. All multicellular species develop from an egg cell through a complicated and intricate process of embryogenesis. Inevitably, the criteria that must be satisfied to ensure the process is successful are highly constraining.

Consider just one embryological criterion that would have to be satisfied in the development of a large vertebrate type of organism from its egg cell. During embryonic development it is essential that all parts of the growing cell mass are supplied at all times with sufficient oxygen and nutrients. This will require a cardiovascular and respiratory system. Clearly, not just any sort of circulatory system will do; only a system which can be continuously transformed and can develop in functional harmony with the growing embryo will suffice. Circulation must be conserved at every stage of development. This problem is of course solved in the case of mammalian development, as every medical student learns, by the expediency of having the system undergo a series of changes from the "first" heart, consisting only of a simple contractile tube, which then undergoes a gradual transformation into an increasingly powerful pumping system until finally it is converted into the efficient four-chambered pump of the adult mammal. One suspects that there may be very few alternative ways of continuously changing a circulatory system from simple tube to four-chambered pump while at the same time maintaining physiological function throughout the process.

In addition to such specific requirements, all developing embryos are constrained greatly by what two authorities refer to as "generic" physical mechanisms. As they point out, these include such diverse physical processes as adhesion, surface tension, and viscosity. Such phenomena may cause morphological rearrangements of cytoplasm, tissue, or extracellular matrix. These can sometimes lead to complex forms such as "micro fingers," and to chemical waves and stripes. They suggest that "major morphological reorganisations in phylogenetic lineages may arise by the action of generic physical mechanisms in developing embryos."[40]

Such physical mechanisms may severely constrain some early events during development such as the formation of the blastula and gastrulation. The diffusion rate of macromolecular morphogens through an embryo may limit the size over which a diffusion gradient can be read during development. This limitation taken in conjunction with the viscoelastic properties of tissues and cells and the small distances over which cells can move in a short space of time is probably the reason why the major morphogenic events, including the cell and tissue movements associated with gastrulation and the specifying of segmentation patterns, occur early in development, when embryos are less than a centimeter across.

Fitness of the Earth

There is a final point to consider regarding the doctrine of plenitude. Only in a uniquely fit environment can a plenitude be actualized. Only in an environment supplied with a reliable and constant source of energy and which is also chemically and physically stable for billions of years and also very diverse could any sort of evolutionary process lead eventually to a manifestation of the full plenitude of life forms. But although these conditions are stringent, the hydrosphere of the earth satisfies them all. For billions of years it has been bathed in a constant radiant energy source (which contains just the type of radiation required for life), and over the vast eons of geological time its chemical and physical stability has been ensured by the various geochemical cycles, including the water and the tectonic cycle. Moreover, although the hydrosphere is remarkably stable, it also contains a great variety of diverse environments—the polar oceans, tropical rain forests, high mountains, hot sandy deserts, arctic tundra, hot springs, the ocean depths, wetlands. That it should be exceedingly stable in terms of its general physical and chemical properties yet at the same time contain very diverse environments is of considerable interest in itself, as this lends further support to the notion of its unique fitness for life. If the earth's hydrosphere had been a monoenvironment, the potential for diverse life forms would have been greatly constrained.

This would seem to be yet another significant coincidence—that the unique environment suitable for carbon-based life, the hydrosphere of a planet of the mass and size of the earth, is also ideally suited for the manifestation of a plenitude of life forms. As we saw in chapter 4, there are convincing grounds for believing that the hydrosphere of the earth (and of any planet of the same size and mass and the same distance from its own sun) is the inevitable end of natural law. In other words, nature would seem to be arranged to generate an environment perfectly fit for the manifestation of what would appear to be the plenitude of all possible carbon-based life forms.

Conclusion

Even though at present we may still be far from having any final answer to the question of plenitude, most of the evidence is nevertheless remarkably consistent with the doctrine. That so many diverse forms of life and basically dissimilar body plans have in fact been actualized during the course of evolution on earth supports the concept that the evolutionary tree of life on earth was generated by direction from a unique program embedded in the order of nature, one that was specifically arranged to generate through a myriad of unique and intricate transformations the fullest possible plenitude of natural biological forms.

Chapter 14

The Dream of Asilomar

In which the challenge posed to undirected evolution by the constraints inherent in complex systems is examined. In complex systems like a watch or a living system, all the subsystems are intensely integrated. Engineering changes in such systems is complex because each change to any one subsystem must be compatible with the functioning of all the other subsystems. Any change beyond a trivial degree is bound to necessitate intelligently directed compensatory changes in many of the interacting subsystems. In this context it is hard to understand how undirected evolution via a series of independent changes could ever produce a radical redesign in any sort of system as complex as a living organism. It is precisely this integrated complexity which provides a major barrier to engineering radical change in living things from viruses to mammals. In the future, if genetic engineers are ever able to radically redesign living systems from proteins to whole organisms, this will only be via intelligently directed changes which will almost certainly necessitate progammed simultaneous change in many of the basic subsystems. Artificial evolution will be per saltum *and not* per *a succession of independent changes. Living organisms not only exhibit an immense integrative complexity but are also immensely complex in terms of the sheer number of unique components they contain. In the case of higher organisms the number of different unique genetic readouts used throughout the life of the organism may approach several billion.*

Three different neurons in the mammalian brain.

Each nerve cell consists of a cell body (c), an axon (a), and dendrites (d). The dendrites receive signals from other nerve cells and transmit them to the nerve cell body. The axons carry the nerve impulses from the nerve cell body to other nerve cells. Nerve cells make between 10,000 and 100,000 connections with other nerve cells. Most axons make highly specific contacts with other nerve cells. Altogether, there are ten trillion nerve cells in the human brain (10^{11}). The total number of connections is about 10^{15}.

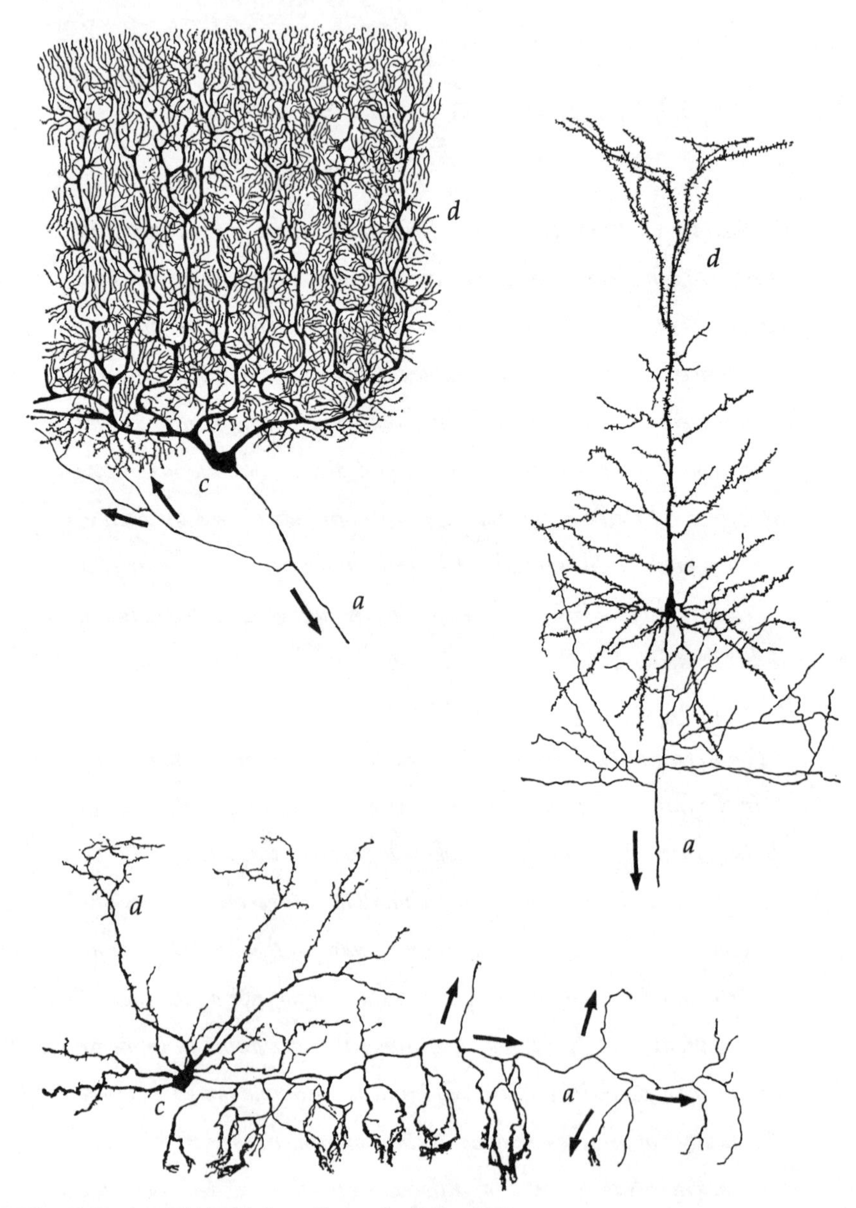

From S. Ramón y Cajal (1909) *System Nerveux,* figs. 8, 9, and 21.)

Breeders habitually speak of an animal's organisation as something plastic which they can model almost as they please . . . [William] Youatt . . . speaks of the principle of selection as "that which enables the agriculturist, not only to modify the character of his flock but to change it altogether. It is the magician's wand by means of which he may summon into life whatever form and mould he pleases." Lord Somerville, speaking of what breeders have done for sheep says, "it would seem as if they had chalked upon a wall a form perfect in itself and then had given it existence."

—Charles Darwin, *On the Origin of Species,* 1859

A handful of sand contains about 10,000 grains, more than the number of stars we can see on a clear night. But the number of stars we can see is only a fraction of the number of stars that are. . . . The cosmos is rich beyond measure: the total number of stars in the universe is greater than all the grains of sand on all the beaches on the planet earth.

—Carl Sagan, *Cosmos,* 1980

Some seventy-five miles south of San Francisco, on the beautiful California coast, beside Monterey Bay, lies the small town of Pacific Grove. It was to this idyllic setting in late February 1975 that 134 of the world's leading biological scientists from the United States and from eighteen other countries gathered for a historic meeting in the Asilomar Conference Center. The meeting was the first example in the history of science of scientists meeting together to discuss the regulation of their activities. The activities which seemed to the participants at the Asilomar Conference so urgently to require regulation concerned experiments using the newly developed experimental techniques that enabled molecular biologists to splice together DNA from two different organisms, creating novel recombinant DNA molecules—that is, molecules of DNA which have never before existed on earth.

Collectively, these new techniques, now generally referred to as genetic engineering, have provided mankind with what is in effect an immense genetic word processor. Just as it is possible with an ordinary word processor to change an individual letter, to delete, to add, or to transpose sections of a text at will, genetic engineering can now be used to create any conceivable DNA molecule, to compose *de novo* any genetic message we may wish and to modify the genetic message in any existing organism in almost any conceivable way. At least, this is true already in the case of microorganisms, and although it is not yet possible to manipulate so readily the genetic text of higher organisms, it seems likely that within a few years such manipulations will also be possible, so that the entire genetic script of any organism will be an open book as accessible and easy to manipulate as the text of an English manuscript using a word processor.

Early Hopes and Fears

The creation of novel recombinant DNA molecules was a cause of widespread concern in the early 1970s. As Nobel laureate Paul Berg pointed out in the now famous "Berg letter" published in the journal *Science* in July 1974:

> Although such experiments [the creation of new recombinant DNA molecules] would facilitate the solution of important theoretical and practical biological problems, they would also result in the creation of novel types of infectious DNA elements, whose biological properties cannot be completely predicted in advance. There is serious concern that some of the artificial recombinant DNA molecules could prove biologically hazardous.[1]

Although the Asilomar Conference was convened to discuss the relatively mundane matter of developing guidelines for researchers using the newly created recombinant DNA molecules, aimed at minimizing the risk of generating a virulent bacterium containing genes harmful to man, the outcome of which might lead, in the words of a *Science* editorial, "to a catastrophe of possibly epidemic proportions,"[2] it was not these immediate practical consequences of genetic engineering that fueled the headlines in the early 1970s.

In those heady days leading up to the Asilomar Conference, the perception was widely shared that the development of recombinant DNA technology represented a historic turning point in the history of science, ushering in a brave new science-fiction era when biologists would indeed possess Darwin's magician's wand to summon up whatever form of life they pleased.

Over the past twenty years articles in glossy magazines, the popular press, and in semipopular and scholarly books with provocative titles such as *Playing God, Man-Made Life,* and *The Ultimate Experiment* have continually proclaimed the theme that fire has finally been stolen from the gods, that man was now creator, that a remodeling of the biosphere was imminent, and that all manner of undreamt-of forms of life were about to leap from the laboratory, conjured up at the whim of the magician's wand.[3]

The scenario of man remodeling the world has not been restricted to the popularizers of science. Even as early as 1972, the distinguished cell biologist Dr. James Danielli, director of the Center for Theoretical Biology at the State University of New York at Buffalo, was widely quoted as claiming that "soon in 20–30 years, but we may well be there in 10 years, scientists will be able to create new species and carry out the equivalent of 10,000,000,000 years of evolution in one year."[4] The scientific community no less than the general public was gripped with what Marie Jahoda described in *Nature* as a sense of "metaphysical awe":

> At the risk of being laughed out of court by these superior minds, I suggest that they were overcome by a metaphysical awe at their own power to fiddle with the very building blocks of life. . . . That metaphysical awe was akin to Oppenheimer's experience in witnessing the first atomic bomb explosion and when it communicated itself to the public, all hell broke loose.[5]

Over the past twenty years the scenario of a biological world remodeled by genetic engineering has been enhanced in the eyes of the general public by a number of widely publicized experiments, some reported in the popular press. For example, *Nature* carried a report in 1982 announcing the first successful transfer of a gene from one species to another.[6] Using a tiny capillary tube culminating in a point one twenty-four-thousandth of an inch in diameter, Richard Palmiter and Ralph Brinster of the University of Washington and the University of Pennsylvania, respectively, injected a DNA molecule containing the rat growth hormone gene plus its promoter sequence (the DNA message which switches the gene on) into the egg cell of a mouse, just after the sperm had penetrated the cell membrane of the ovum. The result of this gene transfer was a so-called transgenic mouse, a mouse containing the gene of another species. Under the alien influence of the rat growth hormone gene, the mouse grew into a giant rat-sized mouse, which gained notoriety in the world press as the "mighty mouse."[7] Another wonder was the sensational "geep."[8] This was a chimera consisting of a mixture of sheep and goat characteristics. The geep was not, strictly speaking, the result of genetic engineering (artificially changing the structure of an organism's DNA), but rather the result of the fusion of embryonic cells from one species with the embryo of another. And during the past year the first successful and widely publicized cloning of a mammal from an adult cell—the sheep Dolly—was announced.

And the wonders are no longer restricted to the laboratory. Since the mid-eighties a number of genetically engineered organisms have been released. One example was the so-called ice-minus microbe, an artificial version of a natural microbe, which is found on the leaves of strawberry plants and which manufactures a protein that acts as a seed for the formation of ice crystals and hence promotes frost formation. Using genetic-engineering techniques, Steven Lindow and Nickolas Panopoulos of the University of California at Berkeley removed the gene which specifies this protein, creating thereby the ice-minus version of the bacterium.[9] In the hope that they would displace their naturally occurring frost-promoting cousins, the ice-minus bacteria were sprayed on strawberry crops in Florida and Southern California.[10] Another artificially engineered organism, created by David Bishop of the Natural Environment and Research Council, Institute of Virology, at Oxford, this time a virus, which naturally infects the pine beauty moth, a serious pest in pine plantations, has been released into the wild. The engineered virus contains new genes, inserted into its DNA coding for insect poisons or insecticides, and has a far more potent destructive effect on the pine beauty moth than the natural virus.

Mere Tinkering

Such events fueled a widespread fear, even hysteria, that the advent of genetic engineering was leading mankind into a genetic hell, a world entirely remodeled, replete with superhumans and all manner of fantastic new types of life. Yet just twenty years later these fears have turned out to be unfounded. The actual achievements of genetic engineering are rather more mundane. The sorts of achievements described above such as the mighty mouse and ice-minus bacteria are far less portentous than they appear. They represent a relatively trivial tinkering rather than genuine engineering, analogous to tuning a car engine rather than redesigning it, an exploitation of the already existing potential for variation which is built into all living systems. An assessment of the achievements of genetic engineering in the agricultural field to 1995 was the subject of a recent *New Scientist* article entitled "Whatever Happened to the Gene Revolution?"[11] As the article points out, although sixty species of plants have been subject to genetic tampering and three thousand field tests of transgenic plants have been conducted worldwide, all of these achievements are relatively trivial and very far from the creation or radical reconstruction of a living organism. Confessed the author of the article, "It isn't what most people expected from genetic engineering."[12]

In fact, genetic engineering does not exist in the usual sense of the word. A mechanical engineer, from his knowledge of the principles of mechanical engineering and of the behavior of materials, can design a suspension bridge on a piece of paper. But no genetic engineer, from his knowledge of the principles of bioengineering and from his knowledge of the behavior and properties of macromolecules, could possibly specify the design of a living system *a priori* and encode the instructions for its assembly in a DNA sequence. Nothing achieved to date by genetic engineering remotely approaches the creation of new living systems from first principles or the radical redesign of an existing organism so that it can compete and survive in a natural environment. The dream has been postponed to some distant future.

Integrative Complexity

The failure to realize the dream of Asilomar, to wave Darwin's magic wand, does not result from our inability to use the DNA word processor. Prometheus in this sense is well and truly unbound. The difficulty in redesigning organisms stems from the fact that living things are immensely complex.

However, the challenge inherent in their complexity resides not primarily in the sheer number of components, which as we shall see below is remarkable enough, but comes from the reality that living systems are such *intensely integrated systems that their components cannot be easily isolated and changed independently.* Consequently, change, even if relatively minor, involves complex compensatory changes. This problem is met with in any attempt to engineer changes in any sort of complex system, such as, for example, a watch, consisting of many richly interconnected component elements.

The watch-organism analogy has of course been one of the most evocative and persistently popular analogies used by philosophers and biologists over the past few centuries.[13] It is useful here, as it very clearly illustrates the challenge of integrative complexity. From a mere cursory examination of the structure of a watch and of the intense functional integration of its various components, it is self-evident that if one cog is to be changed in some way, then if the function of the watch is to be maintained, simultaneous compensatory changes must be made to the entire chain of cogwheels—in effect, the entire watch must be redesigned. It is hard to imagine an object less able to undergo any sort of undirected evolutionary change. Watches are almost infinitely intolerant of any sort of random tinkering or changes in the configuration of any of their components. Change to any part, without intelligently engineered compensatory changes in the other parts, will lead to a complete disruption of the mechanism. The wheels will grind to a halt.

To change or improve the design of a watch or a clock in some way necessitates a total intelligent redesign of the entire system involving compensatory changes to the configuration of nearly all the other components simultaneously. For this reason the historical evolution of watches and clocks occurred in a series of relatively discontinuous steps. Each advance, from the medieval "verge foliot" escapement to the pendulum anchor escapement, was the result of innovative insights on the part of a succession of watchmakers from the twelfth century to the seventeenth, leading to intelligently engineered improvements in watch and clock design. Watches can undergo, and have historically undergone, "directed evolution" but only under the direction of an intelligent engineer. The historical evolution of steam engines, telescopes, car engines, computers, and airplanes conforms to the same discontinuous pattern and for precisely the same reasons.[14]

Before the Darwinian revolution organisms were viewed almost universally as unique wholes, incapable of evolutionary change via a successive series of small independent changes. Just like any other complex system, an organism was "a whole pre-supposed by all its parts," in the view of the early nineteenth-century English poet Samuel Coleridge.[15] This concept can be traced back to the biological philosophy of Aristotle.[16] It was presented forcibly in the first half of the nineteenth century by the leading French comparative anatomist and ardent antievolutionist, Georges Cuvier, whose views profoundly influenced nearly all the leading nineteenth-century biologists in Europe. In Cuvier's own words:

> All the organs of one and the same animal form a single system of which all the parts hold together, act and react upon each other; and there can be no modifications in any one of them that will not bring about analogous modification in them all.[17]
>
> Every organised being forms a whole . . . a peculiar system of its own, the parts of which mutually correspond, and concur in producing the same definitive action, by a reciprocal reaction. None of these parts can change in form, without the others also changing.[18]

According to this Cuvierian view, because of the intensity of the functional integration, no aspect or component subsystem is isolated or independent. Consequently, any major change in any component subsystem will require of necessity a whole train of *simultaneous* compensatory changes of a highly specific kind in all or many of the interacting components to preserve the functional integration upon which the viability of the system depends. *Thus, gradual change resulting from the accumulation of a succession of minor independent changes is impossible.*

One of the most recent and very scholarly presentations of this principle was given by Stuart Kauffman in his *Origins of Order.* As Kauffman points out, as the number of components in a complex system increases, the constraints against change likewise also increase:

> I believe this to be a genuinely fundamental constraint facing adaptive evolution. As systems with many parts increase both the number of those parts and the richness of interactions among the parts, it is typical that the number of conflicting design constraints among the parts increases rapidly. Those conflicting constraints imply that optimisation can attain only poorer compromises. No matter how strong selection may be, adaptive processes cannot

> climb higher peaks. . . . conflicting constraints are a very general limit in adaptive evolution.[19]

Although since 1859 most biologists eventually came to accept the concept of evolution, the idea that the intense functional integration of all the parts of an organism is bound to act as a severe constraint against biological change has been acknowledged almost universally by specialists in every field of biology.[20] The "constraints problem" is a recurring theme in the thinking of many developmental biologists today who acknowledge that the richness of the interconnectedness of living systems is bound to constrain to a large degree the direction of evolutionary change.[21] The same point was made by the Japanese biochemist Susumo Ohno in a *Nature* article defending his notion of saltational evolution by "frozen accidents": "An enzyme seldom functions alone. More often a number of enzymes are functionally coupled together to constitute one metabolic system. Once a reasonably well-functioning system was established, any drastic change in the kinetic property of one enzyme, without concomitant adjustments in the kinetic properties of functionally coupled enzymes, would tend to be disastrous."[22]

In the realm of gene therapy, where genetic engineering has been intensively applied to viral genomes in an endeavor to create safe vectors for the transfer of "healthy genes" into patients with genetic diseases, the problem of the integrative complexity of biological systems is now being directly encountered.[23] The complex and often daunting necessity of having to engineer coordinated and often complex changes in the genomes of viruses to render them capable of efficient or safe gene transfer is at the heart of the problem of gene therapy and one of the major reasons why progress in this area has been relatively slow.

Even Richard Dawkins, one of the staunchest defenders of Darwinian orthodoxy, admits that constraints do exist and would be bound to restrict or channel evolutionary change to some degree.[24] In the context of the above discussion it is ironic that Richard Dawkins should have chosen the title *The Blind Watchmaker* for his recent best-seller. Of all analogies, that of a watch is perhaps the least apt for arguing the case for Darwinism. A watch is the very archetype of a complex integrated system that cannot undergo change other than by intelligent direction because of the stringent demands for simultaneous compensatory change.

The question of how such intensely integrated systems as organisms can undergo continuous change in some part or subsystem without the need for "intelligent compensatory changes" is sidestepped in all discussion of undirected models of evolutionary change. Invariable Darwinian arguments artificially isolate a particular component or organ, such as the eye, from the immensely complex system in which it is embedded. Conveniently isolated from the constraining functional interconnections between organ and organism, it is relatively easy to envisage some organ or structure undergoing gradual change via a long series of hypothetical transitional forms. Thus, Darwinian explanations often appear to be superficially plausible.

For example, it is easy to envisage changing the word MAN via MAT and SAT to CAT via a series of single letter changes when the words are considered in isolation. But to make the same sequence of letter changes while at the same time preserving the meaning of any sentence in which they occur is far more complex, necessitating a number of additional "compensatory changes" to the other words in the sentence.

MAN →	MAT →	SAT →	CAT
THE	THE	HE	THE
MAN	MAT	SAT	CAT
LOOKED	LOOKED	DOWN	SAT
ANGRY	GOOD		DOWN

If an eye or any other organ is to be changed gradually from one state to another via a series of intermediate states, then this of necessity will involve compensatory changes in its biological context.

Dawkins's claim in the concluding paragraph of *The Blind Watchmaker*—"provided we postulate a sufficiently large series of sufficiently finely graded intermediates, we shall be able to derive anything from anything else"—[25] is unrealistic not only because of the functional constraints problem, but also because there are several cases where there are biophysical barriers to particular transformations, and in such cases, no matter how many intermediates we might like to propose, there is simply no gradual route across.

For example, all viral capsids are either cylinders or icosahedrons. The reason is purely biophysical. These are the only two stable forms that can be built up by stacking together a single subunit—the icosahedron and the cylinder. There is no intermediate series of stable forms leading from the cylindrical-shaped viral capsid to the icosahedron. Physics forbids it. Another quite different example regards the two alternative positions of the nervous system within the animal phyla. In invertebrates the nervous system is ventral—situated along the underside of the organism—while in the vertebrates, it is dorsal—situated along the back. No group of organisms exist which have their nervous system on the side, midway between the front and back. It was always hard to imagine how an asymetric intermediate arrangement with the nervous system on the right or left sides could be adaptive. The developmental genetic evidence now suggests there never were such organisms. Recent discoveries have revealed, for example, that the gene specifying which part of the embryo will be dorsal and which ventral is the same in both vertebrates and invertebrates. This in turn suggests that the gene may have suddenly switched its meaning during evolution, causing what was previously ventral to become dorsal. Intermediacy between dorsal and ventral may in effect be excluded, because of the Boolean logic of developmental genetic systems.

Beanbag Genetics

At the beginning of the century, shortly after the rediscovery of Mendel's laws, there was a brief period, the so-called era of beanbag genetics, when individual genes were considered to act separately and independently during development. There was believed to be a simple one-to-one relationship between individual genes and character traits. Given that it was already well established that embryogenesis involved complex phenomena, including highly ordered tissue and cell interactions, it is, from this distance, somewhat difficult to understand how this gross simplification could have had any appeal or how geneticists could have envisaged that a gene could have influenced a particular trait independently of other genes.

Despite the initial appeal of the beanbag concept, as knowledge of genetics and development progressed during the first decades of the century, it became increasingly apparent that most components of an organism are generated by the interaction of very many genes and that most genes influenced more than one character. Just how many components may be affected by the same gene is illustrated by the mutation called "wingless" in domestic fowl. The mutations in this gene cause developmental abnormalities in a wide variety of systems: the wings do not develop properly, the downy cover is underdeveloped, there are multiple abnormalities in the kidney and urinary tract, and the lungs and air sacs are missing.[26] Many other examples could be cited.

Rather than viewing genes as independent entities influencing individual separable character traits, the modern view that has emerged over the past sixty years was summarized by the great evolutionary biologist Ernst Mayr: "Every character of an organism is affected by all genes and every gene affects all characters."[27]

As the molecular structure and organization of the gene in higher organisms was worked out over the past twenty years, one of the reasons the great majority of genes affect diverse characters became clear—because many of the basic functional units in living things are made up of combinations of different gene products—just as words are combined into sentences. (See figure below.) It is generally combinations of gene products which form the key molecular complexes involved in directing and coordinating the development of higher organisms, determining when and where a cell will divide, whether or not a cell will differentiate into another cell type, which genes will be expressed in a particular cell at a particular stage of development, as well as forming the unique cell surface markers by which a cell reads its position in the embryo and signals to other cells its identity. Because the various combinations or complexes in which a particular gene functions may play entirely unrelated biological roles, mutations in any such gene will inevitably lead to a complex pattern of mutation, in which individual anomalies develop without any apparent physiological connection and seem to be distributed in a random manner among very diverse structures and processes.[28]

The fact that many genes are elements in complex combinations which play diverse roles influencing many different aspects of development implies

Gene combination and its consequence, pleiotropy.

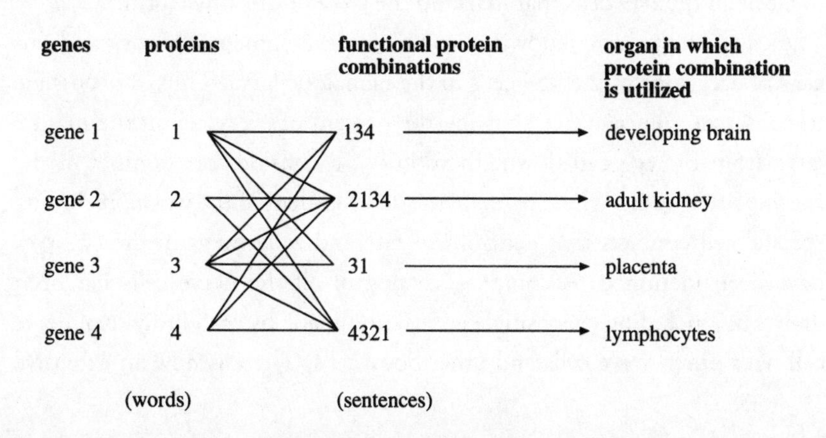

that the process of development is not genetically compartmentalized. Particular processes and organs are not specified by particular sets of genes. Nowhere in the organism is there a set of genes restricted to "making" the brain, an eye, or a leg. No structure or process or organ is genetically isolated; the same genes are involved to a greater or lesser degree in the specifications of all organs, structures, and processes. In terms of the gene interactions which underlie development, higher organisms are indeed richly interconnected systems.

The Genetics and Development of the Nematode

There is no more remarkable illustration of the interconnectedness of living systems than the development of the tiny nematode worm *C. elegans,* which is currently the best understood in terms of its molecular genetics and development of all the multicellular life forms.

The nematodes are tiny cylindrical unsegmented wormlike organisms generally about 1 millimeter long. There are a vast number of different species of nematodes, perhaps as many as 400,000. Normally, they go unnoticed, but they are exceedingly plentiful. A spadeful of garden soil may contain more than 1 million individual nematodes. Most are parasites. The human body hosts at least fifty different parasitic species.

Because they are so small and develop very quickly, and because all their organs and cells can be observed so easily with a microscope, they are an ideal organism for studying in detail the complex mechanisms which direct the development of higher organisms. To exploit the advantages of these tiny worms, the Cambridge biologist Sydney Brenner commenced a long-term project in 1963 to completely describe, cell-by-cell, the entire development of all the 959 cells that make up the larva of this tiny worm.[29]

As a result, we now know the complete developmental history and lineage of every one of the 959 cells in the nematode larva. Thus, it is possible to construct a lineage tree showing the descent of every cell in the mature larva from the egg cell down. In addition, a considerable number of the specific interactions which regulate the development of the worm, including specific cell contacts that control cell fate and movement in the embryo, have been identified. A complete catalog of all the nerve cells has been drawn up, including every single connection made by each individual nerve cell with other nerve cells and other body cells. There is now an extensive catalog of mutants that affect virtually every aspect of the organism's biology, including virtually every mutation that disrupts patterns of development.

As this project got under way, it soon revealed that the assembly of the nematode organs during development was remarkably nonmodular. It is impossible to isolate any part or organ in the nematode and treat it as an independent developmental entity.

Before the detailed analysis of the development of the nematode had been carried out, it might have been predicted that each organ would form separately, perhaps from a separate set of genetic instructions or a separate clone of cells. It turned out, however, that for each organ—the pharynx, the intestine, the nervous system, and the sex organs—the cells composing that organ were not related in any way in terms of lineage. The development of the pharynx of the nematode illustrates this nicely. The cells making up the pharynx are derived from eight completely different cell lines scattered throughout the lineage tree. The same is true of all the organ systems of the nematode; in each case the cells making up a particular organ are completely unrelated in terms of their lineage.

Another curious aspect of the development of the nematode and one that would never have been predicted is that although the organism is bilaterally symmetrical—that is, its left and right halves are mirror images of each other—the equivalent organs and cells on the right- and lefthand sides of the body of the larva *are not derived from equivalent cells in the embryo.* In other words, identical components on the right and left sides of the body are generated in different ways from different and nonsymmetrically placed progenitor cells in the early embryo and have therefore lineage patterns which are in some cases completely dissimilar. This is like making the right and left headlight on an automobile in completely different ways and utilizing completely different processes.

Even individual cells of the same cell type in any one organ, such as, say, the muscle cells, gland cells, or nerve cells of the pharynx, are also derived from different lineages. For example, one particular cell progenitor of the pharynx gives rise to muscle cells, interneurons, gland cells, and epithelial cells. Another progenitor gives rise to muscle and gland cells.

Altogether, there is not the slightest trace of compartmentalization or modularity or logical hierarchy in the assembly of the nematode. The finding was astonishing and completely unpredicted. Before the assembly of the nematode was worked out, "there had been," as a recent editorial in *Science* put it, "a persistent expectation among molecular biologists that the guiding themes of development would somehow be encoded in a genetic program somewhat analogous to the sequential encoding of a protein . . . and Brenner himself expected 'that the nematode would throw light on the logical structure of the program.' . . . Those involved in the project had believed that 'the cells were going to be powers of two, amusing mathematical symmetries, and so on.'" The editorial continues: "One persistent view . . . is . . . that organisms must be partitioned in some kind of molecular fashion, based on anatomical structures, physiological systems or developmental pathways. This modular organisation is then thought to be the basis for molecular genetic representation."[30]

Just as there was not the slightest modularity in organ assembly, the genes likewise showed no sign of a modular order. There was not the slightest evidence for genetic modules—that is to say, the existence of groups of genes involved in the specification of particular parts of the organism, say, the nervous system or the pharynx. No matter which part or organ of the nematode one examined, its development was intimately interconnected with the development of practically all the other organs and parts of the organism.

Brenner acknowledges that the idea that genes might be arranged into functional modules is appealing for both its feasibility for control and its access to ordered evolutionary change. Large computer programs are organized in this manner in order to localize the effects of change: "Thus by analogy it is argued that genes are also arranged in closed logical packets allowing changes to take place in one subsystem without affecting the others. But the analogy has not been borne out."[31]

So tightly "joined" are all the components during nematode development that it would seem a flagrant affront to common sense to presume that any radical change engineered in the structure of any one component, say, the pharynx, would not necessitate compensatory balancing changes in virtually the entire organism. It is safe to say nothing remotely like a radical redesign of the nematode will be achieved for many years to come.

Although the development and genetics of the nematode may be an extreme example of "interconnectedness," as knowledge of the genetic basis of development has advanced over recent decades, the same interconnectedness has been found to some degree in the development of all higher organisms. It may not always be quite as intense, but there is in many instances the same lack of any genuinely independent modules, either genetic or developmental. And it is indisputable that the majority of genes in all higher organisms influence more than one functional or developmental system.

Redundancy

Another very intriguing aspect of development in higher organisms which has become increasingly apparent over the last ten years, and which is bound to impose additional constraints against any sort of bit-by-bit undirected change, is the use of partially or totally redundant components to buffer organisms against random mutational error and ensure reliability, particularly during development. As one authority points out: "The idea that redundancy may be quite common in cell and developmental biology has its origin in Spemann's (1938) idea of double assurance, a term taken from engineering."[32]

The strategy of using several different means to achieve a particular goal, where each of the individual means is sufficient by itself to achieve the goal, is used in all manner of situations to guarantee that the goal will always be achieved, even if one or more of the means fail. Missiles, for example, are often guided to their targets using a number of different automatic guidance systems, including ground-based radar, map matching, inertial guidance, following a graded signal (heat-seeking). Even if one fails, the missile will still home in unerringly on its target. Reliability of information storage on computer discs is increased by encoding the information in two or more different ways. The functional reliability of complex machines such as aircraft and particularly space vehicles invariably involves the use of redundant components.[33] The space shuttle's on-board inertial guidance system, which it uses during boosting into orbit and during reentry, consists, according to the *McGraw-Hill Encyclopedia of Science and Technology*, of "*five redundant computers* and three inertial measurement units. Dual star trackers are used for periodic realignment in space. . . . A radar backup system is provided for safety during launch and landing."[34] (My emphasis.) Another instance where redundancy is exploited to increase reliability is in human and animal navigation, where most often a number of different and individually redundant clues are followed to minimize the risk of navigational error, which might accrue from following only one type or set of clues.[35]

It now appears that a considerable number of genes, perhaps even the majority in higher organisms, are completely or at least partially redundant. One of the major pieces of evidence that this is the case has come from so-called gene knockout experiments, where a gene is effectively disabled in some way using genetic-engineering techniques so that it cannot play its normal role in the organism's biology. A classic example of this came when a gene coding for a large complex protein known as Tenascin-C, which occurs in the extracellular matrix of all vertebrates, was knocked out in mice, without any obvious effect. As the author of a paper commenting on this surprising result cautions: "It would be premature to conclude that [the protein] has no important function . . . [as] it is conserved in every vertebrate species, which argues strongly for a fundamental role."[36] The protein product of the *Zeste* gene in the fruit fly drosophila, which is a component of certain multiprotein complexes involved in transcribing regions of the DNA, can also be knocked out without any obvious effect on the very processes in which it is known to function.[37]

The phenomenon of redundant genes is so widespread that it is already acknowledged to pose something of an evolutionary conundrum. Although in the words of the author of one recent article, "true genetic redundancy ought to be, in an evolutionary sense, impossible or at least unlikely,"[38] partially redundant genes are common. As another authority comments in a recent review article: "Arguments over whether there can be true redundancy are moot for the experimentalist. The question is how the functions for partially redundant genes can be discovered given that *partial redundancy is the rule.*"[39] (My emphasis.)

And it seems increasingly that it is not only individual genes that are redundant, but rather that the phenomenon may be all-pervasive in the development of higher organisms, existing at every level from individual genes to the most complex developmental processes. For example, individual nerve axons, like guided missiles or migrating birds, are guided to their targets by a number of different and individually redundant mechanisms and clues.[40] The development of the female sexual organ, the vulva, in the nematode provides perhaps the most dramatic example to date of redundancy exploited as a fail-safe device at the very highest level. A detailed description of the mechanism of formation of the nematode vulva is beyond the scope of this chapter; suffice it to say that the organ is generated by means of two quite different developmental mechanisms, either of which is sufficient by itself to generate a perfect vulva.[41]

It seems increasingly likely that redundancy will prove to be universally exploited in many key aspects of the development of higher organisms, for precisely the same reason it is utilized in many other areas—as a fail-safe mechanism to ensure that developmental goals are achieved with what amounts to a virtually zero error rate. A very high degree of redundancy in the specification of the development of higher organisms is almost certainly not in the least bit gratuitous, but rather of necessity. *Probably no system remotely as complex as a higher organism could possibly function without a large measure of redundancy in many or even every aspect of its design.*

Now, this phenomenon poses an additional challenge to the idea that organisms can be radically transformed as a result of a succession of small independent changes, as Darwinian theory supposes. For it means that if an advantageous change is to occur, in an organ system such as the nematode vulva, which is specified in two completely different ways, then this will of necessity require simultaneous changes in both blueprints. In other words, the greater the degree of redundancy, the greater the need for simultaneous mutation to effect evolutionary change and the more difficult it is to believe that evolutionary change could have been engineered without intelligent direction. Redundancy also increases the difficulty of genetic engineering, as it means that the compensatory changes that must inevitably accompany any desired change must be necessarily increased.

The Complexity of Proteins

The same fantastic interconnectedness and lack of clear modularity which is manifest in the nematode is seen at every level of biological design right down to individual protein molecules. In the late fifties, as the first three-dimensional structures of proteins were worked out, it was first assumed that each amino acid made an individual and independent contribution to the 3-D form of the protein rather in the same way that beanbag geneticists earlier in the century had envisaged that each gene had an independent effect on the organism's phenotype. This simplifying assumption flowed also from the concept of proteins as "molecular machines," which implied that their design should be like that of any machine, essentially modular, built up of a combination of independent parts each of which made some unique definable contribution to the whole.

It soon became apparent, however, that the design of proteins was far more complex than was first assumed. The contribution of each individual amino acid to the configuration of a protein was not straightforward but was influenced by subtle interactions with many of the other amino acids in the molecule. The situation was further complicated by the finding that most proteins could tolerate considerable changes to their amino acid sequence and still maintain the same three-dimensional form and functional properties. After thirty years of intensive study it is now understood that the spatial conformation adopted by each segment of the amino acid chain of a protein is specified by a complex web of electronic or electrochemical interactions, including hydrogen bonds and hydrophobic forces, which ultimately involve virtually every other section of the amino acid chain in the molecule. *It might be claimed with only slight exaggeration that the position of each one of the thousands of atoms is influenced by all the other atoms in the molecule and that each atom contributes, via immensely complex cooperative interactions with all the other atoms in the protein, something to the overall shape and function of the whole molecule.*

Proteins are thus very much less modular than ordinary machines such as watches, which are built up from a set of relatively independent modules or compartments. Remove the cog from a watch, and it still remains a cog; remove the wheel from a car, and it remains a wheel. The parts of a watch do not determine the configuration of the whole watch, nor does the whole configuration of the watch determine the conformation or properties of any of the parts. Proteins are far more holistic than any machine yet built.

Even today protein engineering in the true sense of the word—the specification of entirely new amino acid sequences that will fold into biological proteins with novel functions—has not in any realistic sense commenced. Although it is the Holy Grail of the protein engineer, in the words of Jennifer Van Brunt, "the obstacles are staggering."[42] Max Perutz considers the goal "utopian."[43]

However, attempts to engineer changes in the structure and function of existing proteins are advancing rapidly.[44] Increasingly ambitious transformations are now being attempted. The most complex transformation engineered to date was recently achieved by a group of researchers at Yale who changed one protein conformation into another quite different structure, a major feat which required changing twenty-eight amino acids in the initial starting protein. The way the Yale researchers carried this out is instructive. It was not achieved Darwinian fashion, bit by bit. In fact, all the necessary twenty-eight changes in the amino acid sequence were first worked out theoretically by applying structure prediction algorithms, model building, etc., and then the new amino acid sequence was synthesized in a bacterial expression system. In other words, all the changes were made simultaneously and the conformational change was engineered per saltum in Cuvierian fashion.[45] The magician's wand was waved, but it obeyed the logic of Cuvier, not Darwin. The magic was per saltum, and not via cumulative selection.

No doubt future advances will one day make it possible to radically modify or redesign existing functional proteins on a routine basis. However, from what is now known of the integrative complexity of proteins, and from the work going on in this field, it is clear that any significant engineered change in the fundamental design of a protein—say, a basic change in the folding configuration of part of the polypeptide chain—will necessitate many complex simultaneous changes throughout the molecule to preserve Christian Anfinsen's "harmonic chords of interaction consistent with biological function."[46]

Minor Changes

Despite the rich interconnectedness of their components, living systems are, at levels from organismic to molecular, modular to some extent. Indeed, no sort of evolution would be possible, either directed or undirected, if there was no modularity whatsoever in organisms. My point is not that there is no modularity in any sense—this would be absurd—but that the functioning of the vast majority of identifiable modules, whether an individual gene or a particular developmental cascade, is bound to be so highly constrained that a succession of undirected mutations is unlikely to result in major adaptive advance. This is not to say that organisms are unable to undergo any degree of adaptive evolution via a succession of independent changes. This again would be absurd. The evolution of the finches of Galápagos is a classic example of completely undirected yet adaptive biological change. Evolutionary change of this sort is often referred to as microevolution.

Evidence that organisms are in general intensely constrained against change, while at the same time capable of undergoing minor functional changes in certain directions, is provided by the phenomenon of mutation. The fact that the vast majority of all mutations which have some detectable influence on the functioning of the organism are deleterious suggests that each functional living system is indeed enormously constrained to adaptive changes along only a tiny fraction of all the possible evolutionary trajectories available to it. However, the fact that a small number of mutations are beneficial suggests that there must be at least a degree of modularity and that not every change requires compensatory changes. Not every change need therefore be "intelligently engineered."

Despite the evidence that organisms can undergo microevolutionary change and their components are clearly not quite as constrained as are the cogs of a watch, there is also no doubt that throughout the twentieth century, with each advance in knowledge, the design of living things has been revealed to be increasingly less and less modular and to increasingly approach the watch model or even the holistic nonmodular ideal of Coleridge and Aristotle.[47] This is particularly true of advances made over the past three decades in studies of the molecular genetics of development. Just as the complexity of living things in terms of the sheer number of unique adaptive components has grown relentlessly, so too has their integrative complexity. The studies of the nematode are graphic testimony to this.

The Necessity for Direction

Although organisms are modular to a degree, the modules are intensely interconnected. And in some cases, like the nematode, the integrative complexity exhibits a qualitatively different order of complexity to that realized in any man-made system or machine. Indeed, the order manifest in the development of the nematode or in the chaotic tangle of amino acid chains in a protein is the *very antithesis of the modular order of the kind required by nondirected evolutionary models.* It is an order lacking hierarchy, compartment, division, regularity, rules, grammar, program, symmetry, or logic.

The design of living systems, from an organismic level right down to the level of an individual protein, is so integrated that most attempts to engineer even a relatively minor functional change are bound to necessitate a host of subtle compensatory changes. It is hard to envisage a reality less amenable to Darwinian change via a succession of independent undirected mutations altering one component of the organism at a time.

Perhaps one day organisms will be radically transformed by genetic engineering, and mankind will at last wave Darwin's magic wand. However, given the complexity of life, this will only be by "intelligent" design. In other words, it will be *directed.* A desired change will be selected. The genes or modules involved in its specification will be identified. The changes necessary in these key components will be worked out. Also, the number of compensatory changes necessary in other interacting genetic systems will be identified. And so the process will continue until all the necessary changes have been documented and the engineering project commenced. It all seems plausible. But if the only way we can conceive of artificial evolution is through coordinated change brought about by intelligent direction, then surely the possibility that the process of natural evolution was similarly engineered can hardly be discounted. It may be also that just as the historical development of clocks and other machines has been discontinuous by necessity, so the course of evolution may have also been saltational, at least to some degree, by necessity.

Toward a Third Infinity

Integrative complexity is only one aspect of the complexity of living systems. They are also immensely complex in terms of the sheer number of unique components they contain. In almost every field of biology, as knowledge advances, complexity in terms of the number of unique components grows. Where one gene carried out a particular function ten years ago, now there are a hundred. Where once, not so long ago, there was one cell type in the retina, now there are fifty. Where there was once one neurotransmitter in the brain, now there are hundreds. Biology is now caught up in an ongoing complexity revolution, which is surely one of the most extraordinary events in the history of modern science. The phenomenal nature of the complexity revolution that currently pervades every field of biology is increasingly a source of comment among researchers in various fields. The general reaction is one of amazement at the ever-greater depths of complexity revealed as biological knowledge advances.[48] The process of endless complexification is acknowledged in the amusing titles of some recent review articles in scholarly journals: "With Apologies to Scheherazade: Tails of 1001 Kinesin Motors,"[49] and "1002 Protein Phosphatases?"[50]

Combinatorial Mechanisms

There is no tale of complexification more astonishing in modern twentieth-century biological science than the remarkable story of the gene as it has unfolded over the past twenty years. Only twenty years ago the gene was a relatively simple section of the DNA molecule about a thousand bases long which contained an encoded message for one protein. This was copied into an RNA molecule. The sequence of the mRNA was then decoded by the translational machinery into the amino acid sequence of a protein. Hence, each gene coded for one unique protein. How relatively straightforward it seemed.

However, during the seventies it was shown that in higher organisms the genes are split into noncontiguous sections in the DNA called exons (letters, in terms of our language analogy). And subsequent research revealed that different combinations of the exons of an individual gene are often combined in many different ways to produce different mRNA molecules, which are translated into many different proteins, each having slightly different functional properties (words). In other words, via this combinatorial device each gene could code for many different proteins. On top of this, during the eighties it became apparent that many proteins function in large multiprotein complexes—in other words, in combination with other proteins (sentences).

The total number of combinations that can be generated even from only a few genes by these combinatorial mechanisms is enormous. Recently, one researcher in this field calculated that the 25 G protein genes (proteins involved in transmitting chemical signals across the cell membrane) are probably combined into as many as five thousand different G protein complexes.[51] Another case is that of the sarcomere, which is the contractile unit in the muscle cell. In vertebrates this organelle is produced by the assembly of seven major contractile proteins, each encoded by a multigene family with a minimum of four members. The authors of an article in *Annual Review of Biochemistry* entitled "Alternative Splicing" point out:

> Assuming an average of 5 genes per family, in combination they have the capacity to generate $5^7 = 78{,}125$ different sarcomeric types. This potential is significantly increased by alternative splicing. To date, more than 3 different exons or pairs of exons are known to be alternatively spliced in sarcomeric contractile genes, raising to *many billions* the number of sarcomeres potentially produced by this limited set of genes. . . . In fact, it is very likely that this maximum potential is never realised because not all genes are concurrently expressed in the same cell. However, even the limited subsets expressed in different muscle types and developmental stages have the potential to generate an impressive number of sarcomeric types.[52] [My emphasis.]

But this is not the end of the story. The G proteins, for example, are always associated in higher-order complexes—in terms of our linguistic analogy, paragraphs—with other elements in the cell membrane, such as the so-called serpentine receptors. There are a large number of these receptors, probably as many as a thousand in the olfactory system alone.[53] To provide the organism with the vast diversity of systems described by Simon as an "apparently bewildering complexity of receptors, heterotrimeric G proteins, and effectors" consisting of "myriads of receptors, isomorphic families of transducers, and multiple effectors" that are "differentially distributed in space and time" and which "interact to generate the appropriate response in different cells"[54] must of necessity require a very large number of unique combinations of these various elements. The possibility that the total number of unique combinations of G proteins and serpentine receptors alone is on the order of a million does not seem out of the question.

Combinatorial strategies are also probably being utilized in the generation of the vast diversity of unique neuronal cell types and axonal branching patterns in the nervous systems of higher organisms. (A few of the different cell types in the mammalian nervous system are shown on page 322.) In a recent paper in the journal *Neuron,* S. L. McIntire and coauthors comment: "Many aspects of the outgrowth of particular neuronal types might be determined by combinations of genes that also function in the outgrowth of other neuron types. In other words, much of the specificity of axonal outgrowth might result from the action of unique combinations of broadly expressed molecules."[55] And this is now a widely held view in neurobiology. Combinations of different neurotransmitters, which are the substances which transmit the nerve impulse from one nerve cell to another, may also be used to generate neuronal diversity in the vertebrate brain. Again, the potential number of biochemically unique cells that could be generated by using combinations of transmitters is very large. The awesome implications inherent in this particular combinatorial expansion is conveyed in the adjectives used in this passage taken from a paper presented at the 1990 Cold Spring Harbor Symposium:

> The *immense* variety of neuronal phenotypes . . . is apparent in considering just the process of chemical transmission. There are approximately 12 known classical neurotransmitters and more than 30 neuropeptides [newly discovered neurotransmitters] thus far identified, and individual neurons simultaneously synthetise, store, and secrete one or more classical transmitters in addition to three or more neuropeptides. The transmitters and peptides are expressed in *an exceedingly large number* of different combinations in different parts of the nervous system. . . . The magnitude of this problem becomes clear if one cal-

culates the number of possible combinations if a neuron is to produce 2 transmitters out of a possible 12 and 3 peptides out of a possible 30. There are *267,960 different potential phenotypes* in this example.[56] [My emphasis.]

Toward a Measure of Complexity

That living things are the most complex objects of which we have any experience is universally accepted. The question naturally arises: Just how complicated an object is a living organism, say, a mammal like a human being? How might we measure it?

A number of approaches might be adopted in attempting to compute the complexity of a higher organism. Perhaps the most conceptually straightforward is simply to count the total number of unique functional gene combinations utilized during the entire process of development, from the fertilization of the egg cell to the final adult form. In the case of a machine, such as a clock, by counting all the unique adaptive components—that is, each component which plays some unique functional role—we can derive some very approximate measure of the complexity of the clock. Similarly, we could count the number of unique sentences in a book. This would again give us a crude estimate of the complexity of the book. Such an approach would provide very much of an underestimate, as the integrative aspect would not be measured.

To count the entire complement of unique gene combinations (each specifying a unique functional multiprotein complex) that conspire together during the development of an organism—a process which in the case of a human results in the production of an organized whole of about 10^{14} cells—would be a titanic task indeed. Such a labor of Hercules would involve enumerating all the gene readouts specifying every single unique molecular switch, device, component, or control signal, etc., from the molecular to the organismic level. Included in this "infinite inventory" would be all the unique molecular signals or devices which cause a particular cell or group of cells to divide at a particular time and place in the embryo throughout the entire period of development; all those myriads of unique signals that turn on a particular gene or groups of genes and cause an individual or a group of cells to start manufacturing a new protein or set of proteins at particular times and positions during development; the entire infinity of molecular markers that tag the surfaces of cells, conferring on them their unique "identity tags" by which cells are recognized by their neighbors and by which cells smell, taste, and touch their ever-changing surroundings in the embryo and by which they also navigate from point to point. Even for a Hercules, this is a hopeless task. The total could easily approach a number in the range of a trillion, or 10^{12}.

The Brain

One aspect of development which could require a vast number of unique gene combinations is the development of the nervous system. This is an area of biology that has always conjured up visions of infinity. Estimates of the total number of connections in the human brain have been possible for more than a century since the famous Spanish neurologist Santiago Ramón y Cajal developed staining techniques that revealed the finest branches of the neuronal dendritic tree. This technique revealed that each cell may make up to 10,000 connections with other neurons. The brain of man, for example, contains about 10^{11} nerve cells, which make between 10,000 and 100,000 connections with other cells, making a total for the whole brain of about 10^{15}, or 1 quadrillion connections. There are certainly more connections in the brain than there are cells in the body.

At present, we cannot answer with any certainty the question as to how many different cells or connections in the human brain might be uniquely specified. But we can still make a tentative guesstimate. We have seen there are 10^{11}, or 100 billion, neurons in the human brain and each connects with about 10,000 other neurons via its dendritic and axonal branches, making a total of about 10^{15} connections. Assuming identifiable subsets of neurons in the human brain contain about 100 cells, this would still mean 1 billion unique, genetically determined cell types (and this is probably an underestimate, as the number of neurons in an identifiable cluster is probably closer to 10 than 100).

But even if we assume that uniquely identifiable classes of neurons in the human brain contain as many as 100 neurons and if we assume that only one-thousandth of the connections are specified—that is, only 10 connections per neuronal equivalence class (again, almost certainly an underestimate)—this would still give 10^{10}, or 10 trillion, uniquely specified connections in the human central nervous system.

To generate such an immense number of unique connections must require unimaginable numbers of cell surface markers and cell signal transducers to guide the movements of the embryonic cells and axons as the order of the brain emerges during development. Even if we can only guess at the total number of uniquely determined connections in the brain, we can be sure that the number of unique genetic readouts involved in the process is phenomenally great and could easily be on the order of 10^{12}.

Attempting to visualize a billion neurons, each a tiny nanoscale navigator, preprogrammed with a unique set of maps and the ability to match each map, at a defined and preprogrammed time, with the unique configuration at a series of unique sites in the ever-changing terrain of the developing brain, all homing in, unerring, toward their target, brings us indeed to the very edge of an "infinity" of adaptive complexity. The unimaginable immensity of "atomic maps," "molecular charts," "nanotimepieces," and other nanodevices used by this eerie infinity of nanorobots which navigate the ocean of the developing human brain, building as far as we can tell the only machine in the cosmos that has genuine understanding, is far greater than that of all the maps, charts, and devices used by all the mariners who ever navigated the oceans of earth, far more even than all the stars in our galaxy, more than all the days since the birth of the earth.

Although neither the number of unique gene combinations utilized in living things nor the number of specified connections in the brain can be estimated with any degree of certainty at present, it is nevertheless clear that *life has been revealed by modern biology to be a thing of phenomenal complexity,* that in terms of the number of unique components utilized in their construction, living things transcend the complexity of our own artifacts by very many orders of magnitude.

And what we have seen already may be only a fraction of what is. In the case of the brain, for example, only 10 percent of its mass is made up of neurons; the rest is composed of the so-called glial cells. There are at least ten times as many glial cells as neurons. For most of the past century these cells have been cast in a supporting role and considered to be a relatively homogeneous mass of relatively inert cells playing little role in the functioning of the brain. But in conformity with the principle of complexification, recent research is revealing signs that this mass of a trillion cells may have important roles to play in information processing in the brain.[57] Similarly, only 1 percent of the genomes code for proteins. All of the complexities of gene expression discovered to date and discussed above involve only this 1 percent;

the remaining 99 percent does not code for protein and has as yet no known function. Further, it has been presumed that it carries no genetic information, and that its role, if any, is merely supportive, like the glial cells in the brain. But again, the first signs are now emerging which suggest that there may well be genetic functions embedded in this vast nonprotein-coding mass of DNA sequences.

The possibilities of informational processing by glial cells and the possibility of genetic functions in the nonprotein-coding DNA could potentially increase the complexity of the nervous system and the genome, respectively, by many additional orders of magnitude.

Then there are the even more radical possibilities, such as the proposal by Stuart Hameroff discussed in chapter 10, that the microtubules in the cytoplasm may be used for computing, providing each individual cell with potentially enormous computing ability. The discovery of entirely new levels of biological order, such as subcellular computing devices, quite unimaginable in terms of our current thinking, could well reveal that the complexity we have uncovered to date is only an infinitesimal fraction of the whole.

The Third Infinity

The concept of the infinite has always been invested with a special kind of reverential awe. It is beyond human understanding. For centuries the church considered it heretical to attribute the infinite to anything but God and consigned Giordano Bruno to the stake in the Roman Piazza del Campo dei Fiori in 1600 for claiming the universe had no end. For the ancient Norsemen the infinite was a challenge too great for any hero to overcome. Even Thor could not drink the sea. It was the unfathomable complexity of the Labyrinth of King Minos that doomed the Athenians and delivered them to the Minotaur.

The infinite stirs many emotions. At the same time, it fascinates, it perplexes, it inspires, it dwarfs, and it terrifies. Who, as a child, asked Rudy Rucker in his *Infinity and the Mind,* "did not lie in bed filled with a slowly mounting terror while sinking into the idea that the universe goes on for ever and ever."[58] Things touched by it are somehow magical or transcendental, reflecting another reality, the realm of the gods. The infinite or measureless are associated with the mysterious in Coleridge's famous poem *Kubla Khan:* "Where Alph, the sacred river, ran / Through caverns measureless to man / Down to a sunless sea." The association of the infinite with things mysterious and beyond understanding is also illustrated by the appearance of the mathematical symbol for infinity, the lemniscus (∞), on Tarot cards in the seventeenth century, shortly after its adoption by mathematicians.

The infinity of the cosmos is an infinity of the very large, the infinity we peer into when we look up at the night sky through a telescope. Another is the infinity we peer into down a microscope, the infinity of the inconceivably small, of the atom and of subatomic particles. But perhaps the most extraordinary infinity of all, conjuring up feelings of paradox and awe every bit as irresistible as those conjured up by contemplating the immensity of the starry heavens or the minuteness of the atom, is the infinity now emerging in modern biology—an infinity of the unimaginably complex. This "third biological infinity" is every bit as awe-inspiring as the previous infinities of the cosmos and the atom.

In effect, modern biology has revealed to us a watch, a watch with a trillion cogs!—a watch which wonderfully fulfills William Paley's prophetic claim in this famous section from his *Natural Theology; or Evidence of the Existence and Attributes of the Deity, Collected from the Appearances of Nature,* published in 1800, that "every indication of contrivance, every manifestation of design, which existed in the watch, exists in the works of nature; with the difference, on the side of nature, of being greater and more, and that in a degree which exceeds all computation."

Chapter 15

The Eye of the Lobster

In which the challenge to undirected Darwinian evolution posed by some complex biological adaptations, is examined. Included is the eye of the lobster, the eye of the scallop, the marsupial frog, and the avian lung. It is argued that such remarkable adaptations are prima facie *evidence that the mechanism of evolution must have involved more than Darwinian processes. The problem of preadaptation is briefly considered in the context of the evolution of the avian lung and the human brain. The question is raised as to whether the evolution of all adaptations such as those very remarkable examples on isolated islands is the result of a built-in plan or whether such cases may be the result of an inherent emergent inventive or creative capacity possessed by all living systems to a small degree.*

The eye of the lobster.

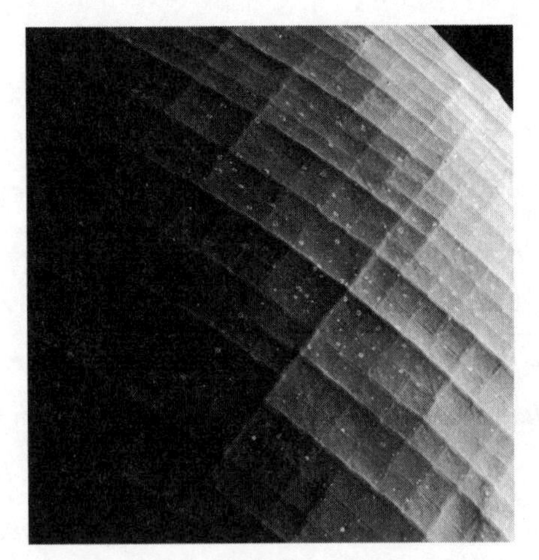

In considering the Origin of Species, it is quite conceivable that a naturalist, reflecting on the mutual affinities of organic beings, on their embryological relations, their geographical distribution, and other such facts, might come to the conclusion that species had not been independently created, but had descended, like varieties from other species. Nevertheless, such a conclusion, even if well founded, would be unsatisfactory, until it could be shown how the innumerable species inhabiting this world have been modified, so as to acquire that perfection and co-adaptation of structure which justly excites our imagination.

—Charles Darwin, *On the Origin of Species,* 1859

Among the most persistent challenges to the Darwinian model of evolution are those many types of complex and unusual adaptations whose evolution is very difficult to account for in terms of a gradual accumulation of successively advantageous changes. The literature of biology is full of examples. The challenge arises because evolution by natural selection can only occur via functional intermediates. Consequently, to get from A to Z by natural selection each step on the path—A to B, B to C, etc.—must be advantageous, and this imposes very stringent constraints on permissible evolutionary paths. Darwin himself spent two chapters of the *Origin* attempting to explain how the origin and evolution of what he called "organs of extreme perfection" may be plausibly accounted for by a gradual accumulation of minor undirected changes. Richard Dawkins has attempted the same in his recent book *Climbing Mount Improbable.* However, despite this "Darwinian apologetic," many biologists have remained unconvinced, finding the "explanations" offered either implausible to some degree or too vague and general to be subjected to critical detailed scrutiny. Moreover, many of these adaptations can be very plausibly accounted for in terms of directed evolution, a possibility which can no longer be dismissed a priori, given the new emerging teleological worldview.

The Eye of the Lobster

There are a great many different types of eye in the animal kingdom, and some are based on very different principles from the well-known vertebrate camera eye. Two of the most remarkable types of eyes found anywhere in the living world are those of the rock lobster and its relatives and of the common scallop.

One of the most striking features of the lobster eye which is immediately obvious even on superficial inspection is that the facets of the eye are perfect squares (see page 352). It is very unusual to meet with perfectly square structures in biology. As one astronomer commented in *Science*: "The lobster is the most unrectangular animal I've ever seen. But under the microscope a lobster's eye looks like perfect graph paper."[1]

The reason for the square facets is that the lobster visual system is based on reflection. Each unit or cell in the eye is a tiny tube of square cross section and plane parallel mirrored sides. The length of each cell is about twice the width. Light which enters the units is reflected off the mirrored sides of the cell wall to points of focus on the retina. The optics of the system were first worked out by Michael Land of Sussex University.[2] The diagram below illustrates the basic principle.

The optical system of the reflector eye of the lobster.

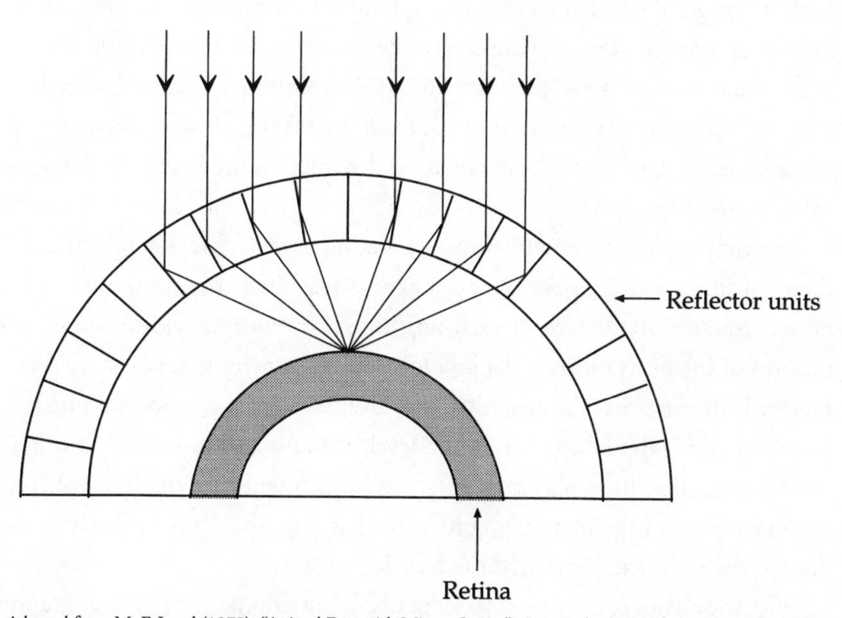

Adapted from M. F. Land (1978), "Animal Eyes with Mirror Optics"; © 1978 by Scientific American, Inc. All rights reserved.

This unique optical system is found in only one group of crustaceans, the so-called long-bodied decapods, which include the shrimps, the prawns, and lobsters. The great majority of crustaceans, and indeed of all invertebrates, have refracting eyes which are based on a completely different design. In these eyes each unit contains a small lens which refracts or bends the light onto the focus on the retina. Moreover, the units are hexagonal or round.

It is generally supposed that the reflecting eye must have evolved from a refracting eye, which are far more common among the Crustacea and of a fundamentally simpler design. The fact that during the development of the lobster—as it changes from the planktonic larval form, which swims in the surface waters of the sea, to the adult form that lives on the sea floor—the larval refracting eye is transformed into the adult reflecting eye would tend to support the idea that evolution followed the same path from refraction to reflection.

However, just what selection pressures may have been responsible and through what intermediate states the reflecting eye evolved is a mystery. The transformation is puzzling because it is very difficult to imagine how the units in some transitional eye—halfway between a hexagon and a square, halfway between a lens and a reflecting surface—could form a better image than the original refracting eye. Consequently, it is difficult to see how those halfway, intermediate eyes would have been selectively advantageous in an evolutionary sense. This is critical, because evolution by natural selection can only follow an evolutionary route from A to B if each step taken on the route to B is adaptively advantageous and confers some increased survival value on the organism.

The mystery is deepened because reflecting eyes of this design can only focus an image if the optics of the system is just right: the units are nearly perfect squares; the length of each unit must be twice its width; the inner surfaces of the units must be flat and mirrored. Deepening the mystery even further is the fact that the remarkable ability of these eyes to form an image is only used by the lobster when the level of illumination is low. At other times, when the illumination is brighter, adjustments are made involving the movement of the opaque pigment so that only those rays directly incident to the units can be seen at the retina.

Why should an organism drop its perfectly functional refracting eyes and start out on the hazardous journey to reflection? Refracting eyes provide organisms with excellent image-forming capabilities, as witness the flight of the dragonfly. Many crustacean cousins of the lobster—crabs, for example—which occupy the same ecological niche as the lobster and have the same predatory lifestyle—have refracting eyes and obviously survive quite well in the same level of illumination.

One would have thought that in the fitness landscape of crustacean eyes the route to an improved or perfected "refraction eye" must be far shorter, less complex, and far more probable than the extraordinary journey from refraction to reflection, from hexagon to square, from curved surfaces to flat surfaces, from lens to mirror, and so forth.

The Eye of the Scallop

The lobster eye is not the only eye that poses this sort of challenge. Another remarkable type of reflecting eye is found in the common scallop, the pecten.[3]

The scallop has about thirty small eyes along the fringe of each mantle, the name given to the two fleshy portions of the scallop which protrude from between the shells. Each eye is small, being only about 1 millimeter across. Each scallop eye contains a large lens in front of the retina. The retina is situated just behind the back of the lens. The back of the eye is formed into a hemispherical reflector surface, and behind this is a layer of brown reflecting pigment. The image is formed as shown by reflection from the hemispheric surface onto the retina.[4]

The eye of the scallop.

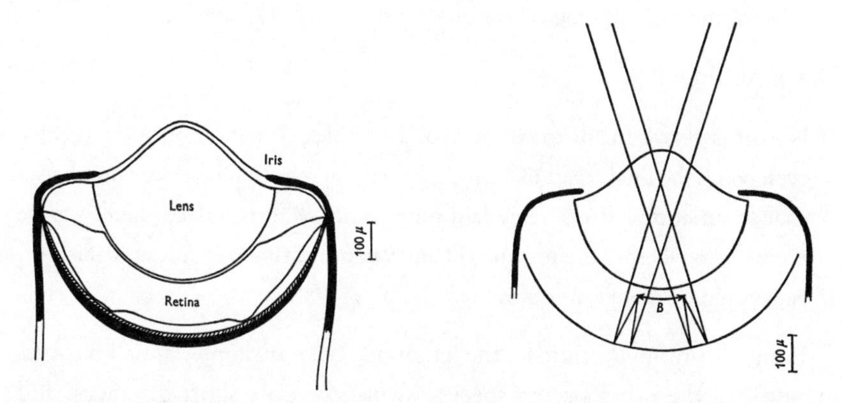

From M. F. Land (1965), "Image Formation by a Concave Reflector in the Eye of the Scallop *Pecten Maximus,*" *Journal of Physiology* 179:138–153, fig. 1, p. 138, and fig. 5, p. 147. Reproduced by permission of Michael F. Land.

The visual system is remarkably complex and sophisticated. Amazingly, the scallop has not one but sixty of these tiny image-forming eyes. What is so striking is the apparent gratuity of the whole system. Why has such a simple organism evolved such a complex image-forming eye?

Long before its optical system was understood, the pecten eye was raising evolutionary problems. William J. Dakin, a professor of zoology at Liverpool University before the Second World War and one of the first to describe the eye in detail, had problems accepting the Darwinian explanation. According to Darwin in chapter 6 of the *Origin:*

> When we reflect on these facts . . . with respect to the wide, diversified and graduated range of structure in the eyes of the lower animals; and when we bear in mind how small the number of living things must be in comparison with those which have become extinct, the *difficulty ceases to be very great* in be-

> lieving that natural selection may have converted the simple apparatus of an optic nerve . . . into an optical instrument as perfect as is possessed by any other member of the Articulate Class. [My emphasis.]

Commenting on this section, Dakin remarks:

> This is an optimistic view of the problem rather than evidence; it is a view to which I find it very difficult to subscribe in so far as the eyes of the Pectinidae are concerned. Indeed after a careful comparative study of the visual organs of the invertebrates one finds greater difficulty in accepting the principle of natural selection as the dominant factor in their origin than is the case with any other of their morphological features.[5]

Dakin continues:

> Now it is very difficult to conceive of a complex structure, complex as these eyes, being the final result of a sifting by natural selection of a large number of chance variations, stress being laid on external factors. Indeed there is grave doubt as to whether the presence of any variations that might lead to such organs could have any survival value.[6]

Even within the Pectinidae, the lethargic *Pecten maximus,* which has eyes identical to the other pecten species, swims for only short distances, and that infrequently. Another challenge to the plausibility of selectionist explanations of the level of development and diversity of eyes within the bivalves is the curious case of *Spondylus.* As Dakin points out, this genus does not swim but is attached to rocks, yet its eyes are as developed as the actively swimming *Pecten* species. Dakin concludes that the size and complexity of the eyes in Pectinidae cannot be explained by natural selection:

> Whatever may have been the *origin* of the eyes of the *Pecten* group I do not hold that utility explains their evolution . . . in the case of Pectinidae there is no evidence of the elimination of types with less complex eyes as unfit . . . in view of the diverse conditions existing in the Lamellibranchs there is no evidence that a reduction in the efficiency of the eyes of *Pecten* would lead to unfitness. . . . We cannot escape from the conviction that in one particular series of bivalves, all intimately related genetically, a distinct type of visual organ arose, independent of other visual organs, and that apart from adaptation, and apart from utility or advantageousness, it attained a certain extraordinary complexity.[7]

Although some recent studies suggest that the eye of the scallop has some selective advantages over the far simpler eyes of its bivalve cousins,[8] the fact remains that these related species survive perfectly well without the assistance of what is for an organism as simple as a mollusc a fantastically sophisticated eye. All the scallop's close relatives have either small hollowed-out photosensitive pits, or slightly more advanced but still very primitive camera-type eyes consisting of a lens positioned at the front of a spherical hollow lined with photosensitive cells. This is basically the standard molluscan type of eye found in such familiar organisms as the garden snail and the slug. Image-forming eyes within the Mollusca are all variations on the camera theme. The most spectacular are those of the octopus and squid, which are comparable with the vertebrate eye in many ways.

While it is possible to envisage a series of camera-type eyes leading from a photosensitive pit to the sophisticated camera eye of the octopus, it is far more difficult to envisage a transitional series of functional eyes leading from a photosensitive pit to the reflecting image-forming eye of the modern scallop. The fitness landscape of molluscan eyes would seem to lead inexorably downhill from a light-sensitive pit toward the camera type of eye.

The reflecting eyes of the lobster and scallop are only the tip of a vast iceberg of similar examples of adaptations that are difficult to explain in terms of a gradual process of cumulative selection.

The Marsupial Frogs

The marsupial, or egg-brooding, frogs are a group of about sixty species which live in the rain forests and mountains of South America. Although the adults are indistinguishable from ordinary frogs, the development of the egg and the method of reproduction are utterly different. In many species of marsupial frogs the mother carries the developing eggs in a special pouch or brood chamber on her back. The eggs lie in the fluid-filled chamber in intimate contact with the mother's tissue through an organ which resembles the mammalian placenta. Through this organ, mother and embryo exchange gases, fluids, nutrients, and wastes. The jelly layer surrounding the marsupial frog egg is very thin, so as to permit the easy transfer of materials to and from the developing embryo.

But even that is not the most remarkable aspect of the marsupial frog reproductive system. It is the very great difference in the way the marsupial frog egg develops that is so astounding. The details of this were worked out in the 1970s by the Ecuadorian biologist Eugenia M. del Pino.[9] Instead of resembling an ordinary developing frog egg, in the marsupial frog the entire organization of early embryogenesis has been radically transformed and resembles the embryonic development of a mammal. The early cell divisions and the most fundamental processes have been completely transformed. Given the very fundamental nature of early embryogenesis and the great number of genes involved, it is clear that the evolutionary route from the standard pattern to that of the marsupial frog must have involved a complex set of genetic and developmental changes.

The device of incubating the eggs on the mother's back is obviously adaptive in the tropical rain forest, where there is intense competition for breeding sites. Nonetheless, many tree frogs adopt the far simpler strategy of laying their eggs in a small pool of water on a leaf in the forest canopy or even just letting the rain carry the eggs to the forest floor on the chance they may reach a pool of water. So the incubation method is one strategy among many, vastly more complex, but apparently no more effective, than the far simpler methods.

Again, as in the two cases cited above, it is hard to believe that any sort of unguided evolutionary mechanism would have realized such an unusual adaptive end. Moreover, changing the basic organization of an embryo would appear to be far more radical than any of the other changes cited above. And again, for this to happen under the agency of natural selection, each one of the individual steps along the evolutionary route must have been selectively advantageous.

The same phenomenon is seen among the embryos of sea urchins. As Jeffrey Levinton commented recently in another *Scientific American* article: "the embryos of sea urchins are fantastically diverse. The larvae of closely related species have radically different forms—some are adapted for a long life of swimming and feeding on plankton, whereas others are suited for a short nonfeeding period while they are dispersed by currents. These . . . specialisations entail monumental differences in the developmental patterns of the larvae and even in the parts of the embryo used to form adult structures."[10] Yet, as Levinton comments, amazingly, the adult members of these species are virtually indistinguishable. It is fantastically difficult to envisage how an accumulation of tiny selectively advantageous changes could have caused such enormously dramatic changes in embryonic development in the case of organisms virtually indistinguishable in their adult forms.

The Avian Lung

Another spectacular adaptation in a group of very familiar organisms is the avian lung. In all vertebrates except birds the air is drawn into the lungs through a system of branching tubes which finally terminate in tiny air sacs, or alveoli, so that during respiration the air is moved in and out through the same passage (the avian lung has already been briefly described in chapter 13). In the case of birds, however, the major bronchi break down into tiny tubes which permeate the lung tissue. These so-called parabronchi eventually join up again, forming a true circulatory system, so that air flows in one direction through the lungs.

This unidirectional flow of air is maintained during both inspiration and expiration by a complex system of interconnected air sacs in the bird's body that expand and contract in such a way as to ensure a continuous delivery of air through the parabronchi. The existence of this air sac system has necessitated a highly specialized and unique division of the body cavity of the bird into several compressible compartments. No lung in any other vertebrate species is known which in any way approaches the avian system.

Just how such a different respiratory system could have evolved gradually from the standard vertebrate design without some sort of direction is, again, very difficult to envisage, especially bearing in mind that the maintenance of respiratory function is absolutely vital to the life of the organism. Moreover, the unique function and form of the avian lung necessitates a number of additional unique adaptations during avian development. As H. R. Dunker,[11] one of the world's authorities in this field, explains, because first, the avian lung is fixed rigidly to the body wall and cannot therefore expand in volume and, second, because of the small diameter of the lung capillaries and the resulting high surface tension of any liquid within them, the avian lung cannot be inflated out of a collapsed state as happens in all other vertebrates after birth. In birds, aeration of the lung must occur gradually and starts three to four days before hatching with a filling of the main bronchi, air sacs, and parabronchi with air. Only after the main air ducts are already filled with air does the final development of the lung, and particularly the growth of the air capillary network, take place. The air capillaries are never collapsed as are the alveoli of other vertebrate species; rather, as they grow into the lung tissue, the parabronchi are from the beginning open tubes filled with either air or fluid.

The avian lung brings us very close to answering Darwin's challenge: "If it could be demonstrated that any complex organ existed, which could not possibly have been formed by numerous, successive, slight modifications, my theory would absolutely break down."[12]

The avian lung is more efficient than the mammalian because of a special countercurrent mechanism whereby the blood flows throughout the lung in the opposite direction to the flow of air. This allows the blood to take up more oxygen and deliver more carbon dioxide than is possible in, say, the lung of a mammal. As Knut Schmidt-Nielsen points out, it is because of the higher efficiency of the avian lung that "birds have been seen in the high Himalayas flying overhead at altitudes where mountain climbers can barely walk without breathing oxygen."[13] But while this adaptation is clearly advantageous to an eagle soaring in the mountains and to advanced modern birds capable of fast powered flight, it is difficult to believe that it would have been so advantageous in the last common ancestor of all modern birds which lived about 100 million years ago and which was presumably nothing like as accomplished a flier as its modern descendants. Here it is hard not to be inclined to see an element of foresight in the evolution of the avian lung, which may well have developed in primitive birds before its full utility could be exploited.

The idea that an adaptation like the avian lung might evolve before its full utility can be exploited is perfectly consistent with a directed model of evolution.

The Evolution of the Human Brain

There are many other cases of this phenomenon. Perhaps the most celebrated and well-known example is the case of human intelligence. Many have commented on the striking fact that our intellectual capabilities, especially our capacity for abstract mathematical thought, upon which the whole enterprise of science is ultimately based, seems vastly in excess of any conceivable intellectual needs of the small tribe of hunter-gatherers who lived in Africa some 200,000 years ago and were the last common ancestors of all modern humans. What selection pressures on the ancient plains of Africa gifted mankind with musical ability, artistic competence, the capacity for profound abstraction, and ultimately the ability to comprehend the entire cosmos from which we sprang.

Commenting on the evolutionary conundrum posed by our intellectual capabilities in his recent book *The Mind of God,* Paul Davies reminds us "that the success of the scientific enterprise can often blind us to the astonishing fact that science works," and he continues:

> What is remarkable is that human beings are actually able to carry out this code-breaking operation, that the human mind has the necessary intellectual equipment for us to "unlock the secrets of nature."
>
> The mystery in all this is that human intellectual powers are presumably determined by biological evolution, and have absolutely no connection with doing science. Our brains have evolved in response to environmental pressures, such as the ability to hunt, avoid predators, dodge falling objects, etc. . . . John Barrow is also mystified: "Why should our cognitive processes have tuned themselves to such an extravagant quest as the understanding of the entire universe? . . . None of the sophisticated ideas involved appear to offer any selective advantage to be exploited during the pre-conscious period of our evolution. . . . How fortuitous that our minds (or at least the minds of some) should be poised to fathom the depths of nature's secrets."[14]

We might add also, how fortunate it is that we are a terrestrial organism of about the size we are, breathing oxygen in a biosphere containing ample quantities of combustible carbon in the form of wood, which burns so gently and controllably. Only these unique conditions provide an intelligent life form with the ability to handle fire and hence provide access to chemistry and eventually to scientific knowledge. How fortunate also that the speed of nerve conduction is 120 meters per second, providing higher organisms like ourselves, despite our relatively large size, with the ability to carry out fine motor manipulation and hence the ability to handle fire and engage in an exploration of the world.

How very fortunate indeed that evolution should have gifted a mind so fit for the scientific enterprise in a physical form so fit to that same unique end long *before* that enterprise was undertaken.

Can Organisms Direct Their Own Evolution?

Although it seems implausible that such complex adaptations could have resulted from any sort of undirected process, the question arises as to whether or not the direction was built into nature from the beginning or was the result of other secondary phenomena. While it is possible to believe that adaptations such as the eye of the lobster and the avian lung might have been programmed into nature from the beginning, in the case of many other types of adaptations the case for classifying them as natural forms is far less certain.

It is particularly hard to believe, in the case of the many complex adaptations restricted to remote islands, that their blueprint is an integral part of the biocentric design of nature. That they are fundamentally contingent seems to be far more likely. One need only consider the extraordinary morphological and behavioral adaptations of the various species of weta, a type of giant ground cricket found in certain restricted regions of New Zealand, sometimes small islands only ten hectares in area, to see the difficulties of explaining all biological adaptive design in terms of a cosmic blueprint. In some species of tusked wetas, the males have tusks protruding from the front of their jaws which they use in ritual jousting matches with other males. Each dominant male tree weta guards a territory containing a harem of breeding females, and the jousts, as in the case of stags or seals, are basically aimed at preventing another male from gaining access to the harem. As if this were not remarkable enough, the pattern of the call of each individual male tree weta appears to be distinctive; each, as it were, has its own acoustic signature.[15]

In the case of such remarkable and complex adaptations restricted to small isolated geographical regions, it is not easy to envisage them as being preprogrammed into the order of nature and being part of a "grand design." The question naturally arises, how did such adaptations come about? If neither natural selection nor any other sort of undirected evolutionary mechanism seem plausible, then could they conceivably have been the result of the activities of life itself operating via some as yet undefined type of inventiveness inherent in all life? Consideration of the many exotic and complex adaptations that grace the living world led many French biologists in the first half of this century, such as Lucien Cuénot,[16] to propose that life must possess to some degree an autonomous creative ability. As he puts it: "The terms finality, adaptation, organisation have only meaning within the world of living matter. . . . One cannot disregard the analogy between the tools or the artifacts of man . . . and the tools or systems of organised beings; accordingly there must be an analogy between the intelligence and reason of man . . . and some property of living matter. I don't dare to call this intelligence or reason so I call it the inventive power."[17]

The possibility that some degree of adaptive evolution may be the result of an inherent emergent inventive capacity possessed by all living things cannot be ruled out. Such a capacity may be manifest in human consciousness and intelligence as well as in a general capacity for adaptation possessed by all living things. In such a view human creativity, rather than being a thing apart from nature, a gift imparted uniquely to mankind by God, becomes instead a profoundly natural phenomenon shared perhaps in some small degree by all life forms on earth.

For Anglo-Saxon biologists such as myself, schooled within a strictly mechanistic view of life, a view which has, by and large, been adopted by all English-speaking biologists, the possibility of life being more than mechanism, possessing some sort of emergent creative capacity, seems to beckon toward some mystical obscurantist cul-de-sac. However, there are many intriguing phenomena cited in the works of well-known authors such as Rupert Sheldrake and Lyall Watson which suggest that life may be more than our current science admits.[18] And certainly the phenomenon of emergence is itself encountered throughout the natural world. We cannot predict the properties of water, or the speed of nerve conduction, from quantum mechanics. Nor can we predict the social behavior of bees, ants, or indeed any organism from observing the behavior of an individual in isolation. It is because of the problem of emergence that the laws of biology cannot be predicted from the laws of physics and why evidence drawn only from physics and astronomy is bound to be insufficient to secure the biocentricity or anthropocentricity of the cosmos. Moreover, as discussed in chapter 12, there exists the possibility that emergent self-organizing phenomena were involved in the origin of life and the origin of evolutionary innovation.

In the context of a teleological view of the cosmos, even if much of the overall order of organic nature was determined from the beginning, it is surely conceivable that the Creator, to paraphrase Darwin in the last paragraph of the *Origin of Species,* could have gifted organisms not only with the capacity for growth, reproduction, inheritance, and variability,[19] but also with a limited degree of genuine autonomous creativity so that the world of life might reflect and mirror in some small measure the creativity of God.

Conclusion:

The Long Chain of Coincidence

In which it is concluded that after four centuries of spectacular increases in scientific knowledge there is still no direct evidence for the existence of any sort of life other than the carbon-based form with which we are familiar on earth. Moreover, evidence which suggests that the laws of nature are specifically adapted for life as it exists on earth is continually increasing as scientific knowledge grows. The main constituents of life such as water and the carbon atom, and environmental conditions such as sunlight and the hydrosphere of the earth, give every appearance of being ideally and uniquely fit for their biological roles. Although the proposition that the cosmos is a uniquely prefabricated whole with life as it exists on earth as its end and purpose cannot be proven, it is easy to refute. Demonstrate the existence of any one example of a life superior to our own or even of an individual constituent, such as water, which is less than ideally tailored for its biological role, and the whole teleological scheme collapses. It is concluded that the anthropocentric presumption has not only stood the test of four centuries of scientific advance, but it increasingly makes more sense of the cosmos as a whole than does any other competitor theory.

"Ammonia! Ammonia!"

In a universe whose size is beyond imagining, where our world floats like a mote of dust in the void of night, men have grown inconceivably lonely. We scan the time scale and the mechanisms of life itself for portents and signs of the invisible. As the only thinking mammals on the planet—perhaps the only thinking animals in the entire sidereal universe—the burden of consciousness has grown heavy upon us. We watch the stars, but the signs are uncertain. We uncover the bones of the past and seek for our origins. There is a path there, but it appears to wander. The vagaries of the road may have a meaning however, it is thus we torture ourselves. . . .

So deep is the conviction that there must be life out there beyond the dark, one thinks that if they are more advanced than ourselves they may come across space at any moment, perhaps in our generation. Later, contemplating the infinity of time, one wonders if perchance their messages came long ago, hurtling into the swamp muck of the steaming coal forests, the bright projectile clambered over by hissing reptiles, and the delicate instruments running mindlessly down with no report.

—Loren Eiseley, *The Immense Journey,* 1957

When I first wrote my treatise about our system, I had an eye upon such principles as might work with considering men, for the belief of a deity, and nothing can rejoice me more than to find it useful for that purpose.

—Isaac Newton, *Principia,* 1687

The defenders of the anthropocentric faith in medieval Europe knew very little, indeed virtually nothing, about the natural world. In a very real sense it was not just the geographical surface of the earth that was largely terra incognita to medieval man but the whole realm of natural phenomena. The nature of such ordinary phenomena as lightning, clouds, stars, infectious diseases, and volcanoes were absolutely mysterious. Moreover, nearly every medieval conception about the natural world was erroneous. The earth was thought to be flat and men could fall off the edge; the alchemists believed that the base metals could be turned into gold; miracles were considered an everyday occurrence; astrology was widely accepted; diseases were thought to be curable by using various herbal and folk remedies, some of which were more harmful than the disease; and mental illness was considered to be the result of demon possession. Given the almost complete absence of any knowledge of the natural world, given also the erroneous nature of so many other far less audacious beliefs, it is all the more remarkable that the most presumptuous of all their beliefs, the central axiom of the Christian faith, on which the whole of medieval civilization was based, has stood the test of time and the critical scrutiny of four centuries of science.

The apparent demolition of the anthropocentric cosmos came about mainly through two great revolutions in thought. The first arose from the discoveries in astronomy in the early seventeenth century, which revealed a cosmos vastly larger than and utterly different in structure from the simple, closed geocentric system conceived of by the medieval theologians, one in which it was increasingly difficult to imagine any teleological place for man and in which there was ample room for many worlds, perhaps inhabited by intelligent beings very dissimilar to ourselves. The second blow that appeared to provide the final *coup de grâce* came in the mid nineteenth century with the rise of Darwinism, which provided what appeared to be an entirely plausible explanation for the apparent adaptive design of living things in terms of a completely undirected process—natural selection. According to this view, all manifestations of life, including our own species, are, in Brian Appleyard's words, "accidents of deep time and chance."[1] With Darwin the process of disenchantment and the secularization of the western mind was completed. However, although these revolutions were based on the findings of science, they were not strictly scientific in essence, but rather philosophical extrapolations from the findings of science which seemed to the scientific community in the seventeenth and nineteenth centuries eminently reasonable.

Other Worlds and Alien Life

Although the existence of a plethora of inhabited worlds full of exotic types of life has always been pure conjecture—no one had actually observed a Jovian, a Martian, or indeed any other extraterrestrial—yet the feeling that the cosmos was inhabited, perhaps by a diverse phantasma of alien types of life and intelligence, arose irresistibly in the minds of the seventeenth-century astronomers as they gazed into the immensity of the sky. This same deep conviction that there is, in Loren Eiseley's words, "life out there beyond the dark" has persisted right down to the present day whenever men have contemplated the enormity of space and time. The concept of extraterrestrial life pervades current culture, both popular and serious, as witnessed in science-fiction movies and television shows such as *Star Wars* and *Star Trek* and in the various serious research projects aimed at detecting radio signals from extraterrestrial civilizations.

The challenge inherent in the existence of extraterrestrial life of a very alien type is self-evident. The claim that the cosmos is specifically designed for the form of carbon-based life that exists on earth can only be seriously defended if the type of life that exists on earth is a unique and peerless phenomenon—the only possible instantiation of the phenomenon compatible with the laws of nature. The existence of extraterrestrial life, especially intelligent life, designed on very different principles to that on earth, would imply that the carbon-based life as it exists on earth is not unique or special in any way and that no one type of biology or intelligence is particularly ordained or prefabricated into the cosmos. Earth's own manifestation of life and intelligence becomes by this doctrine only one possibility among many, interesting to ourselves, no doubt, *but not exceptional in any way and certainly not unique.* And humankind is but one possibility among many equally probable alternatives. If there are indeed exotic forms of extraterrestrial life, including species possessing intelligence vastly greater than our own and perhaps possessing a technology centuries in advance of our own, then one may just as well argue that the cosmos is adapted for any one of these alien life forms such as Christiaan Huygens's long-necked, round-eyed extraterrestrials or Carl Sagan's imaginary gaseous Jovian beings, the size of immense airships, which he named Floaters in his best-seller *Cosmos*[2] (see below), rather than for *Homo sapiens.* Of course, there was no direct evidence for extraterrestrial life or intelligence in the early decades of the seventeenth century. It was not the possible existence of extraterrestrial life in itself that punctured the teleological assumption, but the possible existence of extraterrestrial life of an utterly alien kind. This was the crux of the challenge of the infinity-of-worlds doctrine—the idea of a plurality of extraterrestrial life based on all manner of exotic and utterly alien chemistries. The underlying antipathy between the doctrine of many worlds and the notion of man's uniqueness and his teleological centrality was succinctly captured by Johannes Kepler:

> If there are globes in the heaven similar to our earth, do we vie with them over who occupies a better portion of the universe? For if their globes are nobler, we are not the noblest of rational creatures. Then how can all things be for man's sake? How can we be the masters of God's Handiwork?[3]

However, *no fact* had been uncovered by astronomers in the seventeenth century which compelled belief in other worlds or alien life. The fact that the earth is not the spatial center of the universe has no longer any meaning today, because the cosmos revealed by twentieth-century astronomy has itself no spatial center. Ironically, our relatively peripheral position on the spiral arm of a rather ordinary galaxy is indeed rather fortunate. If we had been stationed in a more central position—say, near the galactic hub—it is likely that our knowledge of the universe of other galaxies, for example, might not have been as extensive. Perhaps in such a position the light from surrounding stars could well have blocked our view of intergalactic space. Perhaps astronomy and cosmology as we know these subjects would never have developed.

There are in fact only two ways to show that the cosmos is not uniquely fit for life as it exists on earth and specifically for the existence of complex higher forms of carbon-based life such as our own species. First, the *straightforward or empirical way*—by directly discovering another type of life based on a completely different chemistry or even more dramatically by directly contacting or discovering a race of beings as intelligent as ourselves but of profoundly dissimilar design and biology to our own. Such discoveries would render hopeless any attempt to interpret the universe as peculiarly fit for life, a designed whole with life and man as its fundamental purpose and goal. Second, there is the *theoretical or indirect way*—by showing from our knowledge of science that nature is at least theoretically fit for the existence of other types of life and other types of intelligence. In other words, that the laws of nature give no indication of their having been arranged specifically for our type of carbon-based life or for the biology of higher forms of life such as ourselves.

The Failure of the Empirical Way

As far as the first way is concerned, it is an extraordinary fact that even as late as 1600—more than a century after the age of exploration had begun with the discovery of the Americas in 1492 and the opening up of the trade routes to Asia around the southern tip of Africa in the early decades of the sixteenth century—it was still conceivable that alien intelligent life very dissimilar to *Homo sapiens,* perhaps even nonmammalian or perhaps even based on an exotic type of biology, might exist in some unexplored region of the globe. In other words, direct demolition of the anthropocentric thesis was still possible by more extensive exploration of our own planet. Myths and ancient legends lent some credence to such a possibility. The legend of the Christian Kingdom of Prestor John beyond the Muslim lands to the east or in Africa and the legends of lost civilizations like Atlantis fueled speculation that the Portuguese caravels might bring back news of an utterly alien civilization in terra incognita. The radical possibility of a highly intelligent nonhuman species extant on earth, members of a civilization perhaps far in advance of sixteenth-century Europe, could not therefore be discounted in 1600. The southern hemisphere had hardly been penetrated, and there was a widespread belief in a great southern land, Terra Australis Incognita, as extensive as Eurasia somewhere east of Africa and south of the Indonesian archipelago. Moreover, the center of the newly discovered continents of North and South America, as well as Central Africa, were still largely unexplored. Such mysterious regions might conceivably have harbored advanced civilizations peopled by nonhuman exotic beings quite unrelated to the biblical Adam and perhaps even of a biology unfamiliar to the Christian Europeans and other cultures of the known world. On the maps of the fifteenth and sixteenth centuries, the interior of these continents are often peopled with a phantasmagoria of strange beasts and semihuman beings. Such imaginings bear witness to the challenging possibility, very real at the time, of intelligent beings very dissimilar to ourselves in the distant unexplored regions of the globe.

Another unexplored realm which held out the possibility of puncturing directly the anthropocentric presumption was the past history of our own planet. Serious scientific exploration of the past through the development of paleontology and archaeology only began in the early nineteenth century. Discovering evidence of nonhuman intelligent activity on the earth before man—for example, complex artifacts in ancient sediments—would pose every bit as significant a challenge as the discovery of extraterrestrial intelligence or St. Augustine's antipodean man. But just as the geographical explorations after 1500 failed to find any nonhuman intelligent races extant in some distant region of the earth, the exploration of the past since 1800 has similarly turned up not the slightest evidence for the existence of intelligent activity on the earth before the advent of man. As far as we can tell, the only life that ever graced our planet was the carbon-based type that is extant today, and no intelligence other than that of *Homo sapiens* has ever left its imprint on the earth in all the 4 billion years that the planet has been capable of sustaining life.

Of course, the earth has, ever since the early seventeenth century, been seen as a mere speck of dust in the immensity of the macrocosm, and it has always been in the stars that the possibility of finding evidence of alien life and intelligence based on a radically different design to our own has seemed the most likely.

Although the concept of extraterrestrial life has been deeply embedded in western culture since the early sixteenth century, it has become particularly popular in the past few decades with the advent of the space age. Almost everyone takes it for granted that the cosmos is teeming with all manner of life forms and that humanity, in Carl Sagan's words, is merely "one voice in the cosmic fugue."[4]

Of the planets in our solar system, Mars has always been considered as one of the more likely candidates for extraterrestrial life. Although the unmanned Mariner spacecraft that flew by Mars in the 1960s and early 1970s and extensively photographed the surface found no evidence of "canals" or any other signs of an extant or vanished civilization, the possibility of there being microscopic life could not be excluded. Consequently, the discovery of at least some microscopic types of life in the Martian soils was very much a possibility when the Viking spacecraft touched down on the Martian surface in the summer of 1976. It carried a vast array of sophisticated instruments for measuring the composition of the atmosphere and seismic activity, for analyzing the chemical composition of the soils, and perhaps most significant of all, a biology lab to carry out experiments designed to test for the presence of life in the Martian soils.

With the imminence of the Viking touchdown on Mars, the summer of 1976 was the culmination of an exciting phase of space exploration. From the successive Mariner missions during the 1960s and 1970s, a fascinating picture of the red planet had emerged. And it was one that was not incompatible with the existence of simple types of microbial life.

Who can recall the first images sent back by *Viking 1* from the plains of Chryse Planitia, in July 1976, and not also recall the mounting excitement of the possible discovery of extraterrestrial life. In the summer of 1976 all mankind stood as they had with Columbus in 1492 and with Cook in 1788 on the shore of a new terra incognita. Moreover, just as was the case with the exploration of "new worlds" on earth, the exploration of this Martian new world could have provided the long feared *coup de grâce* to the man-centered view of the world. When *Viking 1* touched down in July 1976, it was of course already unlikely, judging from the previous observations of the Mariner missions, that any intelligent race would be found on Mars. However, the possibility of the discovery of a completely alien type of life on a planet so close to earth and the very first extraterrestrial world to be examined by humans would have greatly increased the probability of extraterrestrial intelligent life based on completely different principles to those observed on Earth and would have seriously threatened the presumption that the laws of nature are uniquely arranged for life as it exists on Earth.

Even before Viking, the increasing possibility that the planet might prove lifeless was definitely greeted with relief in many conservative Christian quarters. Echoing the fears of the inquisitors of Bruno and Galileo four centuries before, they were, in Sagan's words: "unmistakably relieved. Finding life beyond the earth—particularly intelligent life, although this is highly unlikely on Mars—wrenches at our secret hope that man is the pinnacle of creation. Even simple forms of extraterrestrial life may have abilities denied to us. The discovery of life on some other world will, among many things, be for us a humbling experience."[5]

Although that experience may now be upon us, it may not be as humbling as Sagan implies. Undoubtedly, the evidence of primitive life, detected in a Martian meteor by NASA scientists, would represent, if confirmed, the first genuine scientific evidence of extraterrestrial life and be a defining moment in human history, a moment of great philosophical significance. But it does not threaten in any sense the main claim of this book—that the cosmos is uniquely fit for the specific type of life that exists on Earth. As stressed above, although the discovery of extraterrestrial life *based on dissimilar principles and chemistry* to that on Earth would overthrow immediately the idea that the laws of nature are uniquely arranged for life as it exists on Earth, the discovery of extraterrestrial life *similar to or even indistinguishable from* that on earth tends to reinforce the conception of the cosmos as uniquely fit for life as it exists on Earth. Although the results are only preliminary, a confirmation would imply that ancient Martian life was carbon-based and very closely resembled the earliest life forms on Earth.

If there is, or once was earthlike life on Mars, this would suggest, as planetary systems are probably quite common (see chapter 4), that there may be life similar to that on Earth throughout the cosmos. The discovery also opens up the possibility of there being life on some of the other planets in our solar system and possibly on some of their fifty-four moons.

But even if there is primitive life on Mars or elsewhere in the solar system, it still seems unlikely that advanced life forms, and especially intelligent life, exists anywhere at present in our solar system. The space explorations of the latter half of the twentieth century would seem to have finally laid to rest the visions of generations of dreamers, who from Giordano Bruno to Percival Lowell have envisaged intelligent inhabitants on our near neighbors in space. So while the NASA results open up the possibility of finding in our solar system simple alien life forms based on a different chemistry to our own, the possibility of finding intelligent life based on exotic chemistries is remote.

Of course, outside the orbit of Pluto, outside our solar system, there still lies the unexplored, unimaginable vastness of the cosmos, containing 100 billion galaxies, each containing 100 billion stars. Although the number of stars with planets is unknown, it is now widely assumed that perhaps the majority of stars have planetary systems and that planets rather like Earth may be the inevitable result of the same cosmic processes which lead to the formation of stars. For the most part, this immense terra incognita will remain forever truly incognita to humankind. The planets circling the most distant stars will never be directly explored by man. No spacecraft will ever sniff or taste the soils of a world in another galaxy. Even the Andromeda galaxy, one of the closest, is nearly a million light years away.

Since we cannot seriously contemplate the direct physical exploration of solar systems beyond our own, it would seem that the only hope we have in the near future of obtaining direct evidence from distant regions of the cosmos of sentient extraterrestrial life dissimilar to ourselves is if they contact us, either by physically visiting Earth or by sending a spacecraft advertising their existence and informing us of their biological design—just as our own Voyager spacecraft is presently carrying information about life on Earth deep into interstellar space. Alternatively, they might beam across the void some type of signal capable of traversing the cosmic immensity, perhaps a radio message.

The fact remains that to date no artifact has ever been discovered on Earth which might be interpreted as a "Chariot of the Gods." No structure or piece of machinery has ever been found that might have been constructed by aliens here on earth, or left by ancient star travelers from another world. Invariably, the many intriguing and puzzling phenomena offered as evidence of an alien presence, such as the Nazca lines in the Atacama Desert, corn circles in southern England, the peculiar round stones of Guatemala, the headgear of the Aztec gods, which are often put down at first to the work of aliens, prove to have far more mundane explanations. Similarly, nothing in the solar system, including the supposed Martian Sphinx, the Martian pyramids, or the moons of Mars, provides any evidence of extraterrestrial visitations to our remote corner of space. All such apparent artifacts are either natural objects, or can very easily be interpreted as such. If our solar system was ever inhabited or visited by intelligent beings, they left long ago and left no sign that we can read.

Nor have any radio or optical signals ever been detected that provide evidence of extraterrestrial intelligence. Occasionally, astronomers pick up an unusual signal from some previously unremarkable point in the sky, and for a moment there stirs a faint hope that it might be an intelligent signal from another world. Despite the lack of any evidence, the concept retains a potent force. As Sagan confesses, "There is something irresistible about the discovery of even a token from an alien world."[6] Yet no matter how deeply ingrained the concept of extraterrestrial intelligence, to date neither the direct exploration of Earth nor of our nearest planetary neighbors have provided the slightest evidence for intelligent life beyond the earth. The empirical way has in effect failed. The concept of the cosmos as uniquely fit for our existence and the presumption of our teleological centrality remain intact. After four centuries of ceaseless searching, Earth has yielded no clue, the heavens have remained eerily silent, and even Mars now threatens to disappoint. We still have at present no direct empirical evidence that *the laws of physics might permit the existence of life or of intelligent beings designed along principles fundamentally different from those governing life on Earth.*

The Failure of the Theoretical Way

Curiously, while the sixteenth-century mariners were vastly extending man's knowledge of the earth's surface, their early scientific contemporaries Galileo and Kepler were initiating the far more consequential exploration of the world through science. For in a very real and profound sense, science is an exploration of the cosmos. The analogy is close. This is because the laws of nature are universal and apply everywhere in space and time. Consequently, scientific discoveries on earth in effect carry us to the most distant galaxy and allow us to peer into every corner of the cosmos. Through the abstract eyes of science we can see further and explore realms to which no spacecraft will ever travel. Through science we can know the whole cosmos. The physical and chemical properties of water and carbon dioxide, the characteristics of electromagnetic radiation, the viscosity of silicate rocks—all these are the same in the galaxy of Andromeda as they are on earth. Through science, we can travel in theory to any part of the universe, to any alien planetary surface, and we know that at a particular temperature and atmospheric pressure, oxygen could exist there as a gas or that visible light will activate chemical reactions, that water will expand if cooled below 4°C.

Science has in effect opened up another, theoretical way to disprove the anthropocentric assumption; to provide a convincing blueprint of an alternative form of life, perhaps based on carbon and water or perhaps based on a completely exotic type of chemistry.

Various attempts have been made to envisage life based on exotic chemistries. Gerald Feinberg and Robert Shapiro in their *Life Beyond Earth* discussed in outline a variety of alternative fluids in which an alien life might be based. Water, they conjectured, could perhaps be replaced by liquid ammonia in a low-temperature world. As they point out: "Ammonia may form an ocean when water is hard ice, a kind of rock material in fact, rather like silica in our granite. . . . Ammonia has a good capacity as a reservoir for heat. It dissolves salts and supports reactions between charged substances such as acids and alkalis."[7] Another possible fluid which could serve as the medium of an alternative type of life discussed by Feinberg and Shapiro might be an oily mixture of hydrocarbons that would be liquid at a much lower temperature than ammonia. In a world containing life in oil, the seas would be in effect a giant oil slick, composed of a mixture of hydrocarbons of lower boiling point than those usually carried by oil tankers on earth. They name their hypothetical oil world Petrolia. Yet another possible biosphere considered was one based in liquid silica at a temperature of 1,000°C.[8] In addition to alternative types of chemical life, they also considered the more radical possibility of nonchemical life based on physical rather than chemical ordering systems. They speculate, for example, that a type of electronic organism—a plasmobe—"composed of patterns of magnetic force together with groups of moving charges in a kind of symbiosis" might be possible in the center of stars, where intense pressures and temperatures exclude any sort of molecular or chemical life.[9]

Although most authorities see fluids as the ideal medium for life, some authors, such as Fred Hoyle in his book *The Black Cloud,* have envisaged gaseous life.[10] One of the more exotic scenarios of life in a gaseous medium very different to its earthly manifestation was described by Carl Sagan in *Cosmos:*

> On a giant gas planet like Jupiter, with an atmosphere rich in hydrogen, helium, methane, water and ammonia, you could also be a floater, some vast hydrogen balloon pumping helium and heavier gases out of its interior and leaving only the lightest gas, hydrogen; or a hot air balloon, staying buoyant by keeping your interior warm, using energy acquired from the food you eat. . . . A floater might well eat preformed organic molecules, or make its own from sunlight and air. . . . Salpeter and I imagined floaters kilometers across, larger than the greatest whale that ever was, beings the size of cities. . . . The floaters may propel themselves through the planetary atmosphere with gusts of gas, like a ramjet or a rocket. We imagine them arranged in great lazy herds for as far as the eye can see.[11]

As our knowledge of chemistry is still largely empirical and we cannot predict with any certainty the behavior of chemicals and chemical systems in conditions vastly dissimilar to those with which we are familiar, we cannot rule out completely any of the possibilities raised by Feinberg and Shapiro or even exclude such fantastic possibilities as floaters.

But the fact remains, as Feinberg and Shapiro themselves concede, that no relatively well-developed model providing a detailed account of an exotic type of life has ever been presented. As biologist Norman Horowitz remarks: "It is not possible to evaluate these proposals because they have not been developed in detail. And in the absence of detailed models, all such speculation must be viewed with scepticism."[12] Horowitz's point, however, is critical, because the design of a self-replicating chemical system is bound to pose design problems involving the necessity to satisfy a good many relatively stringent and often conflicting criteria. These will only be revealed on attempting a detailed blueprint. How does alien life originate? How does it evolve? How do alien intelligent beings do science? How did they develop a technology? The questions are endless and unanswered. In our own carbon-based system in many instances, as we have seen, these criteria are in effect already satisfied in the existence of what appear to be a prearranged string of mutual adaptations in many of its basic constituents. It seems more than likely that in developing a detailed blueprint of an exotic biology, it would turn out that in many key instances design problems would prove insurmountable without built-in solutions analogous to those life-facilitating coincidences upon which our own form of life depends.

Despite the optimism of the gurus in the fields of exobiology, artificial life, and nanotechnology, no even remotely detailed blueprint for an alternative feasible self-reproducing system has been worked out. Although the possibility cannot be completely excluded, from the evidence now available it seems increasingly unlikely that life can be realized in any other material system in our cosmos.

Four centuries after the scientific revolution science has provided no significant evidence that any alternative life is possible. The second, theoretical means of disproving the anthropocentric presumption, by showing that the laws of nature are *not specifically fit* for life as it exists on earth, has *failed.* Just as the explorations of the sixteenth-century mariners and all subsequent explorations since have failed to bring back direct evidence to threaten the unique status of life and our own species, so also the scientific exploration has found no token of another life, no shred of evidence for something other than ourselves or of our type of life as it exists on earth. On the contrary, science has revealed a universe stamped in every corner, riven in every tiny detail, with an overwhelmingly and all-pervasive biocentric and anthropocentric design.

Fitness for Life's Being

We may not have final proof that the cosmos is *uniquely* fit for life as it exists on earth—because the possibility of alternative life cannot yet be entirely excluded—but there is no doubt that science has clearly shown that the cosmos is *supremely* fit for life as it exists on earth. For as we have seen, the existence of life on earth depends on a very large number of astonishingly precise mutual adaptations in the physical and chemical properties of many of the key constituents of the cell: the fitness of water for carbon-based life, the mutual fitness of sunlight and life, the fitness of oxygen and oxidations as a source of energy for carbon-based life, the fitness of carbon dioxide for the excretion of the products of carbon oxidation, the fitness of bicarbonate as a buffer for biological systems, the fitness of the slow hydration of carbon dioxide, the fitness of the lipid bilayer as the boundary of the cell, the mutual fitness of DNA and proteins, and the perfect topological fit of the alpha helix of the protein with the large groove of the DNA. In nearly every case these constituents are the only available candidates for their biological roles, and each appears superbly tailored to that particular end.

If these various constituents—water, carbon dioxide, carbonic acid, the DNA helix, proteins, phosphates, sugars, lipids, the carbon atom, the oxygen atom, the transitional metal atoms and the other metal atoms from groups 1 and 2 of the periodic table, sodium, potassium, calcium, and magnesium—did not possess precisely those chemical and physical properties they exhibit in an aqueous solution ranging in temperature from 0°C and about 75°C, self-replicating carbon-based chemical machines would be impossible. And it is not only microorganisms that the cosmic design has "foreseen." Many of the properties and characteristics of life's constituents seem to be specifically arranged for large, complex, multicellular organisms like ourselves. The coincidences do not stop at the cell but extend right on into higher forms of life. These include the packaging properties of DNA, which enable a vast amount of DNA and hence biological information to be packed into the tiny volume of the cell nucleus in higher organisms; the electrical properties of cells, which depend ultimately on the insulating character of the cell membrane, which provides the basis for nerve conduction and for the coordination of the activities of multicellular organisms; the very nature of the cell, particularly its feeling and crawling activities, which seem so ideally adapted for assembling a multicellular organism during development; the fact that oxygen and carbon dioxide are both gases at ambient temperatures and the peculiar and unique character of the bicarbonate buffer, which together greatly facilitate the life of large air-breathing macroscopic organisms.

In short, science has revealed a *vast chain of coincidences which lead inexorably to life* on earth—not just microbial life but all life on earth, including large, air-breathing organisms like ourselves—a chain of adaptations which leads from the dimensions of galaxies, through the physical conditions in the center of stars to the heat capacity of water and the atom-manipulating capacities of proteins, and on eventually to our own species and our ability to comprehend the world. From the inertial resistance we encounter when we move our hand, determined by the mass of the most distant stars, to the radioactive heat in the earth's interior which drives the great tectonic system, thus ensuring a continual replenishing of the vital elements of life—all nature, every facet of reality, is bound together into one mutual self-referential biocentric whole.

What is so particularly impressive and so highly suggestive about these life-giving adaptations is that what at first sight seem to be very trivial aspects of the chemistry and physics of a particular component turns out to be of critical significance for its biological role. Many examples have been cited in earlier chapters, including the decrease in the viscosity of the blood when the blood pressure rises, which increases the blood flow to the metabolically active muscles of higher organisms; the anomalous thermal properties of water, which buffers both the planet and organisms against massive swings in temperature; the curious but critical fact that the hydration of carbon dioxide is quite slow, which prevents a fatal acidosis in the body of higher organisms in anaerobic exercise; the curious fact that it is base sequences in the major groove of the DNA which provide the electrostatic variability that can be recognized by an α helix, and so forth.

It is important also to recall that the vital mutual adaptations are in the essential nature of things and are not the product of natural selection. This was also stressed by Henderson: "Natural selection does but mould the organism without truly altering the primary qualities of environmental fitness."[13] These are antecedent to the existence of life. The precise fit between the α helix and the large groove of the DNA are given by physics; the relationship long predated life. Similarly, the life-giving anomalous expansion of water below 4°C and on freezing and its low viscosity are given by physics. It was given before the first cell appeared in the primeval ocean. The fact that hydrogen bonds and other weak bonds have sufficient strength to hold proteins and DNA in "metastable" conformations at ambient temperatures; the fact that the majority of organic compounds are relatively stable below 100°C; the fact that the reactivity of oxygen, the only feasible terminal oxidant for carbon, is relatively unreactive below 50°C; the fact that the solubility of oxygen in water, the unique matrix for life is relatively low; the fact that carbon dioxide is a gas, that bicarbonate has such excellent buffering capabilities—all these unique coincidences are in effect laws of nature, universals no less than the constants of physics. Commenting on Henderson's arguments, the great biologist Joseph Needham stressed the same point: "Since the properties of water and the . . . elements antedate the appearance of life . . . they can be regarded philosophically as some sort of preparation for life. Purposiveness, then, exists everywhere, it permeates the whole universe. . . . Restricted teleology melts away in the immensity of that discussed by Lawrence Henderson."[14]

Fitness for Becoming

The evidence that life's becoming is also built into nature, presented in the second part of the book, is admittedly not as convincing as the evidence presented in the earlier chapters. But it is consistent with the possibility. The curious equality of mutation rates and evolutionary substitution rates and the just as curious uniformity of protein evolution which have caused endless discussion over the past twenty years have not proved easy to reconcile with Darwinian explanations. And although in no sense can either of these two phenomena be claimed as evidence for design, they are suggestive of something more in the evolutionary process than purely random mutation. Again, the current picture of the origin of life is also compatible with the concept of a uniquely ordained path from chemistry to the cell. The growing evidence that evolution is jumpy and that major evolutionary transformations have occurred rapidly is again suggestive. The more saltational the course of evolution, the easier it is to envisage it as being the result of a built-in program. The enormous diversity of the pattern of life on earth may not represent a full plenitude of all life forms, but it appears to approach closely this ideal. The very great complexity of life, and especially its quite fantastic holistic nature, which seems to preclude any sort of evolutionary transformations via a succession of small independent changes, is perfectly compatible with the notion of directed evolution. The ease with which the evolution of the very many complex adaptations such as the eye of the lobster and the avian lung can be explained in terms of design lends further support to the notion of directed evolution.

The Argument for Design

The strength of any teleological argument is basically accumulative. It does not lie with any one individual piece of evidence alone but with a whole series of coincidences, all of which point irresistibly to one conclusion. It is the same here. Neither the thermal properties of water, nor the chemical properties of carbon dioxide, nor the exceptional complexity of living things, nor the difficulties this leads to when attempting to give plausible explanations in Darwinian terms—none of these individually counts for much. Rather, it lies in the summation of all the evidence, in the whole long chain of coincidences which leads so convincingly toward the unique end of life, in the fact that all the independent lines of evidence fit together into a beautiful self-consistent teleological whole. The evolutionary evidence is similar; it compounds. In isolation, the various pieces of evidence for direction, the speed of evolutionary change, the fantastic complexity of living things, the apparent gratuity of some of the ends achieved, are perhaps no more than suggestive, but taken together, the overall pattern points strongly to final causes.

Reinforcing further the teleological position is the fact that its credibility has relentlessly grown as scientific knowledge has advanced throughout the past two centuries. In the early part of the nineteenth century when chemistry was just beginning, the only biocentric adaptations that Whewell was able to cite in his Bridgewater Treatise as evidence for a biocentric design were a few of the thermal properties of water. But by Henderson's time in the first decade of the twentieth century, while ironically the last vestiges of teleology were being exorcised from mainstream biology, advances in physiological and organic chemistry had revealed an additional and highly significant series of mutual adaptations in life's constituents which provided for the first time a significant body of evidence consistent with the view that our own carbon-based life is unique and that the laws of nature are specifically tailored to that end. Once again, during the past fifty years advances associated with molecular biology have, as we have seen, revealed yet another set of unique mutual adaptations at the heart of life in key constituents such as DNA and protein. And over the same period advances in cosmology and astrophysics have indicated that the overall structure of the universe and the constants of physics seem also to be fine tuned for our existence.

Note also that theories or worldviews are most often accepted not because they can explain everything perfectly but because they make sense of more than any competitor does. Evolution was accepted in the nineteenth century not because it explained everything perfectly but because it accounted for the facts better than any other theory. Similarly, the teleological model of nature presented here is far more coherent and makes far more sense of the cosmos than any currently available competitor. The idea that the cosmos is a unique whole with life and mankind as its end and purpose makes sense and illuminates all our current scientific knowledge. It makes sense of the intricate synthesis of carbon in the stars, of the constants of physics, of the properties of water, of the cosmic abundance of the elements, of the existence throughout the cosmos of organic matter, of the fact that the two adjacent planets Earth and Mars appear so similar, that the atom-building process continues to uranium. No other worldview comes close. No other explanation makes as much sense of all *the facts.*

Falsification

It is sometimes claimed by critics of the design hypothesis that the universe is bound to look as if it is designed for our existence because we could only be here if the universe was adapted for our existence. There is obviously an element of truth in this line of argument, for indeed the universe must be adapted to some degree for our existence if we are to be here as observers. However, the conclusion to design is not based on evidence that the laws of nature are adapted to *some degree* for life but rather on the far stronger claim that the cosmos is *optimally adapted* for life so that every constituent of the cell and every law of nature is uniquely and ideally fashioned to that end.

It is important to note that the design hypothesis is in fact very easy to refute. Any number of observations could potentially refute the claim that the cosmos is uniquely fit for life and overthrow the whole teleological conception of nature. If the laws of nature had permitted another type of life comparable in every way to our own (a possibility that now seems very remote), then the entire argument would collapse. In the seventeenth century it was the mere likelihood that the cosmos might contain a plurality of alternative worlds, some perhaps superior to our own, which threatened the whole anthropocentric worldview. In a sense, for the seventeenth-century scientific community the design hypothesis had been *refuted empirically.* When Kepler, for example, contemplated the immensity of the heavens, he could not imagine a connection between the newly revealed vastness of the cosmos and life on earth.

But the design hypothesis can, of course, be refuted by far less dramatic evidence than the discovery of an exotic type of extraterrestrial life. For example, the discovery of an alternative liquid as fit as water for carbon-based life, or of a superior means of constructing a genetic tape, better than the double helix, of alternatives superior to oxidation, superior to proteins, superior to the bilayer lipid membrane, to the cell system, to bicarbonate, to phosphates, and so on. The creation of a machine with an intellectual capacity superior to that of man would also effectively demolish the argument that the universe is contrived with mankind and human intelligence as its primary end. Just one clear case where a constituent of life or a law of nature is evidently not unique or ideally adapted for life, and the design hypothesis collapses.

If refutability is the hallmark of a scientific theory, as claimed by Karl Popper, then the hypothesis that "the cosmos is uniquely fit for life" might be classed as a perfectly ordinary scientific hypothesis. According to Popper, no hypothesis can be finally proved. All we can do with the hypothesis is attempt to refute it. The longer it resists our attempts, the better the theory. Consistency with the facts is the best we can hope for, even in the most powerful scientific theory. By the criteria of the Popperian philosophy of science, the design hypothesis is demonstrably an exceedingly robust hypothesis. Even the most rigorous skeptic must surely concede that the hypothesis is at least consistent with the facts.

The traditional way of avoiding the teleological conclusion is of course the classic "appeal to chance" which, given an infinite period of time, will generate even the most improbable result. This has been used throughout western history to counter the argument for design. This strategy was first used by the atomists in classical times. It was used specifically for this purpose by Lucretius in his great poem *De rerum natura:*

> For verily not by design do the first beginnings of things station themselves each in his right place, occupied by keen sighted intelligence, . . . but because after trying motions and unions of every kind, at length they fall into arrangements, such as those out of which this our sum of things has been formed. . . .[15]

The same line of reasoning was used by David Hume in his famous *Dialogues on Natural Religion,*[16] and again very recently by the Oxford chemist P. W. Atkins in his book *The Creation* to avoid concluding to design.[17]

According to this tradition, the order of the cosmos is ultimately a matter of chance and we need seek no further explanation. This line of reasoning is of course impossible to refute. This is not a strength but rather a serious defect. For in terms of the Popperian view of knowledge, irrefutable claims are outside the realm of science. Whatever else the appeal to infinity may be, it is not a scientifically refutable hypothesis. Whether we can consider it seriously as a valid strategy to avoid the teleological conclusion is therefore open to debate. Whatever may be the philosophical status of the design hypothesis, unlike the appeal to infinity, it is certainly capable of refutation. But as this book has shown over and over again, new evidence which could potentially have refuted the hypothesis has only ended up confirming it.

I believe that Henderson's verdict on the chance hypothesis—that "the mind baulks at such a view" and "there is no greater [improbability than] that these unique properties should be without some cause uniquely favourable to the organic mechanism"[18]—is the only verdict compatible with a strictly scientific and commonsense approach.

Whether one accepts or rejects the design hypothesis, whether one thinks of the designer as the Greek world soul or the Hebrew God, there is no avoiding the conclusion that the world *looks* as if it has been uniquely tailored for life: it *appears to have been designed.* All reality *appears* to be a vast, coherent, teleological whole with life and mankind as its purpose and goal.

Are We Alone?

If the cosmos is uniquely fit for the being and becoming of our type of life, then earthlike life should arise and flourish on any planet where the conditions are similar to those on Earth. Moreover, the evolution of life on any such planet should result in many life forms similar to those generated during the evolution of life on Earth. If the evolutionary constraints are very stringent, then extraterrestrial life forms may be as identical to those on earth as Australian marsupials are to the placentals elsewhere.

A key question is how many earthlike planets might exist? From the evidence of our own solar system, which contains at least one other planet—Mars—that is very similar in many ways to the Earth and may even have had oceans and glaciers at times in the past, earthlike planets cannot be that rare. There are about 10^{11} stars in a galaxy and a total of 10^{11} galaxies in the universe, which makes 10^{22} stars in all. Since most stars probably have solar systems, the number of planetary systems is simply unimaginable. Even if only one solar system in a billion were like our own, this would still imply that there might be 10^{13} planets like Earth.

Because we cannot judge how likely the route to life, it is difficult to know how many would harbor life. The existence of life on Mars would provide strong evidence that the route to life is highly probable. But even if the route to life is somewhat tortuous and requires stringent conditions, given the number of planets, it is hard not to be inclined to the view that a vast number of planets do in fact harbor life.

As to the question "How many planets might harbor intelligent life?", again it is difficult to judge. To get from a single cell to *Homo sapiens* has taken about 4 billion years on Earth. This is a significant fraction of the entire history of the universe. Moreover, our planet has been struck many times by meteors, which have on several occasions nearly wiped out all life. If the Cretaceous meteor, credited by many with having eliminated the dinosaurs, had been only a few kilometers bigger in diameter, it could easily have boiled the oceans, effectively wiping out all life, not just the dinosaurs. Perhaps only a few planets are lucky. Perhaps on only a few does life survive for so long.

Then there is the difficulty of knowing how stringent the evolutionary constraints may be in moving from a cell to *Homo sapiens.* Perhaps the cosmos abounds in trees of life, but perhaps only one in a million trees grows at all like our own on Earth to reach the higher primates and eventually a species like *Homo sapiens.* Perhaps only in a cosmos of the size and age of our own is there a probability of one that *Homo sapiens* will arise. Science cannot yet say.

Conclusion

In the discoveries of science the harmony of the spheres is also now the harmony of life. And as the eerie illumination of science penetrates ever more deeply into the order of nature, the cosmos appears increasingly to be a vast system finely tuned to generate life and organisms of biology very similar, perhaps identical, to ourselves. All the evidence available in the biological sciences supports the core proposition of traditional natural theology—*that the cosmos is a specially designed whole with life and mankind as its fundamental goal and purpose, a whole in which all facets of reality, from the size of galaxies to the thermal capacity of water, have their meaning and explanation in this central fact.*

Four centuries after the scientific revolution apparently destroyed irretrievably man's special place in the universe, banished Aristotle, and rendered teleological speculation obsolete, the relentless stream of discovery has turned dramatically in favor of teleology and design, and the doctrine of the microcosm is reborn. As I hope the evidence presented in this book has shown, science, which has been for centuries the great ally of atheism and skepticism, has become at last, in these final days of the second millennium, what Newton and many of its early advocates had so fervently wished—the "defender of the anthropocentric faith."

Epilogue

It is not sufficient any longer to listen at the end of a wire to the rustlings of a galaxy: it is not enough even to examine the great coil of DNA in which is coded the very alphabet of life. These are our extended perceptions. But beyond lies the great darkness of the ultimate Dreamer, who dreamed the light and the galaxies. Before act was, or substance existed, imagination grew in the dark. Man partakes of that ultimate wonder and creativeness. As we turn from the galaxies to the swarming cells of our own being, which toil for something, some entity beyond their grasp, let us remember man, the self-fabricator who came across an ice age to look into the mirrors and magic of science. Surely he did not come to see himself or his wild visage only. He came because he is at heart a listener and a searcher for some transcendent realm beyond himself.

—Loren Eiseley, *The Unexpected Universe,* 1970

If the scientific evidence is consistent with the traditional teleological view of the cosmos as an anthropocentric whole with our own race, *Homo sapiens,* as its fundamental end and purpose, then this surely raises the further possibility that the acquisition of scientific knowledge itself and even the actual historical sequence of scientific and technical discoveries which led to modern science and eventually to the revelation of our centrality may also have been prearranged in the nature of things.

A hint that this might be so is the fact mentioned previously that our intellectual capacity and particularly our mathematical and aesthetic sense seem remarkably adapted for the scientific enterprise. There is, for example, the very curious fact which has fascinated philosophers since Plato—that the order of nature is mathematical and that it is primarily because the laws of nature can be described in simple mathematical forms that nature is inherently decipherable. As we saw in chapter 11, Aristotle was also struck by the correspondence between the human mind and the natural world.[1]

But in itself the mutual fitness of the human mind and the underlying structure of nature, although a necessary condition for the scientific enterprise, is clearly insufficient. For no matter how intelligent, no matter how much the deep structure of reality is fit for scientific understanding, no matter how much the human brain is adapted for the scientific enterprise, science is impossible without sophisticated tools, and these in turn are impossible without metals, and these in turn are impossible without wood. For without wood there can be no fire, and without fire, no metallurgy, and without metals, no technology and certainly no science. A race restricted, like the Eskimo, to a treeless tundra would be unable to develop any sort of sophisticated technology no matter how intelligent or manually dexterous.

In short, if the scientific enterprise is to succeed, the environment must also be fit. The major breakthroughs which ultimately made science possible—particularly the discovery of fire and the discovery of metals—have been possible only because of certain unique environmental conditions on the surface of the earth. Fire, which is the oxidation of carbon, is a remarkable phenomenon in itself. It is virtually the only known chemical reaction that is relatively nonviolent while at the same time generating vast quantities of heat. The controlled fire utilized by early man is only possible in turn because the oxygen level in the earth's atmosphere is approximately 20 percent and because of the availability of trees, which provide the carbon in the form of wood. Without trees there would have been no fire and certainly no metallurgy, and human advancement would have been forever frozen somewhere between the Stone Age and the civilization of the Aztecs. But metallurgy is itself possible only because of the existence of various geophysical processes which concentrate metal atoms in certain types of rock, because these metal-bearing ores are relatively common, and because the heat required to extract the metals is not prohibitively high and thus within the easy reach of a wood fire. Without all these favorable coincidences there would be no metallurgy, and no technologically sophisticated society would ever have developed.

The historical path that led humanity from the Stone Age hunter-gatherer to the dawn of science may have had its vagaries, but its general direction was not a matter of contingency. For fire must come before metallurgy and metallurgy before wheels (for wheels can only be easily manufactured with metal tools). Thus, the American Indians, who had only a rudimentary knowledge of metals, never developed the wheel. Again, the route from villages, via the development of agriculture and the domestication of animals, to the construction of cities, a route taken by societies in both the Old World and the Americas, is of necessity. The specific temporal sequence from the initial manufacture of stone tools via the control of fire to the discovery of metals and the development of metallurgy to the manufacture of the wheel and so on down to the scientific revolution was largely of necessity. Even the succession from copper to iron was of necessity, determined by the relative ease of extraction of copper compared with iron.

It is often claimed that chance played a large part in the process of discovery. But this is only true in the immediate sense. The classic case of Hans Christian Oersted's chance finding, during a student lecture, that an electric current in a metal wire held over a magnet could cause the magnet to move, one of the great discoveries in nineteenth-century science[2] and which, by indicating a link between electricity and magnetism, led eventually to the dynamo, electric motors, and the whole modern twentieth-century world of electrotechnology, occurred only because metal wire was freely available, because metals are conductors of electricity, and because magnetism is a natural phenomenon which had been known and used since before classical times. It seems more than likely that the discovery would have been made in the early nineteenth century if not by Oersted then by at least one of the many other physicists who were interested in the connection between electricity and magnetism. Mendel's laws were discovered independently on two occasions within the space of forty years. Differential calculus was discovered simultaneously in the seventeenth century by Newton and Liebnitz.

The fact that the scientific revolution occurred in Europe in the sixteenth century was again hardly a matter of chance. The scientific revolution could only have occurred in a technologically advanced society, skilled in the use of metallurgy, capable of constructing complex tools and artifacts, having a knowledge of mathematics, carrying out trade and commerce in manufactures. In other words, a society very similar to that which existed in Europe in the sixteenth century. Although the relative importance of the various factors which led to the advent of science in Europe at that time are controversial, there is general agreement among most historians (not only among Marxists) that given the level of technological development and the chronic shortage of labor, the rapid growth of a mercantile class anxious to expand trade and apply labor-saving technologies, the political liberty consequent on the breakdown of the feudal state, and the rediscovery of classical science and rationalism, the advent of science was virtually inevitable.

There is, I believe, every justification for viewing our planetary home with its oxygen-containing atmosphere, large land masses covered in trees, with its readily available and well scattered metal-bearing rocks as an ideal and perhaps unique environment for the use of fire and the development of metallurgy and ultimately the emergence of a technologically advanced complex society similar to that of sixteenth-century Europe. Any social animal endowed with our mental and physical attributes, with the capability of language, the ability to preserve and transmit knowledge across several generations and an innate sense of curiosity would be bound to follow the same progressive path. From the camp fires of our Neolithic ancestors we were led inevitably to copper, to iron, to toolmaking, to the wheel, to the lathe and eventually to the level of technological development of sixteenth-century Europe and from there to the scientific revolution.

And so it would seem that that same handful of about thirty atoms, which seem so perfectly fit for the construction of self-replicating chemical systems, which have in other combinations during the evolution of the earth generated a planetary environment supremely fit for life, have also possessed a myriad of properties in other combinations that have proved of very great utility to man and have facilitated greatly the development of a sophisticated technology, without which science would have been impossible and the revelation of our centrality forestalled. And so in perfect concordance with teleology, we are led to view the emergence of science and its revelation of the centrality of man as no less the inevitable and determined end of things than our biological emergence and the original emergence of life.

We have come far from the long dawn of the Paleolithic, when our ancestors huddled by wood fires, gazed wonderingly out into the night, divining imaginary messages and meanings in the pattern of the constellations, far from those distant days when a roughly hewn ax head was our only technological accomplishment. We have traveled far from the July night when the Chinese watched in uncomprehending awe a strange new light in the evening sky; we have come far from the birth of science, from the early seventeenth century, when Galileo first glimpsed the rings of Saturn.

But although the journey was long, the route often slow and tortuous, the evidence increasingly suggests that the end was never in doubt, that we followed a path already charted to an end foreseen and that our success was not in the least a matter of contingency. Like pilgrims seeking the source of their own transcendence, we have been drawn along a predetermined path from the discovery of fire to the birth of science to the revelation of our own centrality in the order of nature. We have deciphered the meaning of the constellations, and in science the cosmos has called us home.

Appendix

Miscellaneous Additional Evidence of the Fitness of the Constituents of Life

In which the properties of a variety of important constituents of the cell—sugar, glycogen, lipids, phosphates, acetic acid, and the bicarbonate buffer—are examined and shown to be uniquely fit for the biochemical roles they perform in living systems. Some of the adaptations which enable proteins to recognize particular sequences of DNA are also examined.

1. Sugar and Glycogen

One of the major molecular fuels used in all living systems are the sugars. The figure below shows the chemical structure of the five-carbon sugar arabinose.

Examples of a fully reduced carbon compound (a hydrocarbon), a partially reduced compound (a sugar arabinose), and the fully oxidized carbon compound (carbon dioxide).

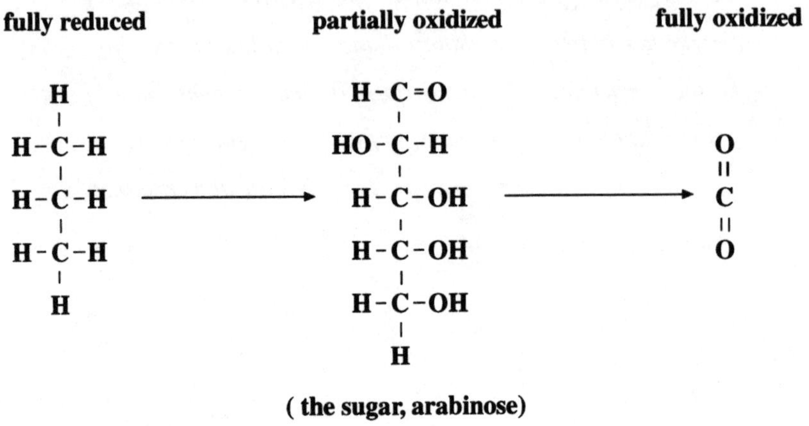

```
fully reduced            partially oxidized              fully oxidized

    H                        H-C=O
    |                          |
  H-C-H                     HO-C-H                            O
    |                          |                              ||
  H-C-H   ------------->     H-C-OH    ------------->         C
    |                          |                              ||
  H-C-H                      H-C-OH                           O
    |                          |
    H                        H-C-OH
                               |
                               H
                      ( the sugar, arabinose)
```

In chemical terms, the sugars represent a halfway stage in the oxidation of carbon between the lipids, $C_{18}H_{36}O_2$ (almost fully reduced carbon) and carbon dioxide, or CO_2 (fully oxidized carbon). The sugars are quite reactive compounds. They have a tendency to oxidize spontaneously, but this reactivity is attenuated in the case of the five- and six-carbon sugars by their characteristic and relatively stable ring-shaped structure. (This structure is known as the hemiacetal structure.) The attenuation of the reactivity because of this hemiacetal ring structure is particularly marked in the case of glucose. As David Green and R. F. Goldberger comment:

> The glucose ring is puckered and has both axial and equatorial groups bonded to the carbon atoms of the ring. The unique structural feature of glucose can now be fully appreciated. In glucose, all the large bulky groups are equatorial. No other six-carbon sugar has this special feature. All other six-carbon sugars have at least one large axial group. This means that glucose is the most stable six-carbon sugar possible.[1]

And as A. E. Needham points out, the stability of the hexoses (the six-carbon sugars) combined with their intrinsic reactivity:

> no doubt accounts for the outstanding biological role of the hexoses. In crystalline form, or in solutions too concentrated to be attacked by living organisms, they decompose very slowly. On the other hand, once activated they oxidise rapidly, even explosively, and are ideal fuels for ready use.[2]

On top of its basic chemical nature, which fits glucose so perfectly for its fundamental biological role as the "current coin" of the cell's energy system, glucose and the other hexoses have the additional very necessary characteristics of being nonvolatile (hence not lost by evaporation), very soluble in water and small enough to diffuse rapidly throughout the cell. On top of this, the breakdown of glucose generates many small molecules of great value for synthetic and other purposes.

The sugars have another property that contributes further to their fitness for their biological role and that is their capacity to polymerize into relatively insoluble complex-branched macromolecules, the polysaccharides, such as starch in plants and glycogen in animals. The molecular characteristics of polysaccharides such as glycogen, shown in the figure below, are perfectly suited to act as a readily available store of glucose to meet the ever-changing needs of the cell for energy.

The fitness of the branched structure of glycogen is commented on by Green and Goldberger:

The molecular structure of glycogen.

The black circles represent sugar molecules, the black lines chemical bonds holding them together.

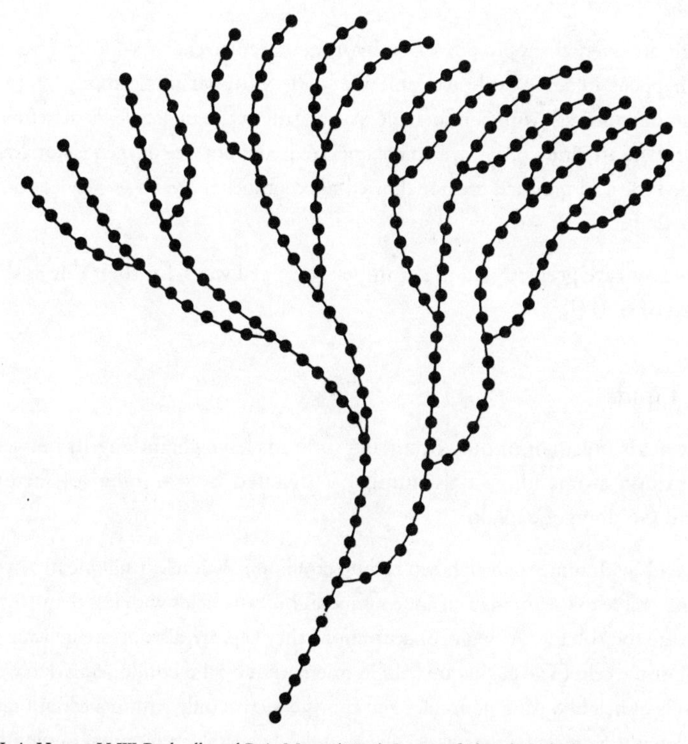

From H. A. Harper, V. W. Rodwell, and P. A. Mayes (1979) *Review of Physiological Chemistry* (Englewood Cliffs, N. J.: Prentice-Hall), fig. 8.24. Reproduced by permission of V. W. Rodwell.

> The structure of glycogen is not just *a piece of fanciful design by nature: it is uniquely suited for the role of glycogen as a storehouse of glucose in the animal organism.* The storage of thousands of glucose molecules in one big macromolecule accomplishes three related purposes. First, being a large molecule, glycogen cannot diffuse across the cell membrane and is thus a stable source of glucose. Second, the storage of many glucose units in a single macromolecule forestalls the osmotic problem which a high concentration of free glucose units would entail for the cell. Finally, the localisation of glucose units within a macromolecule simplifies enormously the logistics, both of commandeering glucose when the concentration of free glucose is low and of storing it when the concentration of free glucose is high. [My emphasis.]

Moreover, as Green and Goldberger continue:

> By virtue of the branched structure, the outer surface of the glycogen molecule presents a high concentration of substrate to the enzymes that regulate glycogen synthesis and breakdown, although in molar terms glycogen is present in very low concentrations in the cell. The lower degree of branching in starch is no disadvantage to plant cells, since plants have a much lower metabolic rate than animals.[3]

There is another characteristic of glucose which contributes further to its biological fitness, and this is seen in the process of glycolysis, where, as Needham points out, energy is released:

> without an external supply of oxygen by intramolecular changes—a mild version of what happens in some explosive molecules. The virtue of this carbohydrate-halting stage in biosynthesis, with considerable oxygen still in the molecule, is therefore obvious for there are situations in which energy is required but free oxygen is not available. Provided the acid products are not allowed to accumulate, glycolysis may continue for long periods.[4]

The hexoses are perfectly fit, in a number of crucial ways, for their role as the primary fuels of the cell.

2. The Lipids

The fatty acids most commonly occurring in lipids have chain lengths between 16 and 18 carbon atoms long. This number is dictated by a number of factors. As Green and Goldberger explain:

> Fatty acids with more than eighteen carbon atoms are so nearly insoluble in water that it is impossible to use them in an aqueous medium. Fatty acids with less than 16 carbon atoms are too soluble. At high concentrations they rapidly disrupt the delicate membranes of the cell. (The chains are held in orientation by the London-van der Walls attractions of neighbouring molecules and this is effective only within a certain range of chain lengths, namely C16–C36, . . . beyond this length the molecules topple and tan-

gle while shorter chains have insufficient attraction.) Thus the exact length of the fatty acids that are selected for use as a source of energy is a compromise between the potentially dangerous, more soluble acids and the unmanageable, less soluble acids. Nevertheless, one may ask why fatty acids are utilised, and not fatty alcohols, or fatty amines? Two chemical advantages of the fatty acids over alcohols or fatty amines are decisive: these are their greater solubility in water and higher reactivity, particularly after thioester formation.[5]

Chain lengths of 16 to 18 carbon atoms are ideal for other reasons as well. Fortuitously, fatty acids of these chain lengths are fluid or near fluid over the temperature range in which most metabolic processes occur in living things. In this section from his *Uniqueness of Biological Materials,* Needham comments on this interesting point:

> Lipid also forms a good cushion and shock absorber, and the viscera of mammals are usually well surrounded by fat. . . . In all cases the lipid is also, if not primarily, a reserve fuel . . . so that cushioning is a gratuitous asset . . . the high viscosity of lipids [compared with water] reduces their rate of flow under shearing forces, and there will be less displacement of the organs of the body during locomotion . . . than if water were the packing fluid. It is of course essential that the lipid should be fluid at body temperature, and this is found to be true. [Cold-blooded animals] have fats of lower melting point, mainly due to a larger percentage of unsaturated fatty acids. . . . Since their body temperature is very near that of the environment, the amount of unsaturated fat increases with latitude within the same species. . . . It appears that the melting point of body fat of [warm-blooded animals] also varies with the distance from the surface. . . . In man the temperature gradient is considerable, the skin being as cool as 24^0C compared with the maximum of 37^0C.[6]

The subcutaneous lipid of warm-blooded animals acts as a heat insulator preventing heat loss. The extreme example of this is the Cetacea with a layer of blubber up to 15 inches thick, which allows it to thrive in the polar seas.[7] It is fortunate for whales that carbon is lighter than oxygen so that most lipids are lighter than water. In fact, the low density of lipids is exploited universally among aquatic life forms to provide buoyancy. Needham comments:

> Oil drops give buoyancy to some Protozoa, siphonophores, embryos of Crustacea and fishes, and other plankton. . . . In adult marine fish the fat content of the body is fairly closely related to the height in the water in which they normally swim.[8]

It seems that the viscosity of just that class of lipids which for other reasons are ideally suited within the cell for constructional, metabolic purposes and energy storage would seem to be also ideally suited for buffering the organs of higher organisms against shearing forces. This is another instance in which the viscosity of one of life's constituents has a critical bearing on its biological role. Previously, we saw that the viscosity of water and ice also have a critical bearing on their respective biological roles.

Because of its hydrophobic properties the lipid bilayer has another characteristic—that of an electrical insulator. This is because, as Needham explains,

> most of the molecule is hydrocarbon, non-ionising and bearing only weak electrostatic charges . . . most electrical activity in living organisms depends on ions rather than electrons and the insulating properties of such lipid structures as the myelin sheath of nerve fibres are very great. They have a high resistance, and lipid therefore also acts as a good dielectric medium for condensers of high potential and low capacity; this type is believed to play a part in the generation of the sudden large changes in electrical potential which accompany the propagation of a nerve impulse. For such condensers a low dielectric constant is required and lipids can satisfy this requirement admirably.[9]

And in addition to its insulating characteristics the lipid bilayer makes another vital contribution to the generation of the membrane potential. As Bruce Hendry puts it in *Membrane Physiology and Cell Excitation:* "the lipid bilayer is vital as it provides the appropriate environment in which membrane proteins can reside and function. . . . The proteins endow the membrane with its specific ionic permeability."[10] Some of these proteins span the membrane. Hendry continues: "Such proteins are clearly well suited to providing a means by which ions may cross the membrane without passing through the bilayer . . . the importance of these membrane proteins for excitability is that they give the membrane the ability to pump ions into or out of the cell and to exhibit selective ion permeabilities."[11]

3. Phosphates

The energy released by oxidation is carried throughout the cell by a special class of energy-rich phosphate compounds. The figure on page 405 shows the structure of adenosine triphosphate (ATP), the main energy-rich phosphate utilized in the cell.

The biochemist F. H. Westheimer, writing in a recent *Science* article, comments that phosphates play a dominant role in the living world:

> Most of the coenzymes are (phosphates). The principal reservoirs of biochemical energy (adenosine triphosphate—ATP, creatine phosphate, and phosphoenolpyruvate) are phosphates. Many intermediary metabolites are phosphate esters, and phosphates or pyrophosphates are essential intermediates in biochemical synthesis and degradations.[12]

However, as Westheimer points out, organic chemists invariably use quite different compounds when carrying out analogous reactions to those carried out by phosphates in living things; for synthetic reactions, chlorides, bromides, iodides, tosylates, triflates, trialkylamines, sulfoxides, and selenoxides among others; for activating molecules for reaction, compounds such as carbodiimides. Chemists need to use reactive intermediates or at least intermediates which react at only moderately raised temperatures. While these compounds may well suit the organic chemist, living things could not tolerate reactive compounds, such as alkyl halides or dialkyl sul-

The chemical structure of ATP.

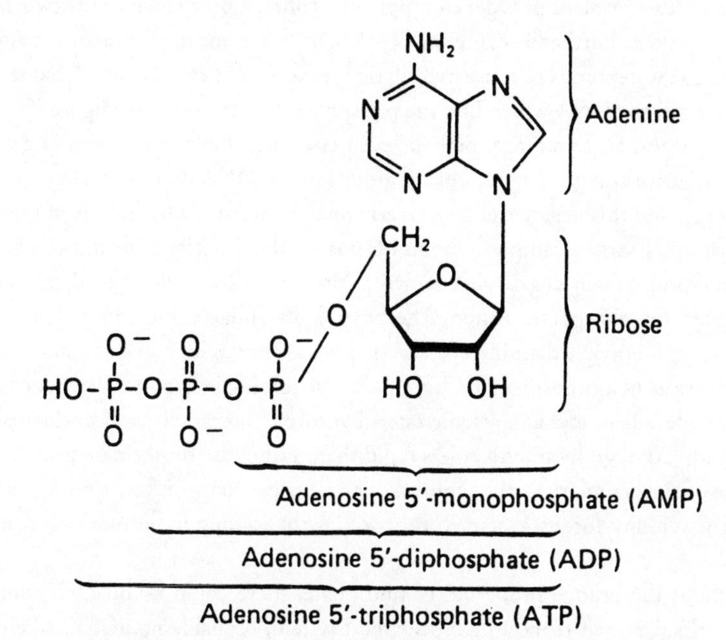

From Robert K. Murray et al., eds. (1996) *Harper's Biochemistry,* 24th ed. (Stamford, Conn.: Appleton & and Lange), fig. 35.12. Reproduced by permission of V. W. Rodwell.

fates, as they would alkylate enzymes and metabolites and inactivate the delicate machinery of the cell.

The energy-rich activating compounds used in the cell are phosphates, such as ATP and GTP. These are far more stable and far less reactive than their equivalents used in organic chemistry, but still sufficiently labile and reactive to fulfill these roles in the cell. In fact, the chemical characteristics of the phosphates are exactly those required of an energy-rich molecule to drive the biochemistry of the cell. Westheimer comments:

> In sharp contrast to [other energy-rich compounds] the phosphoric anhydrides are protected by their negative charges from rapid attack by water and other nucleophiles so that they can persist in an aqueous environment even though they are thermodynamically unstable, and thus can drive chemical processes to completion in the presence of a suitable catalyst (enzyme). This remarkable combination of thermodynamic instability and kinetic stability was noted many years ago by Lippmann, who correctly ascribed the kinetic stability to the negative charges in ATP. A citric acid anhydride would not survive long in water and could not serve as a convenient source of chemical energy.[13]

Interestingly, the phosphates have another key biological role. As we saw in chapter 7, they link together the nucleotide bases in the DNA. As Westheimer points out, in constructing a "tape" from small molecules, the connecting groups must be at least bivalent in order to supply one connection to each of the two adjacent nucleotides. Further, because the DNA helix functions in an aqueous environment and as water tends to cause hydrolytic breakdown of ester bonds—that is, the –0– bonds in the DNA which link the phosphates to the sugars (see figure on page 142)—it is also an advantage, perhaps even essential, that the connecting groups carry a negative charge. The phosphate groups in the DNA do in fact carry a negative charge, and this negative charge greatly retards the rate of hydrolysis of DNA.

What other sorts of simple compounds possess the capacity to form two chemical bonds and remain negatively charged at the same time and might therefore be substituted for phosphate? Among the possible candidates, including citric acid, glutamic acid, ethylenediaminemonoacetic acid, arsenic acid, and silicic acid, none of them are as fit as phosphate to function as the joining group in the genetic tape. For example, silicic acid and arsenic esters hydrolyze too rapidly, and in the case of citric acid, although hydrolysis is less rapid than in the case of silicic or arsenic acid because of the geometry of the molecule, the negative charge is too weak to confer sufficient stability for citric acid to function as the joining molecules in a genetic tape.

Without the unique properties of phosphate, there could be no DNA double helix, perhaps no self-replicating biochemical system, probably no life. Westheimer, commenting on the role of phosphate in linking the nucleotides in DNA, remarks that any such compound must be at least divalent, it must also possess a negative charge, and the charge must be physically close to the two ester bonds to protect them against hydrolysis: "All of these conditions are met by *phosphoric acid and no other alternative is obvious.*"[14] (My emphasis.) No other compounds possess the correct mix of properties to drive the chemical machinery of the cell.

4. Acetic Acid

One individual compound that plays a central role in the metabolic system is acetic acid. It is nature's choice for building many of the larger molecules in the cell. This is again almost certainly no accident, because there is no other obvious candidate available. As Green and Goldberger comment:

> The arithmetic for the synthesis of long chain fatty acids takes the form of the series 2+2+2 . . . rather than 1+1+1 . . . or 3+3+3. . . . The numbers refer to the numbers of carbon atoms in the fatty acid. Formic acid, the one carbon atom acid, can be disqualified on each of several important chemical grounds. An essential metabolic feature of a fatty acid is its terminal methyl group, and this is missing in formic acid. Thus it would be difficult to build up larger saturated molecules by successive condensations of units of formic acid. The three carbon acid is too unreactive chemically for purposes of condensation. The two carbon molecule of choice could be an acid (CH_3COOH), an al-

cohol (CH_3CH_2OH), or an aldehyde (CH_3CHO). Why was the acid selected as the starting point for synthesis? The alcohol is too unreactive and the aldehyde, although reactive, is somewhat unstable. We may conclude that chemical considerations were paramount in the selection of acetic acid as the starting point for the synthesis of larger molecules. *No other compound comes close to acetic acid in respect to properties excellently suited for building molecules by condensation reactions.*[15]

5. Buffers

The acidity of a solution is determined by the concentration of H^+ ion. The greater the concentration of H^+ ions, the more acid the solution. Pure water contains only a very low concentration of H^+ ions—about one H^+ ion to every 200 million molecules of water, or one 10-millionth of a gram of H^+ ions per liter. A common way of expressing the concentration of H^+ ions is as the negative logarithm of the concentration. Therefore, pure water, which has a concentration of H^+ ions of 10^{-7} grams per liter has a pH of 7. The diagram below gives the pH scale and corresponding H^+ ion concentration.[16]

The pH values of various fluids.

	H^+ CONCENTRATION (g/liter)		pH	BODY FLUIDS	SOLUTIONS
↑	10^0	1.0	0		
	10^{-1}	0.1	1		0.1N HCl
				Gastric	
	10^{-2}	0.01	2	Juice	
Acidic	10^{-3}	0.001	3		
					0.1N H_2CO_3
	10^{-4}	0.0001	4		0.0001N HCl
	10^{-5}	0.00001	5		
↓	10^{-6}	0.000001	6	Urine	
				Saliva	
Neutral	10^{-7}	0.0000001	7		Pure Water
				Blood	
↑	10^{-8}	0.00000001	8	Bile	
				Intestinal Juice	
	10^{-9}	0.000000001	9		
	10^{-10}	0.0000000001	10		
Basic	10^{-11}	0.00000000001	11		0.001 NaOH
	10^{-12}	0.000000000001	12		0.1N NH_4OH
	10^{-13}	0.0000000000001	13		0.1N NaOH
↓	10^{-14}	0.00000000000001	14		

From M. Toporek (1968) *The Basic Chemistry of Life;* © 1968 by Appleton & Lange, Stamford, Conn.

At every pH value from 1 to 14 a fraction of water molecules dissociate into H^+ (hydrogen) and OH^- (hydroxide) ions:

$$H_2O = H^+ + OH^-$$

The concentration of these two radicals varies inversely: when the concentration of H^+ ions goes up, the concentration of OH^- ions goes down. The diagram below shows the way the concentration of these two radicals varies over the complete pH range.

Note that the pH at which the concentration of these two charged radicals is minimum is 7. At pH 7 there are about two of these radicals for every 200 million molecules of water, while at pH values markedly lower or higher than 7 there is a dramatic increase in the number of these radicals. For every 200 million molecules of water at a pH of 1, there are 10 million H^+ ions, while at a pH of 14, there are 10 million OH^- ions.

The concentration of hydrogen and hydroxyl ions at various pH values.

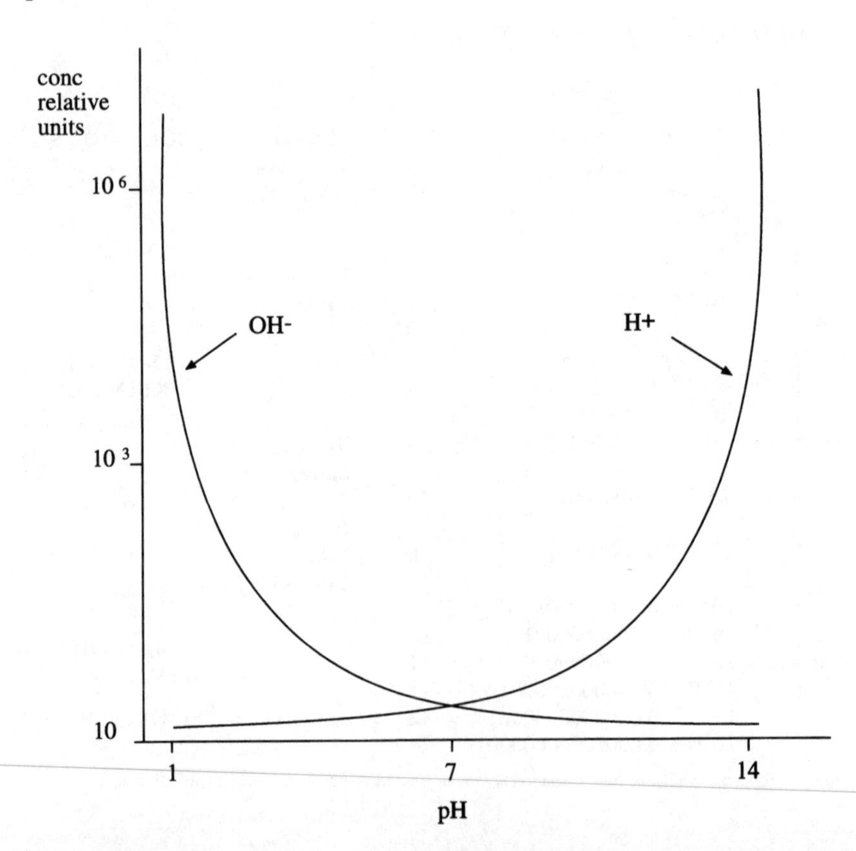

Unfortunately, both these ions are highly reactive, interacting in a variety of ways with practically all of the most important organic chemicals in the cell, with sugars, proteins and with the constituents of DNA and RNA, etc. High levels of either of these two reactive radicals causes the rapid hydrolytic breakdown of most organic compounds and are therefore incompatible with cellular life. And even slight increases in their concentration interfere in all manner of subtle ways with the finely tuned chemical activities of the cell and are therefore inimical to life. This was already well known in Henderson's day, as he remarks:

> The hydrogen ion concentration exerts a marked influence upon the rate of progress of chemical reactions. Thus for example [the hydrolysis of cane sugar] into glucose and fructose is commonly accomplished by warming a solution . . . to which a little acid has been added. . . . This process depends upon the strength of the acid, or, according to the modern view, upon the concentration of hydrogen ions. . . . Reactions of this type in which carbohydrates, fats, proteins, and other substances take part, make up a large if not the largest fraction of all the processes of metabolism. . . . In the body, to be sure, such reactions are under the control of enzymes, but the concentration of hydrogen and hydroxyl ions is not less, rather more important for that reason. Besides retaining their direct influence upon the reaction, the ions exert an influence on the enzymes themselves.[17]

By maintaining the pH of their body fluids close to 7, living systems can carry out their chemical activities with minimal interference from these two highly reactive ions. Consequently, a pH of 7, which is the pH of pure water, is the most fit for life. The further the pH is from 7 (either higher or lower), the greater the deleterious influences of the H^+ and OH^- ions on the vital activities of the cell. Not only are most of the organic constituents of the cell most chemically stable in aqueous solutions around pH 7, but the weak chemical bonds which hold together the complex 3-D architecture of the cell are also most effective at pH values close to neutrality. DNA, for example, which is stabilized by hydrogen bonds, is most stable at around this pH. Those critical activities involving the reduction and oxidation of water that occur in oxidative metabolism and photosynthesis and which involve transitional metal atoms and electron currents are also optimal close to pH 7. Also, the vital properties of the calcium and phosphate ions can only be utilized at pH values near neutrality. Bone cannot be formed at pH values below 5, while at pH levels much above neutrality the insolubility of many calcium compounds, but particularly calcium phosphate, becomes prohibitive.[18]

Life based on the sophisticated manipulation of organic compounds is impossible in a strongly acid or alkaline condition—i.e., at very low or high pH levels. In such conditions the delicate structures of the cell would suffer rapid hydrolysis and breakdown. As Fraústo da Silva and Williams comment: "Virtually all biological cells have a cytoplasm which operates at pH = 7; in almost any environment, it is as if nature had selected a buffer with as low an OH^- and H^+ concentration as possible."[19] The choice of pH 7 is therefore no accident. It is to a large extent of necessity.

To maintain a pH close to 7 and ensure that chemical activities can proceed in optimal conditions necessitates the use of substances known as buffers. Buffers are therefore also an essential part of life's design.

A buffer is merely a compound which tends to resist changes in the level of acidity or H^+ ion concentration in a solution. To understand how a buffer works, in simple terms, we may think of the buffer as having two forms, an *acid form* and an *alkaline, or basic, form.* (Chemists refer to these two forms as the conjugate acid and the conjugate base.) When the level of H^+ ions rises as the solution becomes more acid, the alkaline form of the buffer soaks up the H^+ ions:

$$H^+ + \text{Buffer}^- = \text{Buffer.H}$$
$$\text{(basic form)} \quad \text{(acid form)}$$

Conversely, when the level of H^+ ions falls, the buffer gives off H^+ ions into the solution and the reaction moves to the left.

$$H^+ + \text{Buffer}^- = \text{Buffer.H}$$

Thus, through the activities of buffers, changes in the concentration of H^+ ions can be "buffered."

6. The Bicarbonate Buffer

The major buffer in the body is bicarbonate. Bicarbonate is an excellent buffer mainly because of the volatility of the gas carbon dioxide (CO_2). When acid accumulates and the concentration of H^+ ions increases, bicarbonate (the basic form) combines with the H^+ ions forming H_2CO_3 (carbonic acid—the acid form), which then dissociates into the two absolutely innocuous substances water (H_2O) and carbon dioxide (CO_2). Thus, the reaction below tends to move toward the right, driving up the concentration of CO_2 which leaves the body via the lungs. It is as if, almost by magic, the excess acid (H^+ ions) were simply drawn out of the body in the lungs.

$$H^+ + HCO_3^- = H_2CO_3 = H_2O + CO_2$$

If CO_2 were not a gas which can leave the body, its concentration would relentlessly increase, and eventually the reaction would tend to run in the reverse direction.

The bicarbonate buffer only works because carbon dioxide can be exhaled in the lungs. In effect, because of the volatility of CO_2, the acid is simply drawn out of the body via the lungs. Thus, by respiration the reaction can be moved permanently to the right.

$$H^+ + HCO_3^- = H_2CO_3 = H_2O + CO_2$$

By this wonderfully clever device even a large and potentially fatal accumulation of H^+ ions can be almost instantly removed from the body. Because in effect the H^+

ions are simply breathed out of the body, the speed and ability of this remarkable system to handle increases in the concentration of H^+ ions is unrivaled. However, there is an additional and marvelously adaptive feature of the bicarbonate buffer system. To grasp its significance, first we need to recall that the H^+ ion concentration or acidity of solutions varies over a very large range—from very high in acidic solutions to very low concentrations in alkaline solutions. The range is 14 orders of magnitude or 10^{14} (see diagram on page 407).

There are a vast number of different buffers known to chemists. Each buffer works best at a particular level of acidity, or H^+ ion concentration. One buffer may function optimally in an acid solution with a pH value of 3. Another might work best in an alkaline solution at a pH of 10, another in a solution of pH 5.5 and so on.

At the H^+ ion concentration, or pH, where a particular buffer functions optimally, the basic and acidic forms of the buffer are present in equal concentrations—they are in equilibrium. The optimal pH for a buffer is known as its pK_a value. The pK_a values for a number of buffers are given in the table on page 414.

As we have seen in the case of the bicarbonate system, the acid form is H_2CO_3 (carbonic acid), and the basic form is HCO_3^-, or bicarbonate, and the H^+ ion concentration (pH) at which these two forms are in equilibrium is about 6.4 in pure water and close to 6.1 in blood and other body fluids.

To function as a buffer in the body, ideally the concentration of the acid and basic forms of a buffer should be in equilibrium—i.e., should be present in equal concentrations at the level of acidity in the body fluids, which are close to pH 7. From this it would appear that the bicarbonate system is not ideally adapted to buffer the body fluids because it functions optimally at a pH level of 6.1, which is considerably lower than pH 7. This appears to be an anomaly but it isn't.

First, because of the volatility of CO_2 and the ease with which it can be eliminated from the body, the system can still function with great efficiency even at a pH of 7. In fact, as B. D. Rose points out in his *Clinical Physiology of Acid-Base and Electrolyte Disorders,* calculation shows that because of the ease with which the carbon dioxide (and with it, in effect, the hydrogen ions) can be breathed away, the buffering capacity of the bicarbonate system is in effect increased by between ten and twenty times (compared with an ordinary buffer) and is far more efficient than an ordinary buffer working at its pH optimum.[20]

There is yet another twist to this intriguing story, arising from the fact that the pK_a value of the bicarbonate buffer is approximately 6.1 in the body. It turns out that this pK_a value, rather than being "non-ideal," is in fact perfectly suited to defending the organism from the major challenge to pH homeostasis, which nearly always comes from an accumulation of acids rather than alkalis. The production of acids and the consequent generation of H^+ ions is an inevitable result of the oxidative metabolism of organic compounds because many of the intermediates on the path from sugar to CO_2 are acids. In fact, the first part of the main catabolic pathway involves oxidative rearrangements in the intermediary compounds, which can occur in the absence of oxygen, and when an organism is deprived of oxygen for any length of time, a considerable buildup of acid is bound to occur. Some organisms derive all their energy from this first part of the catabolic pathway; indeed, catabolism in the absence of oxygen, known as anaerobic metabolism, is vital even in organisms such as mammals, which derive most of their energy from the complete oxidative catabolism of carbon to H_2O and CO_2. Moreover, as Henderson noted: "In the main the foodstuffs are neutral substances, but their principal end products, except water, are almost exclusively acid compounds, . . . phosphoric acid, sulphuric acid, uric acid."[21]

Because the pK_a of the bicarbonate system is close to 6.1 in blood and body fluids, the organism is able to maintain a far higher concentration of bicarbonate ion in its body fluids than would be possible if its pK_a was actually a 7, and this in turn allows the organism to maintain a larger reserve of bicarbonate ion to soak up the acids generated during metabolism.

To understand why this is so, recall that at the H^+ ion concentration (pH) at which a buffer works with maximal efficiency, i.e., its pK_a value, the concentration of its acid and basic forms are equal. In the case of the bicarbonate system this occurs at a pH of 6.1.

However, at a pH of 7 the equilibrium is forced to the left and the relative concentration of bicarbonate is about twenty times greater than the concentration of carbonic acid.

This relatively high concentration of bicarbonate is precisely what is required to defend the body from the accumulation of acids arising from the metabolism of food or during anaerobic exertion. That the bicarbonate levels can be maintained at such a high level to soak up any excess acids is only possible because the pH optimum of the bicarbonate buffer is 6.1.

It turns out, therefore, that the bicarbonate system is exquisitely adapted to defend the body against increases in acidity. Because its optimum pH is considerably lower than 7, this allows a buildup of bicarbonate ions in the body to a much higher concentration than would be possible if the pH optimum of the system was 7. This is just what the body needs for defense against the accumulation of acids. However, this advantage can only be exploited in turn because, as we have seen, unlike an ordinary buffer, when the level of acid increases, the bicarbonate base is converted into the neutral gas carbon dioxide, which can be easily excreted from the body.

The bicarbonate buffer system is therefore anomalous on two counts: it functions far from its optimal pH, and on soaking up acid it generates a neutral gas which can be readily excreted from the body. Both these two anomalies are essential to the design of the whole system. If carbon dioxide was not a gas, the system would fail. If the pH optimum of the bicarbonate system was 7, the system would again fail. The system works only because the pH optimum of the system is very close to 6.1 in body fluids and only because the acid form is in equilibrium with the gas carbon dioxide.

Because the pH optimum of the bicarbonate buffer system is 6.1 rather than 7, this raises the amount of bicarbonate available to soak up acid tenfold or more, which, together with the ten-to-twentyfold increase in its efficiency over ordinary buffers consequent on the volatility of CO_2, provides a buffering system of unparalleled efficiency and one wonderfully fit to maintain the hydrogen ion concentration in the body fluids of a large terrestrial organism close to neutrality.

What is really astonishing about this ingenious buffer system is that there are few if any buffers which function optimally close to pH 7 suitable to replace the bicarbonate system in living organisms. (See table below, which lists the pH optima, i.e., the pK_a values, of a number of buffers). Note that of the few candidate buffers that have a pH optimum close to 7 and might conceivably be utilized by living things such as oxalic acid, benzoic acid, etc., nearly all can be excluded for various reasons.

Phosphoric acid has a pH optimum in the right range and is in fact utilized as a buffer by living organisms, but for a variety of reasons only a small proportion of the buffering can be handled by phosphate because many phosphate compounds are insoluble and levels of phosphate much higher than those which occur physiologically would lead to the precipitation of these compounds with various deleterious effects. We are led inexorably to the conclusion that bicarbonate is the only buffer available and, as with so many of the other constituents of life, is also ideally adapted for this very specific and absolutely critical role.

7. DNA Recognition by Proteins

Individual α helices are incapable of tight binding to a unique DNA sequence because the number of weak interactions between an individual α helix and a 4-base-long section of DNA, which is generally about 5 or 6, are insufficient to stabilize the binding. The recognition α helix which fits into the large groove of the DNA is generally about 10 amino acids long. The key recognition interactions usually involve about 3 of the amino acids in the recognition helix and these make a total of about 5 or 6 specific weak interactions, mainly hydrogen bonds with the bases in the DNA.

Strong binding between a protein and DNA requires approximately 10 to 18 weak interactions. Thus, from first principles it would seem that only by using a combination of 2 or more α helices can this number of weak bonds be achieved and a protein be constructed to bind firmly to a particular sequence of the DNA. And this is precisely what is found.

As Suzuki and Yagi point out:

> An α helix can bind to no more than five base pairs because of the curvature of the major groove; it can only access one side of the DNA. To recognize more than five base pairs, two or more helices are used in combination, essentially by relating the two by a twofold symmetry axis or repeating them in tandem. The classic helix turn helix proteins . . . use the symmetric arrangement, while the Zn Finger proteins, use a tandem arrangement.[22]

The ionization constants and pK_a values of a number of buffers.

TABLE AI-3. IONIZATION (DISSOCIATION) CONSTANTS AND pK_a VALUES OF SOME WEAK ACIDS

Conjugate Acid	MW	Conjugate Base	K_a	pK_a (= $-\log K_a$)
Acetic acid	60.05	Acetate^{-1}	1.75×10^{-5}	4.76
Ascorbic acid	176.12	Ascorbate^{-1}	8×10^{-5} (K_{a_1})	4.1 (pK_{a_1})
Ascorbate^{-1}		Ascorbate^{-2}	1.6×10^{-12} (K_{a_2})	11.79 (pK_{a_2})
Benzoic acid	122.12	Benzoate^{-1}	6.30×10^{-5}	4.20
Boric acid	61.84	Borate^{-1}	5.8×10^{-10}	9.24
n-Butyric acid	88.10	Butyrate^{-1}	1.5×10^{-5}	4.82
Carbonic acid	62.03	Bicarbonate^{-1}	4.31×10^{-7} (K_{a_1})	6.37 (pK_{a_1})
Bicarbonate^{-1}		Carbonate^{-2}	5.6×10^{-11} (K_{a_2})	10.25 (pK_{a_2})
Citric acid	192.12	Citrate^{-1}	8.7×10^{-4} (K_{a_1})	3.06 (pK_{a_1})
Citrate^{-1}		Citrate^{-2}	1.8×10^{-5} (K_{a_2})	4.74 (pK_{a_2})
Citrate^{-2}		Citrate^{-3}	4.0×10^{-6} (K_{a_3})	5.40 (pK_{a_3})
3,6-Endomethylene-1,2,3,6-tetrahydrophthalic acid,	182.2		5×10^{-5} (K_{a_1})	4.3 (pK_{a_1})
"EMTA"			1×10^{-7} (K_{a_2})	7.0 (pK_{a_2})
Formic acid	46.03	Formate^{-1}	1.77×10^{-4}	3.75
Fumaric acid	116.07	Fumarate^{-1}	9.3×10^{-4} (K_{a_1})	3.03 (pK_{a_1})
Fumarate^{-1}		Fumarate^{-2}	3.4×10^{-5} (K_{a_2})	4.47 (pK_{a_2})
Glycerophosphoric acid	172.08	Glycerophosphate^{-1}	3.4×10^{-2} (K_{a_1})	1.47 (pK_{a_1})
Glycerophosphate^{-1}		Glycerophosphate^{-2}	6.4×10^{-7} (K_{a_2})	6.19 (pK_{a_2})
Hippuric acid	179.17	Hippurate^{-1}	2.3×10^{-4}	3.64
Hydrocyanic acid	27.03	Cyanide^{-1}	4.9×10^{-10}	9.31
Hydrofluoric	20.01	Fluoride^{-1}	1×10^{-3}	3.00
Hydrogen sulfide	34.08	Hydrosulfide^{-1}	5.7×10^{-8} (K_{a_1})	7.24 (pK_{a_1})
Hydrosulfide^{-1}		Sulfide^{-2}	1.2×10^{-15} (K_{a_2})	14.92 (pK_{a_2})
Hydroquinone	110.11	Hydroquinone^{-1}	1.1×10^{-10}	9.96
N-2-hydroxyethylpiperazine-N′-2-ethanesulfonic acid, "HEPES"	238.3		2.82×10^{-8}	7.55
Itaconic acid	130.10	Itaconate^{-1}	1.46×10^{-4} (K_{a_1})	3.84 (pK_{a_1})
Itaconate^{-1}		Itaconate^{-2}	2.8×10^{-6} (K_{a_2})	5.55 (pK_{a_2})
Lactic acid	90.08	Lactate^{-1}	1.39×10^{-4}	3.86
Maleic acid	116.07	Maleate^{-1}	1.0×10^{-2} (K_{a_1})	2.0 (pK_{a_1})
Maleate^{-1}		Maleate^{-2}	5.5×10^{-7} (K_{a_2})	6.26 (pK_{a_2})
Malic acid	134.09	Malate^{-1}	4×10^{-4} (K_{a_1})	3.40 (pK_{a_1})
Malate^{-1}		Malate^{-2}	9×10^{-6} (K_{a_2})	5.05 (pK_{a_2})
Malonic acid	104.06	Malonate^{-1}	1.4×10^{-3} (K_{a_1})	2.85 (pK_{a_1})
Malonate^{-1}		Malonate^{-2}	8.0×10^{-7} (K_{a_2})	6.10 (pK_{a_2})
2-(N-morpholino)-ethane sulfonic acid, "MES"	195.2		7.06×10^{-7}	6.15
Nitrous acid	47.02	Nitrite^{-1}	4×10^{-4}	3.40
Oxalic acid	126.07	Oxalate^{-1}	6.5×10^{-2} (K_{a_1})	1.19 (pK_{a_1})
Oxalate^{-1}		Oxalate^{-2}	6.1×10^{-5} (K_{a_2})	4.21 (pK_{a_2})
Phenol	94.11	Phenolate^{-1}	1.3×10^{-10}	9.89
Phosphoric acid (H_3PO_4)	98.00	$H_2PO_4^{-1}$	7.5×10^{-3} (K_{a_1})	2.12 (pK_{a_1})
$H_2PO_4^{-1}$		HPO_4^{-2}	6.2×10^{-8} (K_{a_2})	7.21 (pK_{a_2})
HPO_4^{-2}		PO_4^{-3}	4.8×10^{-13} (K_{a_3})	12.32 (pK_{a_3})

From I. H. Segal, *Biochemical Calculations;* © 1968 by John Wiley & Sons, New York. Table *AI-3*. Reprinted by permission of the publisher.

The diagram below illustrates the two different ways in which combinations of helices are used to recognize a DNA binding site.

The symmetrical arrangement exploits the symmetry inherent in the DNA double helix. Note that the sequence looked at from the right is identical to that looked at from the left. As Mark Ptashne puts it in his short monograph *A Genetic Switch:* "a tiny demon standing at the middle of the target sequence facing right and then left would see identical corridors of chemical groups."[23] The proteins which recognize these symmetrical target sequences are invariably dimers consisting of two identical subunits or monomers. One subunit recognizes via its recognition helix the base sequence on the left-hand side of the target sequence, and the other recognizes the identical sequence in the right half of the target.

In the symmetrical arrangement the recognition helix of one subunit might bind to the bases GCAT in the large groove of the DNA on the left side of the target while the helix of the other subunit will bind to the identical bases GCAT in the large groove on the right side of the target. In addition to possessing a recognition

The symmetrical arrangement of two recognition α helices.

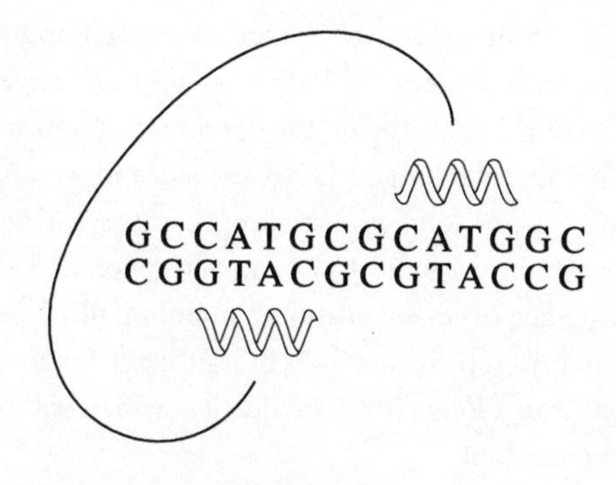

The tandem arrangement of recognition α helices.

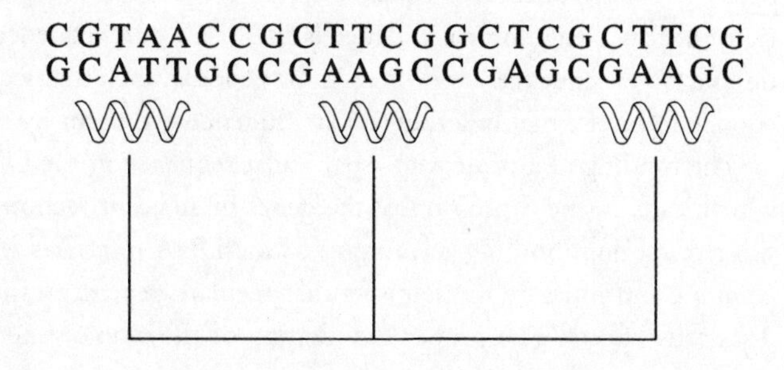

helix which recognizes and binds to the DNA, each subunit also possesses the capacity to recognize and bind to another subunit forming the active dimer. The 2 dimers exhibit cooperativity when binding such that the binding of one subunit enhances the strength of the binding of the other subunit—so the dimer binds more strongly than either of its constituent monomers individually. Thus only 1 identical protein is required to recognize what are in effect 2 different but sterically identical binding sites placed back-to-back in the DNA.

As mentioned above, effective binding between a protein and DNA necessitates about 10 to 18 weak bonds, and the use of 2 or 3 α helices can achieve this number. Much more than about 18 weak bonds and the binding is too strong. This means that the longest DNA sequence that can be recognized by an individual protein recognition system is about 15 to 16 bases long. As Mark Ptashne comments in *A Genetic Switch,* attempting to double the size of the DNA-protein recognition complex will not work. By doubling the size of the DNA sequence and its recognition protein we have increased the specificity of binding by many orders of magnitude. But now we run into another problem—the absolute strength of binding is now so high that the recognition proteins spend most of their time bound nonspecifically to random DNA sequences, i.e., on nonspecific sites. Calculation shows that our hypothetical oversized protein will never find its target sequence during a bacterial cell generation of 2.5 hours. And there is another problem with our oversized repressor. Its binding is so strong that once it has bound it will stay on virtually forever. Because the energy of binding would be so large, the repressor would have to be drastically altered while on the operator to remove it from the operator.[24]

The fact that the longest DNA sequence to which a protein can reversibly bind is about 15 bases long is intriguing because a simple mathematical calculation shows that in the genome of a mammal ("genome" is the term used for the complete DNA sequence of an organism, containing all its genes and hereditary information), which is approximately 3 billion bases long—about one meter in length if pulled out into a long string—sequences which will occur by chance only once are about 15 bases long.

Note that DNA sequences 5 bases long occur by chance once in a sequence about 4^5 or 1,000 bases long, DNA sequences 10 bases long occur by chance once in a sequence about 4^{10} or 1 million bases long, and DNA sequences 15 bases long occur by chance once in a sequence about 4^{15} or 1 billion bases long. The significance of this intriguing coincidence is that if a DNA sequence is to function as a unique target sequence somewhere in the genome, then that sequence must be long enough to ensure that no other similar sequence will occur by chance.

The binding of a protein to a particular sequence in the DNA is the key event which leads to the expression of the genes in adjacent sections of the DNA. Unwanted, random binding of proteins to the DNA interferes with the enormously complex and intricate mechanisms that regulate gene expression and would have disastrous effects. The proper functioning of the whole genetic data bank is absolutely dependent on the recognition proteins, binding strongly *only to their proper target sequences.*

If each target sequence used to label or address particular regions of the genome were not unambiguously unique, then chaos would inevitably ensue. The genome would resemble a filing cabinet with the same labels on different drawers. Hence the necessity that target sequences in the DNA of higher organisms like mammals be 15 to 16 bases long, for only sequences of this length will not occur again in the genome by chance, causing unwanted binding of the recognition protein to a non-target area of the DNA.

The use of the four bases in DNA means that the required length of 15 bases for a unique label can be met even in the 3-billion-base-long genome of a mammal. From these considerations it is clear that in a theoretical mammalian genome 3 billion bases long, constructed out of DNA composed of 2 bases, where unique target sequences would have to be, of necessity, 20 to 30 bases long (which is twice as long as in DNA composed of four bases), given the existing energy levels of weak bonds, functional repressors designed to recognize such superlong sequences would not be feasible, as they would bind irreversibly to the DNA. Clearly, a hypothetical two-base system would therefore be of little utility for protein-DNA recognition, certainly less than the existing four-base system.

A unique target sequence 15 bases long in a theoretical genome constructed out of only 2 bases would occur once by chance in about 130,000 bases—the length of the genomes of many viruses. This means that to use proteins as devices for recognizing unique DNA sequences in a 2-base genetic system we will be restricted to genomes ten thousand times smaller than that of a mammal—too small to construct organisms of any complexity.

Ptashne's comment raises again the significance, discussed in chapters 5 and 8, of the energy levels of noncovalent bonds. If these had been, say, twice as strong, then proteins designed specifically to bind to unique sequences of DNA 10 to 15 bases long—i.e., long enough to function as unique target sequences in the genome—would bind to their target sequences irreversibly and would have been of no utility to the cell. This would mean that the regulation of gene expression, which is dependent on the ability of proteins to recognize and bind reversibly to unique DNA recognition sequences adjacent to a gene, would be greatly constrained.

Here is another case where the actual energy level of the weak interactions are critical to an absolutely vital biological function. Not only are they set just right to confer on proteins their unique "metastability" and their consequent ability to undergo allosteric transitions, but they are also set at precisely the level needed for the reversible binding of proteins to unique target sequences between 10 and 15 bases long in the DNA composed of four bases.

A 2-base genome would be disadvantageous in other ways too. As we have seen in chapter 8, an α helix can feel only 4 contiguous bases in the DNA. As we also saw, a 4-base system provides 256 unique quadruplets for a helix recognition. A 2-base system would provide only 16 different quadruplets for a helical recognition, a number surely far too small to allow anything remotely approaching the sophistication of the genetic regulation possible with the existing system.

We have already considered the protein coding consequences of constructing DNA of 6 instead of 4 bases. As we saw in chapter 7, the use of 6 bases to code for 20 amino acids would necessitate what would appear to be a number of less elegant alternative solutions to the amino acid coding problem. What might be the consequences for protein-DNA recognition of having six bases instead of four?

In a large mammalian genome 3 billion bases long composed of 6 bases, a unique sequence will still need to be about 12 bases long, which would still require the same number of α helices, approximately 2 or 3 for protein recognition. As an α helix can only feel a DNA sequence 4 bases long, the additional 2 bases would provide only a very slight advantage over the existing system. And against any very minimal advantage that might be gained, a 6-base system would require the additional burden and complexity of synthesizing and maintaining a readily available pool of 6 rather than only 4 bases for the construction of DNA.

Notes

Note to the Reader

1. P. C. W. Davies (1982) *The Accidental Universe* (Cambridge: Cambridge University Press). See also J. D. Barrow and F. J. Tipler (1986) *The Anthropic Cosmological Principle* (Oxford: Oxford University Press).
2. Davies (1982); see Preface.
3. P. C. W. Davies (1995) "Physics and the Mind of God," the Templeton Prize Address, *First Things,* August-September, pp. 31–35.
4. P. C. W. Davies (1995) *Are We Alone?* (London: Penguin Books), pp. 70, 85.
5. Ibid.; see p. 25.
6. P. C. W. Davies (1992) *The Mind of God* (London: Penguin), p. 232.
7. F. Hoyle and C. Wickramasinghe (1996) (London: Orino Books); see chap. 4.
8. S. L. Miller and L. E. Orgel (1996) *The Origins of Life on the Earth* (Englewood Cliffs, N.J.: Prentice-Hall); see chap. 15.
9. S. V. W. Beckwith and A. I. Sargent (1996) "Circumstellar Discs and the Search for Neighbouring Planetary Systems," *Nature* 383:139–144.
10. W. Paley (1807) *Natural Theology* (London: Faulder & Son).
11. L. J. Henderson (1958) *The Fitness of the Environment* (Boston: Beacon Press).
12. D'Arcy W. Thompson (1952) *On Growth and Form,* 2nd ed. (Cambridge: Cambridge University Press).
13. G. Wald (1964) "The Origins of Life," *Proceedings of the National Academy of Sciences of the United States of America* (hereafter *Proc. Natl. Acad. Sci. USA*), 52:594–611, see pp. 600–601.
14. A. E. Needham (1965) *The Uniqueness of Biological Materials* (Oxford: Pergamon Press), pp. 9–10.
15. C. F. A. Pantin (1968) *The Relations Between the Sciences,* eds. A. M. Pantin and W. H. Thorpe (Cambridge: Cambridge University Press); see Appendix 1 entitled "Life and the Conditions of Existence," pp. 129–154, and see pp. 148–152. See also C. F. A. Pantin (1951) "Organic Design," *Advancement of Science* 8(29):138–150.
16. S. Kauffman (1995) *At Home in the Universe* (New York: Oxford University Press).
17. Ibid., p. 112.
18. Ibid., p. 92.
19. C. de Duve (1995) *Vital Dust* (New York: Basic Books).

20. Ibid., p. 301.
21. K. E. Yandell (1986) "Protestant Theology and Natural Science in the Twentieth Century," in *God and Nature,* ed. D. C. Lindberg and R. L. Numbers (Berkeley: University of California Press), pp. 448–471; see pp. 468–469.
22. A. Peacocke (1979) *Creation and the World of Science* (Oxford: Oxford University Press). See also A. Peacocke (1993) *Theology for a Scientific Age* (London: SCM Press); and D. J. Bartholomew (1984) *God of Chance* (London: SCM Press).
23. Peacocke (1993), p. 119.

Prologue

1. A. J. Gurevich (1985) *Categories of Medieval Culture* (London: Routledge & Kegan Paul), pp. 57, 61.
2. Ibid., p. 57.
3. From Bacon's *De sapientia veterum,* cited in J. D. Barrow and F. J. Tipler (1986) *The Anthropic Cosmological Principle* (Oxford: Oxford University Press), p. 48.
4. B. V. Subbarayappa (1989) "Indian Astronomy: An Historical Perspective," in *Cosmic Perspectives,* ed. S. K. Biswas, D. C. V. Mallik, and C. V. Vishveshwara (Cambridge: Cambridge University Press), pp. 25–39; see p. 25.
5. W. Theodore de Bary, W. Chan, and B. Watson (1960) *Sources of Chinese Tradition* (New York: Columbia University Press), pp. 518–519.
6. Barrow and Tipler, op. cit., pp. 92–93.

Chapter 1: The Harmony of the Spheres

1. C. Sagan (1985) *Cosmos* (New York: Ballantine Books), p. 192.
2. J. D. Barrow and F. J. Tipler (1986) *The Anthropic Cosmological Principle* (Oxford: Oxford University Press), see chap. 1.
3. H. Ross (1989) *The Finger of God* (Orange, Calif.: Promise Publishing Co.), p. 127.
4. P. C. W. Davies (1982) *The Accidental Universe* (Cambridge: Cambridge University Press), p. 118.
5. Ibid.
6. V. Trimble (1977) "Cosmology: Man's Place in the Universe," *American Scientist* 65: 76–86.
7. Davies (1982), p. 39.
8. J. Boslough (1985) *Stephen Hawking's Universe* (New York: Quill), p. 101.
9. Ibid.
10. For a discussion of the sorts of universe that would result if the constants were different, see J. R. Gribbin and M. J. Rees (1989) *Cosmic Coincidences* (New York: Bantam Books), chap. 10, pp. 241–269. See also Trimble, op. cit.
11. Davies (1982), Preface.
12. S. J. Dick (1982) *Plurality of Worlds* (Cambridge: Cambridge University Press), p. 61.
13. I. B. Cohen (1958) *Isaac Newton's Papers and Letters on Natural Philosophy* (Cambridge: Cambridge University Press), pp. 360–361.
14. P. C. W. Davies (1987) *The Cosmic Blueprint* (London: Penguin), p. 203.
15. S. J. Gould (1989) *Wonderful Life* (New York: Norton), p. 291.
16. Ibid., p. 320.
17. Ibid., p. 323.

Chapter 2: The Vital Fluid

1. I. B. Cohen (1958) *Isaac Newton's Papers and Letters on Natural Philosophy* (Cambridge: Cambridge University Press), pp. 381–382.
2. See A. E. Needham (1965) *The Uniqueness of Biological Materials* (Oxford: Pergamon Press), pp. 9–10. See also N. V. Sidgwick (1937) "Molecules," *Science* 86:335– 340, for discussion of the conditions for life.
3. J. Von Neumann (1966) *Theory of Self-reproducing Automata,* ed. W. A. Burks (Urbana: University of Illinois Press); see p. 82.
4. W. Paley (1807) *Natural Theology* (London: Faulder & Son), p. 401. See also Paley's discussion on p. 406, which illustrates the difficulty encountered at the time in alluding to properties of water that were specially beneficial to life.
5. W. Whewell (1871) *Astronomy and General Physics Considered with Reference to Natural Theology,* 8th ed. (London: Bohn).
6. Ibid., pp. 70–72.
7. Ibid., p. 78.
8. Ibid., p. 79.
9. Ibid., pp. 123–125.
10. L. J. Henderson (1958) *The Fitness of the Environment* (Boston: Beacon Press).
11. J. Needham (1929) *The Sceptical Biologist* (London: Chatto & Windus), pp. 211–218.
12. H. J. Morowitz (1987) *Cosmic Joy and Local Pain* (New York: Scribner), pp. 99–107.
13. Henderson, op. cit.; see Preface.
14. Ibid., pp. 99–100.
15. Ibid., p. 106. See also F. Franks (1972) "Water, the Unique Chemical," in *Water: A Comprehensive Treatise,* vol. 1 (New York: Plenum Press), p. 488.
16. Henderson, op. cit., pp. 86–89.
17. Ibid.
18. Ibid., p. 105.
19. Ibid., p. 106.
20. Ibid., pp. 126–127.
21. A. E. Needham, op. cit., p. 11.
22. Franks, op. cit., p. 20.
23. Henderson, op. cit., p. 111.
24. Franks, op. cit., p. 20.
25. Henderson, op. cit., pp. 112–115.
26. Ibid., pp. 113–115.
27. A. E. Needham, op. cit., p. 23.
28. Henderson, op. cit., p. 79.
29. A. E. Needham, op. cit., p. 12.
30. K. Schmidt-Nielsen (1975) *Animal Physiology* (Cambridge: Cambridge University Press), p. 671.
31. H. Davson and J. F. Danielli (1952) *The Permeability of Natural Membranes* (Cambridge: Cambridge University Press), pp. 51–52.
32. H. Stern and D. L. Nanney (1965) *The Biology of Cells* (New York: Wiley), p. 77.
33. Schmidt-Nielsen, op. cit., p. 671.
34. As M. W. Clark (1948) *Topics of Physical Chemistry* (Baltimore: Williams & Wilkins), p. 128, explains, "The slowness of diffusion over long distances and its great rapidity over short distances is, as the physiologist A. V. Hill who carried out pioneering work on diffusion in the thirties pointed out: 'the basis of the capillary circulation and therewith the whole design of the larger animals.'"

35. Ibid., p. 146.
36. Schmidt-Nielsen, op. cit., p. 143.
37. Ibid., p. 44.
38. M. Reiner (1959) "The Flow of Matter," *Scientific American* 201(6):122–137.
39. G. Ranalli (1987) *Rheology of the Earth* (Boston: Allen & Unwin), p. 71. A. Holmes (1965) *Principles of Physical Geology* (London: Nelson), pp. 61–62.
40. R. C. Weast and M. J. Astle (1980) *CRC Handbook of Chemistry and Physics,* 61st ed., (Boca Raton, Fla.: CRC Press). The densities cited are all taken from this source.
41. Morowitz, op. cit., pp. 152–153.
42. Ibid., 154.
43. A. E. Needham, op. cit., p. 22.
44. Henderson, op. cit., pp. 130–131.

Chapter 3: The Fitness of the Light

1. C. Sagan (1985) *Cosmos* (New York: Ballantine), p. 199.
2. Ibid., p. 199.
3. I. M. Campbell (1977) *Energy and the Atmosphere* (London: Wiley), pp. 1–2.
4. G. Wald (1959) "Life and Light," *Scientific American* 201(4):92–108.
5. K. L. Coulson (1975) *Solar and Terrestrial Radiation* (New York: Academic Press), from fig. 3.1, p. 40.
6. Campbell, op. cit., p. 1.
7. *Encyclopaedia Britannica* (1994), 15th ed., vol. 18, p. 200, fig. 5.
8. Ibid., from fig. 3, p. 198.
9. Ibid., p. 198.
10. Ibid., p. 203.
11. J. A. Maclaughlin, R. R. Anderson, and M. F. Holick (1982) "Spectral Character of Sunlight Modulates Photosynthesis of Previtamin D3 and its Photoisomers in Human Skin," *Science* 216:1001–1003.
12. K. L. Bushaw et al. (1996) "Photochemical Release of Biologically Available Nitrogen from Aquatic Dissolved Organic Matter," *Nature* 381:404–407.
13. K. Ya. Kondratyev (1969) *Radiation in the Atmosphere* (New York: Academic Press), p. 123.
14. T. Fenchel and B. J. Finlay (1995) *Ecology and Evolution in Anoxic Worlds* (Oxford: Oxford University Press). See also N. R. Pace (1997) "A Molecular View of Microbial Diversity and the Biosphere," *Science* 276:734–740.
15. H. B. Barlow (1964) "The Physical Limitations of Visual Discrimination," in *Photophysiology,* ed. A. C. Giese, vol. 2, pp. 163–202; see esp. 187–197, and table on p. 189 and discussion on p. 197.
16. H. B. Barlow (1981) "Critical Limiting Factors in the Design of the Eye and the Visual Cortex," *Proceedings of The Royal Society of London, Series B* 212:1–34; see p. 4.
17. J. R. Brobeck (1979) *Best and Taylor's Physiological Basis of Medical Practice* (Baltimore: Williams & Wilkins), pp. 8–67.
18. M. D. Levine (1985) *Vision in Man and Machine* (New York: McGraw-Hill); see p. 61.
19. S. Duke-Elder (1958) *System of Ophthalmology* (London: Henry Kimpton), pp. 450–451.
20. Barlow (1981), p. 5. See also Brobeck, op. cit., pp. 8–71.
21. M. A. Ali and M. A. Klyne (1985) *Vision in Vertebrates* (New York: Plenum Press), p. 153.
22. Barlow (1981), pp. 4–5.
23. A. Despopoulous and S. Silbernagl (1991) *Color Atlas of Physiology* (New York: Thieme Medical Publishers), p. 306.

24. B. Alberts et al. (1989) *Cell,* 2nd ed. (New York: Garland Publishing); see pp. 1104–1107.
25. Despopoulos and Silbernagl, op. cit.
26. Alberts et al., op. cit.
27. N. Henbest and M. Marten (1983) *The New Astronomy* (Cambridge: Cambridge University Press). M. A. Mitton (1977) *The Cambridge Encyclopaedia of Astronomy* (London: Jonathan Cape). See also J. M. Pasachoff and M. L. Kutner (1978) *University Astronomy* (Philadelphia: W. B. Saunders).
28. *Encyclopaedia Britannica,* vol. 18, p. 203.
29. Ibid., pp. 196–197.
30. Henbest and Marten, op. cit., p. 8.

Chapter 4: The Fitness of the Elements and the Earth

1. J. J. R. Fraústo da Silva and R. J. P. Williams (1991) *The Biological Chemistry of the Elements* (Oxford: Oxford University Press), pp. 3–4. P. A. Cox (1995) *The Elements on Earth* (Oxford: Oxford University Press).
2. Fraústo and Williams, op. cit., p. 5.
3. C. Ponnamperuma (1983) "Cosmochemistry and the Origin of Life," in *Cosmochemistry and the Origin of Life,* ed. C. Ponnamperuma (Dordrecht, Holland: Reidel), chap. 1, fig. 6.
4. J. T. Edsall and J. Wyman (1958) *Biophysical Chemistry,* vol. 1 (New York: Academic Press), chap. 1, fig. 2.
5. Cox, op. cit., pp. 183–186.
6. S. Levy (1993) *Artificial Life* (London: Penguin), pp. 87–120.
7. F. Press and R. Siever (1986) *Earth* (New York: W. H. Freeman), p. 11.
8. Ibid., p. 366.
9. Ibid., p. 367.
10. Ibid., p. 366.
11. Ibid., pp. 11–12.
12. Ibid., pp. 13–14.
13. Ibid., p. 14.
14. Ibid., pp. 367–368.
15. H. Rymer (1996) "Book Reviews," *Nature* 383:684.
16. Ibid., p. 122.
17. Yi-Fu Tuan (1968) *The Hydrologic Cycle and the Wisdom of God* (Toronto: University of Toronto Press), p. 4.
18. Ibid., pp. 4–5.
19. Press and Siever, op. cit., p. 479.
20. W. D. Parkinson (1983) *Introduction to Geomagnetism* (Edinburgh: Scottish Academic Press), pp. 356–357. M W. McElhinney (1973) *Paleomagnetism and Plate Tectonics* (Cambridge: Cambridge University Press), pp. 146–147.
21. N. C. Brady and R. R. Weill (1996) *The Nature and Properties of Soils* (Englewood Cliffs: Prentice Hall), pp. 242, 270.
22. Ibid., p. 241.
23. Ibid., p. 270.
24. C. F. Ugolini and H. Spaltenstein (1992) "Pedosphere," in *Global Biogeochemical Cycles,* ed. S. S. Butcher et al. (London: Academic Press), pp. 123–153; see p. 128.
25. A. Holmes (1965) *Principles of Physical Geology* (London: Nelson), p. 63.
26. J. E. Fergusson (1982) *Inorganic Chemistry and the Earth* (Oxford: Pergamon Press); see chap. 7.
27. J. G. Hering and W. Stumm (1990) "Oxidative and Reductive Dissolution of Minerals," *Reviews in Mineralogy* 23:427–465; see p. 456.

28. R. Siever (1983) "The Dynamic Earth," *Scientific American* 249(3), pp. 30–39.
29. R. A. Berner (1995) "Chemical Weathering and Its Effect on Atmospheric CO_2 and Climate," *Reviews in Mineralogy* 31:565–583; see pp. 567–568.
30. Ibid., p. 570. As Berner points out: "The importance of mountain uplift to weathering of silicates has been emphasized recently. . . . The idea is that uplift results in rugged relief, and cold temperatures at high elevation. The rugged relief enhances physical erosion and the removal of protective covers of highly weathered clay residues, allowing greater exposure of primary silicates to chemical weathering. Cold temperatures result in greater physical weathering due both to freeze-thaw and to the grinding where glaciers are present. The enhanced physical weathering exposes more surface area of the primary minerals to the weathering solutions. In addition, mountains can bring about enhanced rainfall due to orographic effects resulting in greater flushing of rocks by water. All these factors should have brought about greater weathering of silicate minerals during geologic periods when the extent of high mountains was globally more important. An example is the late Cenozoic when the uplift of the Himalayan/Tibetan system occurred."
31. J. E. Lovelock (1987) *Gaia: A New Look at Life on Earth* (Oxford: Oxford University Press); see Preface.
32. Ibid., p. 34.
33. D. Attenborough (1995) *The Private Life of Plants* (London: BBC Books), p. 70.
34. E. Pennisi (1995) "The Secret Language of Bacteria," *New Scientist,* September 16, pp. 30–33.
35. Press and Siever, op. cit., p. 4. See also G. Wald (1964) "The Origins of Life," *Proc. Natl. Acad. Sci. USA* 52:594–611; see pp. 600–601. In Wald's words: "Those conditions almost surely involve a planet somewhat resembling the Earth, of about this size and temperature, and receiving about this quantity of radiation from its sun. To mention a few points of the argument: a much smaller planet could not hold an adequate atmosphere, a much larger one might hold too dense an atmosphere to permit radiation to penetrate to its surface. Too cold a planet would slow down too greatly the chemical reactions by which life arises; too warm a planet would be incompatible with the orderly existence of macromolecules. The limits of temperature are probably close to those at which water remains a liquid, itself almost surely a necessary condition for life."
36. Press and Siever, op. cit., p. 4.
37. Ibid.
38. J. S. Kargel and R. G. Strom (1996) "Global Climatic Change on Mars," *Scientific American* 275 (5):80–85; see p. 80.
39. Ibid., p. 82.
40. A. Henderson-Sellers (1986) "The Evolution of the Earth's Atmosphere," in *The Breathing Planet,* ed. J. Gribbin (Oxford: Basil Blackwell), pp. 19–26; see p. 21.
41. Ibid., pp. 21–22.
42. J. Davies (1995) "Searching for Alien Earth," *New Scientist,* May 13, pp. 24–28. See also *Science* 267 (1995):1273, and G. A. H. Walker (1996) "A Solar System Next Door," "News and Views," *Nature* 382:23–24.
43. Noam Stoker cited by M. Chown, *New Scientist* 155, no. 2091 (1997): 21. See also S. V. W. Beckwith and A. I. Sargent (1996) "Circumstellar Disks and the Search for Neighbouring Planetary Systems," *Nature* 383:139–144.
44. C. Sagan (1980) *Cosmos* (New York: Ballantine); see p. 177.
45. See "Worlds Around Other Stars Shake Planet Birth Theory," in the "Research News" section, *Science* 276:1336–1339.
46. Ibid.; see p. 144.
47. G. W. Wetherill (1995) "How Special is Jupiter?" *Nature* 373:470. See also G. W.

Wetherill (1993) "Our Friend Jove," *Discover,* July, p. 15. See also M. A. Corey (1995) *The Natural History of Creation* (Boston: University Press of America); see p. 69.
48. Wetherill (1993).

Chapter 5: The Fitness of Carbon

1. R. E. D. Clark (1961) *The Universe: Plan or Accident?* (London: Paternoster Press), p. 98.
2. Ibid., p. 119.
3. H. T. Pledge (1966) *Science Since 1500* (London: Her Majesty's Stationery Office), pp. 124–125.
4. W. Prout (1855) *Chemistry, Meteorology, and the Function of the Digestion,* 2nd ed. (London: Bohn), p. 6. Prout was well-known in his day as the first to propose the notion that the atomic weights of all the elements were multiples of the atomic weight of hydrogen, the first to divide foods into sugars, fats, and proteins, and as the discoverer of hydrochloric acid in the stomach.
5. L. J. Henderson (1958) *The Fitness of the Environment* (Boston: Beacon Press), pp. 193–194.
6. N. V. Sidgwick (1950) *The Chemical Elements and their Compounds,* vol. 1 (Oxford: Oxford University Press), p. 490. As Sidgwick explains, the reason for the stability of carbon compounds is that "in the first place the typical four-covalent state of the carbon atom is one in which all the formal elements of stability are combined. It has an octet, a fully shared octet, an inert gas number, and in addition, unlike all the other elements of the group, on octet which cannot increase beyond 8, since 4 is the maximum covalency possible for carbon. Hence the saturated carbon atom cannot co-ordinate either as donor or as acceptor, and since by far the commonest method of reaction is through co-ordination, carbon is necessarily very slow to react and even in a thermodynamically unstable molecule may actually persist for a long time unchanged. More than 50 years ago Victor Meyer drew attention to the characteristic inertness (*Tragheit*) of carbon in its compounds, and there is no doubt that this is its main cause."
7. Ibid.
8. Ibid.
9. G. Wald (1964) "The Origins of Life," *Proc. Natl. Acad. Sci. USA* 52:595–611; see p. 603: "Silicon . . . forms looser, less stable compounds . . . (and) silicon chains . . . are susceptible to attack by molecules possessing lone pairs of electrons, in part because of their more open structure but still more because silicon, a third period element, possesses 3*d* orbitals available for further combination. . . . Silicon however has another fatal disability, its failure to form multiple bonds." As Wald explains, silicon dioxide molecules bond with each other, forming the long polymers which make up quartz, the major constituent of most rocks. Silicon is "fit for making rock," while carbon is "fit for life."
10. Sidgwick, op. cit., p. 490. A. E. Needham (1961) *The Uniqueness of Biological Materials* (London: Pergamon Press), p. 32.
11. J. B. S. Haldane (1954) "The Origin of Life," *New Biology* 16:12–27.
12. Needham, op. cit., p. 30.
13. Ibid., pp. 32–33.
14. Henderson, op. cit., pp. 220–221, 235–237.
15. J. Yudkin (1985) *The Penguin Encyclopaedia of Nutrition,* (London: Penguin), p. 98.
16. T. Hoyem and O. Kvale (1977) *Physical, Chemical, and Biological Changes in Food Caused by Thermal Processing* (London: Applied Science Publishers), pp. 185–201.
17. H. R. White (1984) "Hydrolytic Stability of Biomolecules at High Temperatures and Its Implication for Life at 250°C," *Nature* 310:430–432. See also H. Bernhardt, D. Lude-

man, and R. Jaenicke (1984) "Biomolecules Are Unstable Under Black Smoker Conditions," *Naturwissenschaften*, 71:583–586.

18. S. L. Miller and L. E. Orgel (1974) *The Origins of Life on the Earth* (Englewood Cliffs, N.J.: Prentice-Hall); see chap. 9 on the stability of organic compounds.
19. H. Eyring, R. P. Boyce, and J. D. Spikes (1960) "Thermodynamics of Living Systems," in *Comparative Biochemistry*, vol. 1, ed. M. Florkin and H. S. Mason (New York: Academic Press), pp. 60–62.
20. Clark, op. cit., p. 59.
21. Wald, op. cit., p. 605.
22. L. M. Lederman (1989) *From Quarks to the Cosmos* (New York: Scientific American Library), p. 152.
23. F. Hoyle (1954) "Ultrahigh Temperatures," *Scientific American* 191(3): 145–154.
24. The bonds are: ionic bonds, van de Waals forces, hydrogen bonds, and the hydrophobic force. A detailed description of these bonds can be found in any major textbook of biochemistry.
25. J. Walker (1981) "The Physics and Chemistry of the Lemon Meringue Pie," *Scientific American* 244(6):154–159; see pp. 154–155.
26. G. N. Somero (1995) "Proteins and Temperature," *Annual Review of Physiology* 57:43–68; see p. 61.

Chapter 6: The Vital Gases

1. T. Fenchel and B. J. Findlay (1995) *Ecology and Evolution in Anoxic Worlds* (Oxford: Oxford University Press). See also N. Pace (1997) "A Molecular View of Microbial Diversity and the Biosphere," *Science* 276:734–740.
2. Fenchel and Findlay, op. cit. See also Pace, op. cit. One interesting group of bacteria, the methanogenic bacteria, derive energy from the reduction of carbon dioxide by hydrogen to produce methane, i.e., $CO_2 + H = CH_4 = H_2O$.
3. Fenchel and Findlay, op. cit.; see chap. 2, pp. 62–63, and chap. 5.
4. N. V. Sidgwick (1950) *The Chemical Elements and Their Compounds*, vol. 1 (Oxford: Oxford University Press), pp. 1124–1129.
5. L. J. Henderson (1958) *The Fitness of the Environment* (Boston: Beacon Press), pp. 247–248.
6. J. E. Lovelock (1987) *Gaia* (Oxford: Oxford University Press), p. 71.
7. Ibid.
8. B. Halliwell and J. M. C. Gutteridge (1990) *Methods in Enzymology* 186:1–88; see p. 1.
9. M. J. Green and A. O. Hill (1984) *Methods in Enzymology* 105:1–21.
10. J. Needham (1970) *The Chemistry of Life* (Cambridge: Cambridge University Press), pp. 30–33.
11. L. L. Ingraham (1966) "Enzymic Activation of Oxygen," in *Comprehensive Biochemistry*, ed. M. Florkin and E. H. Stotz, vol. 14 (Amsterdam: Elsevier), pp. 424–446; see p. 424.
12. Sidgwick, op. cit., p. 490.
13. *Van Nostrand's Scientific Encyclopedia* (1995), 8th ed., vol. 2, p. 2321. G. L. Pollack (1991) "Why Gases Dissolve in Liquids," *Science* 251:1323–1330; see p. 1323. A. Krogh (1941) *The Comparative Physiology of Respiratory Mechanisms* (Philadelphia: University of Pennsylvania Press); see p. 12, table 5.
14. Schmidt-Nielsen, op. cit., pp. 676–677. Krogh, op. cit., pp. 5–6.
15. G. N. Ling (1967) "Effects of Temperature on the State of Water in the Living Cell," in *Thermobiology*, ed. A. H. Rose (New York: Academic Press), pp. 5–24; see fig. 2, p. 10. L. Watson (1988) *The Water Planet* (New York: Crown Publishers); see p. 132.

16. I. Fridovich (1976) "Oxygen Radicals, Hydrogen Peroxide, and Oxygen Toxicity" in *Free Radicals in Biology*, ed. W. A. Pryor, vol. 1 (New York: Academic Press), pp. 239–277.
17. Ibid., pp. 239–240.
18. V. B. Mountcastle (1968) *Medical Physiology*, vol. 1 (St. Louis: C. V. Mosby), p. 629.
19. Ibid., p. 631.
20. P. B. Bennett and D. H. Elliott (1969) *The Physiology and Medicine of Diving and Compressed Air Work* (London: Bailliere, Tindall & Cassell); see chap. 22, pp. 508–509.
21. Pollack, op. cit., p. 1324.
22. A. Naqui, B. Chance, and E. Cadenas (1986) "Oxygen Intermediates," *Annual Review of Biochemistry* 55:137–166.
23. Fenchel and Findlay, op. cit., p. 237.
24. A. Henderson-Sellers (1983) *The Origin and Evolution of the Planetary Atmospheres* (Bristol: Adam Hilger); see chaps. 4 and 5, and p. 164.
25. J. B. West (1979) "Mechanics of Breathing," in *The Physiological Basis of Medical Practice,* 10th ed. J. B. Brobeck, ed. (Baltimore: Williams & Wilkins), pp. 636–653.
26. Ibid. V. B. Mountcastle (1968) *Medical Physiology*, vol. 1 (St. Louis; C V. Mosby), pp. 622–626. *Encyclopaedia Britannica* (1994), 15th ed., vol. 26, p. 745.
27. Bennett and Elliott, op. cit.; see chap. 4, pp. 76–109; p. 81.
28. D. R. Lide (1995) *CRC Handbook of Chemistry and Physics,* 76th ed. (Boca Raton, Fla.: CRC Press), pp. 6–17.
29. Lovelock, op. cit., p. 78.
30. J. W. Drake (1970) *The Molecular Basis of Mutation* (San Francisco: Holden-Day), p. 171.
31. A. E. Needham (1905) *The Uniqueness of Biological Materials* (Oxford: Pergamon Press), p. 35.
32. Henderson, op. cit., pp. 139–140.
33. Ibid., p. 153.
34. J. T. Edsall and J. Wyman (1958) *Biophysical Chemistry*, vol. 1 (New York: Academic Press), p. 550.
35. I. B. Cohen (1958) *Isaac Newton's Papers and Letters on Natural Philosophy* (Cambridge: Cambridge University Press), p. 361.
36. Edsall and Wyman, op. cit., p. 554. As the authors point out: "The hydration of CO_2 to H_2CO_3 is a process requiring a rearrangement of the valence bonds, the two C-O bonds of CO_2, 180 degrees apart and 1.15 A long, being transformed to the three C-O bonds of H_2CO_3, approximately 120 degrees apart and not far from 1.3 A long. We shall not attempt to comment here on the details of the electronic rearrangements that must be involved in the process, and indeed little is known of them. It is not surprising however that a process such as this should require an appreciable time, in contrast for example to a process such as the hydration of NH_3 to NH_4OH in which the hydration process simply involves the formation of a hydrogen bond between the unshared electron pair in the ammonia molecule."
37. Henderson, op. cit., p. 141.
38. Ibid., p. 138.
39. Ibid., pp. 266–267.
40. Ibid., p. 272.

Chapter 7: The Double Helix

1. J. Von Neumann (1966) *Theory of Self-Reproducing Automata,* ed. A. W. Burks (Urbana: University of Illinois Press).
2. C. G. Langton (1989) "Artificial Life," in *Artificial Life,* ed. C. G. Langton, Proceedings

of an Interdisciplinary Symposium held in September 1987 in Los Alamos, New Mexico (Redwood City, Calif.: Addison-Wesley), pp. 1–48; see p. 1.

3. C. Schneiker (1989) "Nano Technology with Feynman Machines: Scanning, Tunneling, Engineering, and Artificial Life," in *Artificial Life,* ed. C. G. Langton, Proceedings of an Interdisciplinary Symposium held in September 1987 in Los Alamos, New Mexico (Redwood City, Calif.: Addison-Wesley), pp. 443–500.
4. Ibid., p. 449.
5. K. E. Drexler (1987) *The Engines of Creation* (New York: Anchor Press). See also K. E. Drexler (1992) *Nanosystems: Molecular Manufacture and Computation* (New York: Wiley).
6. A. K. Dewdney (1988) "Computer Recreations," *Scientific American* 258(1):88–91.
7. V. Tartar (1961) *The Biology of Stentor* (London: Pergamon Press); see chap. 7, pp. 105– 134.
8. H. F. Judson (1979) *The Eighth Day of Creation* (New York: Simon & Schuster), pp. 173–175.
9. Ibid., p. 175.
10. F. Crick (1974) "The Double Helix: A Personal View," *Nature* 284:766–769.
11. W. Saenger (1984) *Principles of Nucleic Acid Structure* (New York: Springer-Verlag), p. 8.
12. "News and Views" (1997) *Nature* 389:231–233.
13. G. G. Simpson (1960) "The History of Life," in *Evolution of Life,* ed. Sol Tax (Chicago: University of Chicago Press), pp. 117–180; see p. 135.
14. A. A. Travers (1989) "DNA Conformation and Protein Binding," *Annual Review of Biochemistry* 58:427–452; see p. 428. See also A. A. Travers (1990) "Why Bend DNA?" *Cell* 60:177–180.
15. Saenger, op. cit., pp. 220–241.
16. D. E. Draper (1995) "Protein-RNA Recognition," *Annual Review of Biochemistry* 64: 593–620; see. p. 596.
17. Ibid. See also L. Gold et al. (1995) "Diversity of Oligonucleotides," *Annual Review of Biochemistry* 64:763–797.
18. "News and Views" (1995) *Nature* 376:548. M. Bolli, A. Micura, and A. Eschenmoser (1997) "Pyranosyl-RNA: Chiroselective Self-Assembly of Base Sequences by Ligative Oligomerization of Tetranucleotide –2',3'-Cyclophosphates," *Chemistry and Biology* 4:309–320. A. Eschenmoser (1993) "Towards a Chemical Etiology of the Natural Nucleic Acids," in *40 Years of the DNA Double Helix* (Houston: Robert A. Welch Foundation), pp. 201–235. See also L. E. Orgel (1992) "Molecular Replication," *Nature* 358:203–209.
19. A. Eschenmoser (1997) personal communication. See also A. Eschenmoser, "Towards a Chemical Etiology."
20. A. Rich (1963) "On the Problems of Evolution and Biochemical Information Transfer," in *Horizons in Biochemistry,* ed. M. Kasha and B. Pullman (New York: Academic Press), pp. 103–126; see p. 120.
21. J. D. Bain et al. (1992) "Ribosome-Mediated Incorporation of a Nonstandard Amino Acid into a Peptide Through Expansion of the Genetic Code," *Nature* 356:537–539. See also J. Piccirilli (1990) "Enzymic Incorporation of a New Base Pair into DNA and RNA Extends the Genetic Alphabet," *Nature* 343:33–37.
22. Saenger, op. cit., see chap. 5.
23. Ibid., p. 114.
24. W. Saenger (1997), in a personal communication, summarized the fitness of DNA thus: "The Watson-Crick base pairs are ideally suited for the [biological function] of DNA as (1) they have the same overall dimensions so that a regular double helix can be formed, (2) the hydrogen bonds can be opened and closed at a rate that permits rapid read-out and replication, and (3) the ribose rings of the sugars have sufficient flexibility to permit conformational changes from the A to the B form. If you modify the bases chemically, it

is still possible to form selective base pairs so that the specificity is retained, but you will change the strength of the hydrogen bonds so that the kinetics of read-out and replication will be altered. This is because any chemical modification will change the p*K*-values of the bases so that the hydrogen-bonding strength will be influenced. As to the sugar moieties, the riboses are never planar but have envelope or twist conformations so that the DNA backbone has a certain flexibility. This would be impossible with six-membered sugar rings which are rigid and cannot confer flexibility that is necessary for biological functioning of nucleic acids. . . . one could also speculate on the phosphodiester link that connects adjacent ribose units. It could be replaced by a peptide or a sulphate diester or some other link which, however, is not found. It appears that the negative charge of the phosphate is necessary to maintain the solubility of the nucleic acids, and a certain flexibility and geometry to provide the properties of the nucleic acids."

25. A. Rich, op. cit., pp. 119–120.
26. D. Maxime, M. D. Frank-Kamenetskii, and S. M. Mirkin (1995) "Triplex DNA Structures," *Annual Review of Biochemistry* 64:65–95.
27. D. K. Gifford (1994) "On the Path to Computation with DNA," *Science* 266:993–994.
28. T. Kaeler (1995) "Designing Molecular Components," in *Prospects in NanoTechnology,* ed. M. Krummenacker and J. Lewis (New York: Wiley), pp. 53–66; see p. 65.

Chapter 8: Nanomanipulators

1. J. Monod (1972) *Chance and Necessity* (London: Collins), p. 64.
2. M. F. Perutz (1969) "X-Ray Analysis: Structure and Function of Enzymes," *European Journal of Biochemistry* 8:455–466.
3. Ibid., p. 462.
4. See "News and Views," *Nature Structural Biology* 4 (1997):424–427.
5. J. Watson (1976) *The Molecular Biology of the Gene,* 3rd ed. (Menlo Park, Calif.: W. A. Benjamin), p. 100. Chap. 4 contains a discussion of the role and biochemical significance of weak bonds.
6. Ibid., p. 65.
7. I. Hirao and A. D. Ellington (1995) "Re-creating the RNA World," *Current Biology* 5:1017–1022.
8. G. J. Narlikar and G. Herschlag (1977) "Mechanistic Aspects of Enzymic Catalysis," *Annual Review of Biochemistry,* 66:19–59.
9. G. Stix (1996) "Waiting for Breakthroughs," *Scientific American* 274(4):78–83.
10. Ibid., p. 81.
11. Ibid.
12. Ibid., p. 82.
13. Ibid.
14. Ibid., p. 83.
15. C. O. Pabo and R. T. Sauer (1984) "Protein DNA Recognition," *Annual Review of Biochemistry* 53:293–321; see pp. 313–314. See also Y. Cho et al. (1995) "Crystal Structure of a p53 Tumor Suppressor-DNA Complex: Understanding Tumorigenic Mutations," *Science* 265:346–355; see p. 353.
16. Ibid., p. 314.
17. N. P. Pavletich and C. O. Pabo (1991) "Zinc Finger-DNA Recognition: Crystal Structure of a Zif 268-DNA Complex at 2.1 A," *Science* 252:809–817; see p. 816.
18. M. Suzuki and N. Yagi (1994) "DNA Recognition Code of Transcription Factors in the Helix Turn Helix, Probe Helix, Hormone Receptor and Zinc Finger Families," *Proc. Natl. Acad. Sci. USA* 91:12357–12361. See also "History," *Nature Structural Biology* (1998) 5:100.
19. M. Ptashne (1986) *A Genetic Switch* (Palo Alto, Calif.: Blackwell Scientific Publications).

Chapter 9: The Fitness of the Metals

1. C. Sagan (1985) *Cosmos* (New York: Ballantine); see p. 198.
2. R. J. P. Williams (1985) "The Symbiosis of Metal and Protein Function," *European Journal of Biochemistry* 150:231–248; see p. 232.
3. Ibid., p. 247.
4. E. Frieden (1974) "Evolution of Metals as Essential Elements," in *Protein-Metal Interactions,* ed. M. Friedman (New York: Plenum Press), pp. 1–31; see p. 11.
5. J. J. R. Fraústo da Silva and R. J. P. Williams (1991) *The Biological Chemistry of the Elements* (Oxford: Oxford University Press).
6. Ibid., p. 107.
7. E. Baldwin (1964) *An Introduction to Comparative Biochemistry,* 2nd ed. (Cambridge: Cambridge University Press), p. 81.
8. N. N. Greenwood and A. Earnshaw (1984) *Chemistry of the Elements* (Oxford: Pergamon Press), pp. 1276–1277.
9. Baldwin, op. cit., p. 81.
10. Frieden, op. cit., pp. 20–21.
11. Ibid., p. 19.
12. Ibid.
13. Ibid., p. 22.
14. R. Gennis and S. Ferguson-Miller (1995) "Structure of Cytochrome *c* Oxidase, Energy Generator of Aerobic Life," *Science* 269:1063–1064.
15. Fraústo da Silva and Williams, op. cit., pp. 411–435. See also E. I. Stiefel (1977) "Molybdoenzymes: The Role of Electrons, Protons, and Dihydrogen," in *Bioinorganic Chemistry,* ed. K. N. Raymond (Washington D.C.: American Chemical Society), pp. 353–388; see p. 388 for reasons for the choice of molybdenum.
16. Ibid., p 427.
17. Williams, op. cit., p. 238.
18. Ibid.
19. M. Calvin (1962) "Evolutionary Possibilities for Photosynthesis and Quantum Conversion," in *Horizons in Biochemistry,* ed. M. Kasha and B. Pullman (New York: Academic Press), pp. 23–57; see p. 53. For discussion of the unique properties of Mg in chlorophyll, see also J. Katz (1973) "Chlorophyll," in *Inorganic Biochemistry,* vol. 2, ed. G. L. Eichhorn (Amsterdam: Elsevier), pp. 1022–1066; see pp. 1025–1026.
20. G. Wald (1959) "Life and Light," *Scientific American* 201 (4):92–108. See p. 97.
21. Williams, op. cit., p. 247.
22. Ibid., p. 246.

Chapter 10: The Fitness of the Cell

1. V. Tartar (1961) *The Biology of Stentor* (London: Pergamon Press); see chap. 7, "Regeneration," pp. 105–135.
2. D. Small (1986) *The Physical Chemistry of Lipids* (New York: Plenum Press); see first page of Preface.
3. D. E. Green and R. F. Goldberger (1967) *Molecular Insights into the Living Process* (New York: Academic Press), p. 25.
4. A. E. Needham (1961) *The Uniqueness of Biological Materials* (London: Pergamon Press), p. 77.
5. Ibid., p. 78.
6. J. P. Trinkaus (1984) *Cells into Organs* (Englewood Cliffs, N.J.: Prentice-Hall), p. 53.
7. Ibid., pp. 51–52.

8. Needham, op. cit., p. 78.
9. B. Hendry (1981) *Membrane Physiology and Cell Excitation* (London: Croom Helm), pp. 18–21.
10. Trinkaus, op. cit., p. 69.
11. A. Kotyk and K. Janacek (1977) *Membrane Transport: An Interdisciplinary Approach* (New York: Plenum Press), p. 100.
12. G. I. Bell (1978) "Models for the Specific Adhesion of Cells to Cells," *Science* 200:618–627; see p. 624. J. M. Baltz and R. A. Cone (1990) "The Strength of Non-covalent Biological Bonds and Adhesion by Multiple Independent Bonds," *Journal of Theoretical Biology* 142:163–178. See p. 172.
13. T. P. Stossel (1993) "On the Crawling of Animal Cells," *Science* 260:1086–1094.
14. Ibid., pp. 1086–1087.
15. D. A. Lauffenburger and A. F. Horwitz (1996) "Cell Migration: A Physically Integrated Process," *Cell* 84:359–369. T. Oliver, J. Lee, and K. Jacobson (1994) "Forces Exerted by Locomoting Cells," *Seminars in Cell Biology* 5:139–147. P. A. DiMilla, K. Barbee, and D. A. Lauffenburger (1993) "Mathematical Model for the Effects of Adhesion and Migration on Cell Migration Speed," *Biophysical Journal* 60:15–37.
16. B. Alberts et al. (1989) *The Molecular Biology of the Cell*, 2nd ed. (New York: Garland Publishing), p. 308. See also D. M. Woodbury (1974) "Physiology of Body Fluids," in *Physiology and Biophysics*, vol. 2, ed. T. C. Ruch and H. D. Patton (Philadelphia: W. B. Saunders), pp. 450–479.
17. Ibid.
18. Alberts et al., op. cit., p. 304.
19. *Encyclopaedia Britannica* (1994) 15th ed., vol. 23, p. 650.
20. J. C. Waterlow, P. J. Garlick, and D. J. Millward (1978) *Protein Turnover in Mammalian Tissues and in the Whole Body* (Amsterdam: Elsevier), see chap. 14.
21. H. C. Berg (1990) "Bacterial Microprocessing," *Cold Spring Harbor Symposium on Quantitative Biology* 55:539–544; see p. 539.
22. H. S. Jennings (1962) *Behavior of the Lower Organisms* (Bloomington: Indiana University Press); see pp. 15–18.
23. S. Hameroff (1988) "Molecular Automata in Microtubules: Basic Computational Logic of the Living State, in *Artificial Life*, ed. C. G. Langton (Redwood City, Calif.: Addison-Wesley), pp. 521–553; see p. 543.
24. Ibid.
25. Ibid.
26. C. Pantin (1951) "Organic Design," *Advancement of Science* 8(29):138–150; see p. 149.
27. R. Penrose (1994) *Shadows of the Mind* (Oxford: Oxford University Press), p. 357.
28. J. Brown (1994) "Tell Me Where Consciousness Is Bred," *New Scientist*, Nov. 19, pp. 46–47; see p. 47. Penrose, op, cit., p. 366.
29. D. Bray (1995) "Protein Molecules as Computational Elements in Living Cells," *Nature* 376:307–311.
30. R. F. Goldberger (1967) *Molecular Insights into the Living Process* (New York: Academic Press), pp. 24, 35–36. See also Needham, op. cit., pp. 43–44.
31. F. H. Westheimer (1987) "Why Nature Chose Phosphates," *Science* 235:1173–1178. See also comments in D. E. Green and R. F. Goldberger (1967) *Molecular Insights into the Living Process* (New York: Academic Press), p. 27.
32. G. Wald (1964) "The Origins of Life," *Proc. Natl. Acad. Sci. USA* 52:594–611; see discussion on pp. 607–608.
33. Ibid., p. 608.
34. C. Pantin (1951) "Organic Design," *Advancement of Science* 8(29):138–150; see pp. 143–144.

35. J. E. Gordon (1980) "Biomechanics: The Last Stronghold of Vitalism," *Symposia of the Society for Experimental Biology* 34:1–11; see p. 1.
36. Pantin, op. cit., p. 145.

Chapter 11: Homo Sapiens: *Fire Maker*

1. G. G. Simpson (1967) *The Meaning of Evolution* (New Haven: Yale University Press), p. 288.
2. P. Lieberman (1975) "On the Evolution of Language: A Unified View," in *Primate Functional Morphology and Evolution,* ed. R. Tuttle (The Hague: Mouton Publishers), pp. 501–540; see pp. 504–510.
3. Ibid., pp. 508, 536.
4. Aristotle, *Metaphysics* 1.1,980a21–7.
5. R. D. Martin (1990) *Primate Origins and Evolution* (London: Chapman & Hall), pp. 496–497.
6. R. M. Yerkes and A. W. Yerkes (1929) *The Great Apes* (New Haven: Yale University Press), p. 346.
7. J. Kidd (1952) *The Bridgewater Treatise on the Physical Condition of Man,* 6th ed. (London: Bohn); see chap. 3 on the hand, p. 26.
8. Ibid., pp. 29–31.
9. R. Tuttle (1975) "Knuckle-Walking and Knuckle-Walkers: A Commentary on Some Recent Perspectives on Hominoid Evolution," in *Primate Functional Morphology and Evolution,* ed. R. Tuttle (The Hague: Mouton Publishers), pp. 203–211; see p. 203.
10. J. Huxley (1941) *The Uniqueness of Man* (London: Chatto & Windus), p. 147.
11. K. Schmidt-Nielsen (1975) *Animal Physiology* (Cambridge: Cambridge University Press), pp. 514–515.
12. M. S. Block (1996) "Nanometers and Piconewtons: The Macromolecular Mechanisms of Kinesin," *Trends in Cell Biology* 5:169–175. See also R. A. Crowther, R. Padron, and R. Craig (1985) "Arrangement of the Heads of Myosin in Relaxed Thick Filaments from Tarantular Muscle," *Journal of Molecular Biology* 184:492–439; and B. Alberts et al. (1994) *Cell* (New York: Garland Publishing), p. 851.
13. R. Simmons (1996) "Molecular Motors: Single-Molecule Mechanics," *Current Biology* 6:392–394. See also Block, op. cit. For strength of weak bonds, see Alberts et al., op. cit., pp. 90–92. For energy levels of kJ of myosin cross bridges, see W. F. Harrington (1979) "On the Origin of the Contractile Force in Skeletal Muscle," *Proc. Natl. Acad. Sci. USA* 76:5066–5070. For energy levels of affinity bonds composed of multiple weak bonds, see J. M. Batz and R. A. Cone (1990) "The Strength of Non-covalent biological Bonds and Adhesions by Multiple Independent Bonds," *Journal of Theoretical Biology* 142:163–178; and F. Amblard et al. (1994) "Molecular Analysis of Antigen-Independent Adhesion Forces Between T and B Lymphocytes," *Proc. Natl. Acad. Sci. USA* 91:3628–3632.
14. Simmons, op. cit.
15. A. J. Wilson (1994) *The Living Rock* (Cambridge: Woodhead Publishing); see pp. 10–16.
16. Schmidt-Nielsen, op. cit.; see p. 625, table 13.2.
17. W. A. H. Rushton (1951) "A Theory of the Effects of Fibre Size in Medullated Nerve," *Journal of Physiology* 115:101–122.
18. Schmidt-Nielsen, op. cit.; see p. 53.
19. Ibid.
20. D. W. Sciama (1959) *The Unity of the Universe* (London: Faber & Faber), pp. 118–119.
21. Huxley, op. cit., p. 8.
22. Ibid., pp. 15–16.
23. V. G. Dethier (1964) "Microscopic Brains," *Science* 143:1138–1145.

24. E. J. Slijper (1962) *Whales* (London: Hutchinson); see pp. 245–246.
25. Ibid.
26. W. E. Le Gros Clark (1969) *The Antecedents of Man,* 3rd ed. (Chicago: Quadrangle Books), p. 260.
27. C. Wills (1993) *The Runaway Brain* (New York: HarperCollins); see p. 7.
28. G. A. Shariff (1953) "Cell Counts in the Primate Cerebral Cortex," *Journal of Comparative Neurology* 98:381–400.
29. Ibid.; see p. 263. J. DeFelipe and E. G. Jones (1988) *Cajal on the Cerebral Cortex* (New York: Oxford University Press); see chap. 5 and chap. 7, fig. 34, p. 69.
30. Ibid.; see p. 263.
31. See "News and Views," *Nature* 385 (1997): 207–210.
32. M. Ward (1997) "End of the Road for Brain Evolution," *New Scientist* 153, no. 2066:14.
33. *Nature,* op. cit., p. 210.
34. J. Searle (1987) "*Minds and Brains Without Programs,*" in *Mindwaves: Thoughts on Intelligence, Identity and Conciousness,* ed. C. Blakemore and S. Greenfield (Oxford: Basil Blackwell), pp. 209–233.
35. R. Penrose (1990) *The Emperor's New Mind* (London: Vintage).
36. E. P. Wigner (1960) "The Unreasonable Effectiveness of Mathematics in the Natural Sciences," *Communications on Pure and Applied Mathematics* 13:1–14.
37. P. C. W. Davies (1992) *The Mind of God* (London: Penguin); see p. 157.
38. J. Lear (1988) *Aristotle: The Desire to Understand* (Cambridge: Cambridge University Press), p. 230.

Chapter 12: The Tree of Life

1. N. C. Gillespie (1979) *Charles Darwin and the Problem of Creation* (Chicago: University of Chicago Press); see chap. 5, "Providential Evolution and the Problem of Design."
2. R. Chambers (1969) *Vestiges of the Natural History of Creation* (New York: Leicester University Press), pp. 152–154.
3. Ibid., p. 158.
4. Ibid., pp. 165–167.
5. Ibid., pp. 250–251.
6. Ibid., pp. 163–164.
7. Ibid., Introduction.
8. Ibid.
9. Ibid.
10. A. Koestler (1970) *The Ghost in the Machine* (London: Pan Books), p. 174.
11. Ibid., pp. 174–175.
12. S. J. Gould (1972) "Zealous Advocates," *Science* 176:623–625; see p. 625.
13. R. Dawkins (1986) *The Blind Watchmaker* (London: Longman Scientific), p. 291.
14. B. Rensch (1959) *Evolution Above the Species Level* (New York: Wiley), pp. 57–58.
15. L. S. Berg (1969) *Nomogenesis* (Cambridge: MIT Press), p. 110.
16. Ibid., pp. 118–120. Re: the reduction of the gametophyte, Berg comments: "We may thus trace the entire process of the reduction of the gametophyte, commencing with its flourishing condition in mosses, and proceeding with its gradual decline in the seed ferns until its complete disappearance in the conifers. . . . A definite course of evolution is here strikingly exemplified."
17. Ibid., pp. 121–124.
18. H. Miller (1869) *Footsteps of the Creator,* 11th ed. (Edinburgh: Nimmo), pp. 293–294.
19. N. C. Gillespie (1979) *Charles Darwin and the Problem of Creation* (Chicago: University of Chicago Press), p. 85.

20. Ibid., p. 104.
21. Gillespie, op. cit.; see chap. 5.
22. M. J. Denton (1985) *Evolution: A Theory in Crisis* (London: Burnett Books).
23. M. Behe (1996) *Darwin's Black Box* (Chicago: Free Press).
24. J. Diamond (1992) *The Rise and Fall of the Third Chimpanzee* (London: Vintage); see p. 23.
25. S. Conway Morris (1995) "Book Reviews," *Nature* 376:736.
26. R. P. Harvey (1996) "NK-2 Homeobox Genes and Heart Development," *Developmental Biology* 178:203–216.
27. B. John and G. Miklos (1988) *The Eucaryotic Genome in Development and Evolution* (London: Allen & Unwin), p. 112.
28. G. Bernardi and G. Bernardi (1986) "Compositional Constraints and Genome Evolution," *Journal of Molecular Evolution* 24:1–11; see p. 1.
29. N. Sueoka (1992) "Directional Mutation Pressure, Selective Constraints, and Genetic Equilibria," *Journal of Molecular Evolution* 34:95–114.
30. G. Vines (1982) "Molecular Drive: A Third Force in Evolution," *New Scientist,* December 9, pp. 664–665; see p. 665.
31. We can envisage such a contriving or tampering of the DNA space to be analogous to rearranging the structure of the English lexicon to permit the evolution of a particular word tree, which could grow from a single beginning to include a galaxy of long complex words. As the lexicon is structured at present, most English words over ten letters long are completely isolated and cannot generally be transformed via single letter changes via functional intermediates to nearby words. Consequently, it is impossible to go from a simple starting word such as "a," to reach complex English words more than ten letters long. However, by playing God and restructuring the lexicon we would be able, if we wished, to arrange a vast word tree within the letter space, so that all functional words were clustered together in a highly ordered manner in the space and so that starting from one unique letter string we would be led inevitably by the necessity to move in single-letter steps via functional intermediates to find all functional words and trace out all the branches of the tree.
32. S. Kauffman (1995) *At Home in the Universe* (New York: Oxford University Press).
33. S. Kauffman (1993) *The Origins of Order: Self-Organization and Selection in Evolution* (New York: Oxford University Press).
34. B. Goodwin (1995) *How the Leopard Changed its Spots* (New York: Simon & Schuster).
35. Kauffman (1995), p. 23.
36. Goodwin, op. cit.; see p. 168.
37. Ibid., p. 131.
38. Ibid., p. 114.
39. M. P. Schutzenberger (1996), personal communication.
40. J. W. Drake (1991) "Spontaneous Mutation," *Annual Review of Genetics* 25:125–146. N. Symmonds (1991) "A Fitter Theory of Evolution," *New Scientist,* September, 21, pp. 30–34.
41. Dawkins, op. cit., p. 313. J. Monod (1972) *Chance and Necessity* (London: Collins), p. 114. E. Mayr (1976) *Evolution and the Diversity of Life* (Cambridge: Harvard University Press), p. 32.
42. T. Dobzhansky et al. (1977) *Evolution* (San Francisco: W. H. Freeman), p. 65.
43. Ibid.
44. M. Delbruck (1947) *Cold Spring Harbor Symposium on Quantitative Biology* 11:154.
45. J. Cairns, J. Overbaugh, and S. Miller (1988) "The Origin of Mutants," *Nature* 335:142–145; see p. 145.
46. "Research News," *Science* 277 (1997):176–178.

47. M. J. Denton (1986) *Evolution: A Theory in Crisis* (Bethesda, Md.: Adler & Adler), p. 249.
48. B. Lewin (1988) "Molecular Clocks Turn a Quarter of a Century," *Science* 239:561–563; see p. 561.
49. F. Hoyle and C. Wickramasinghe (1996) (London: Orino Books); see chap. 4.
50. J. Horgan (1993) "In the Beginning," *Scientific American* 264 (2):101–109; see p. 101.
51. Monod, op. cit., p. 135.
52. F. Crick (1981) *Life Itself* (New York: Simon & Schuster), p. 88.
53. Horgan, op. cit., p. 103.
54. Ibid., p. 102.
55. I. Hirao and A. D. Ellington (1995) "Re-creating the RNA World," *Current Biology* 5:1017–1022.
56. Horgan, op. cit.
57. Ibid.
58. Ibid., p. 106.
59. Ibid., p. 102.
60. "News and Views," *Nature* 382 (1996):496–497.
61. Ibid., p. 109.
62. M. M. Waldrop (1990) "Spontaneous Order, Evolution, and Life," *Science* 247:1543–1545; see p. 1543.
63. R. Lewin (1993) "Order for Free," *New Scientist* 137, no. 1860, Supplement on Complexity: pp. 10–11; see p. 10.
64. Ibid., p. 10.
65. S. J. Gould (1996) "War of the World Views," *Natural History* 105:22–33.
66. Ibid., p. 30.
67. N. Eldredge and S. J. Gould (1973) "Punctuated Equilibria: An Alternative to Phyletic Gradualism," in *Models in Paleontology*, ed. T. J. M. Schopf (San Francisco: Freeman, Cooper & Co.), pp. 82–115.
68. R. A. Kerr (1995) "Did Darwin Get it Right?" *Science* 267:1421–1422. For evidence that avian evolution conforms to the same pattern, see A. Feduccia (1995) "Explosive Evolution in Tertiary Birds and Mammals," *Science* 267:637–638.
69. Ibid.
70. R. A. Kerr (1993) "Evolution's Big Bang Gets Even More Explosive," *Science* 261:1274–1275.
71. S. J. Gould (1994) "In the Mind of the Beholder," *Natural History* 103(3):14–23.

Chapter 13: The Principle of Plenitude

1. W. Coleman (1964) *Georges Cuvier, Zoologist* (Cambridge: Harvard University Press), pp. 171–172.
2. A. O. Lovejoy (1953) *The Great Chain of Being* (Cambridge: Harvard University Press); see p. 52.
3. N. Pace (1997) "A Molecular View of Microbial Diversity and the Biosphere," *Science* 276:734–740. See also T. Fenchel and B. J. Findlay (1995) *Ecology and Evolution in Anoxic Worlds* (Oxford: Oxford University Press).
4. E. Florey (1966) *An Introduction to General and Comparative Animal Physiology* (Philadelphia: W. B. Saunders), p. 39.
5. S. Conway Morris (1993) "The Fossil Record and the Early Evolution of the Metazoa," *Nature* 313:219–225. Morris lists seventy-one different phyla in all, twenty-nine of which are extinct; see fig. 2, p. 221.

6. S. J. Gould (1989) *Wonderful Life* (New York: Norton); see chap. 3.
7. B. Rensch (1959) *Evolution Above the Species* (New York: Wiley); see p. 66.
8. A. R. Jones (1974) *The Ciliates* (London: Hutchinson), p. 17.
9. Rensch, op. cit., p. 67.
10. Ibid., p. 59.
11. Ibid.
12. Ibid.
13. Ibid., p. 60.
14. Ibid., p. 61.
15. Ibid., p. 62.
16. Ibid., p. 66.
17. M. F. Land and R. D. Fernald (1992) "The Evolution of Eyes," *Annual Review of Neuroscience* 15:1–29.
18. E. T. Burtt (1974) *The Senses of Animals* (London: Wykeham Publications), p. 115. See also G. Wald (1959) "Light and Life," *Scientific American* 201 (4):92–108.
19. E. A. Newman and P. H. Hartline (1982) "The Infrared 'Vision' of Snakes," *Scientific American* 246 (3):98–107.
20. K. Schmidt-Nielsen (1984) *Scaling* (Cambridge: Cambridge University Press), pp. 1–2.
21. H. J. Morowitz (1966) "The Minimum Size of Cells," in *Principles of Biomolecular Organisation,* G. E. W. Wolstenholme and M. O'Connor (London: Churchill), pp. 446–459.
22. Ibid., p. 456.
23. Schmidt-Nielsen, op. cit., p. 213.
24. Rensch, op. cit., p. 171.
25. K. Schmidt-Nielsen (1975) *Animal Physiology* (Cambridge: Cambridge University Press); see section on insect respiration, pp. 61–68. See also A. D. Imms (1964) *A General Textbook of Entomology,* pp. 133–150; see p. 144 on diffusional constraints. The tracheal system and the size of insects is also discussed in R. McNeil Alexander (1971) *Size and Shape* (London: Edward Arnold), pp. 21–22.
26. Schmidt-Nielsen (1984); see discussion on pp. 204–208.
27. Ibid., p. 207.
28. Ibid., pp. 207–208.
29. Ibid., p. 10.
30. Ibid., p. 205.
31. J. D. Curry (1970) *Animal Skeletons* (London: Edward Arnold). See pp. 5–8 for a discussion of the comparative merits of exo- and endoskeletons. An exoskeleton, for example, is far superior to an endoskeleton in regard to failure by buckling and bending, the sorts of stresses likely to be encountered by a small organism. However, an endoskeleton is far more resistant to external forces of impact because the soft tissues can absorb a great amount of energy without serious damage whereas the hard, stiff exoskeleton is unprotected. See also Schmidt-Nielsen (1984), pp. 52–53: "The entire kinetic energy is absorbed on impact, and for a fast-moving large animal, the forces impacting on a hard exoskeleton are likely to cause local failure."
32. S. A. Wainwright et al. (1976) *Mechanical Design in Organisms* (London: Edward Arnold). On Page p. 239 the authors note: "Vertebrates must have an endoskeleton . . . but sclerotised cuticle has a great disadvantage—it is extremely resistant to enzymes . . . invertebrates overcome this problem by shedding their cuticle as they grow . . . bone on the other hand, although stiff and seemingly immutable, is in a dynamic state of erosion and deposition the whole time; therefore remodelling necessitated by growth is easily brought about."
33. Ibid., pp. 49–51.
34. J. Huxley (1941) *The Uniqueness of Man* (London: Chatto & Windus), pp. 134–135.

35. Ibid., p. 141.
36. Alexander, op. cit.; see sec. 5.2 on animal legs.
37. S. J. Gould (1981) "Kingdoms Without Wheels," *Natural History* 90 (3):42–48.
38. Ibid., pp. 47–48.
39. Schmidt-Nielsen (1984), p. 212.
40. S. A. Newman and W. D. Comper (1990) "Generic Physical Mechanisms of Morphogenesis and Pattern Formation," *Development* 110:1–18; see summary, p. 1.

Chapter 14: The Dream of Asilomar

1. P. Berg et al. (1974) "Potential Biohazards of Recombinant DNA Molecules," *Science* 185:303.
2. "News and Comments," *Science* 190 (1975):1175.
3. J. Goodfield (1977) *Playing God* (London: Hutchinson); see pp. 4–5.
4. Ibid., pp. 58–59.
5. M. Jahoda (1982) "Once a Jackass," *Nature* 295:173–174.
6. J. G. Williams (1982) "Mouse and Supermouse," in "News and Views," *Nature* 300:575. See also I. Wilmut, J. Clark, and P. Simons (1988) "A Revolution in Animal Breeding," *New Scientist* 119, no. 1620:56–59.
7. A. Wyke (1988) "A Survey of Biotechnology," *Economist,* April 30, pp. 5–10; see p. 7.
8. Ibid., p. 17.
9. Ibid., p. 6.
10. Ibid., p. 20.
11. K. Schmidt (1995) "Whatever Happened to the Gene Revolution?" *New Scientist* 145, no. 1959:21–25; see p. 22.
12. Ibid.
13. G. Cuvier (1854) *Animal Kingdom* (London: W. Orr). See p. 18: "Life then presupposes organisation in general, and the life proper to each being presupposes the organisation peculiar to that being, just as the movement of a clock presupposes the clock."
14. K. E. Drexler (1987) *Engines of Creation* (Garden City, N.Y.: Anchor Books). K. E. Drexler (1988) "Biological and Nanomechanical Systems: Contrasts in Evolutionary Capacity," in *Artificial Life,* ed. C. G. Langton (Redwood City, Calif.: Addison-Wesley), pp. 501– 519; see pp. 509–510.
15. P .C. Ritterbush (1972) "Organic Form: Aesthetics and Objectivity in the Study of Form in the Life Sciences," in *Organic Form: The Life of an Idea,* ed. G. S. Rousseau (London: Routledge & Kegan Paul), pp. 25–59.
16. W. Coleman (1964) *Georges Cuvier, Zoologist* (Cambridge: Harvard University Press); see Introduction, p. 2, and pp. 38–43.
17. G. Cuvier (1812) *Reserches sur les ossements fossiles de quadrupèdes, Discours préliminaire,* English trans. by R. Kerr (1813), entitled *Essay on the Theory of the Earth* (Edinburgh and London), pp. 94–95. Also referred to in Coleman, op. cit., p. 108.
18. G. Cuvier (1829) *Revolutions of the Surface of the Earth* (London: Whittaker, Treacher & Arnot), p. 60.
19. S. A. Kauffman (1993) *The Origins of Order* (Oxford: Oxford University Press); see pp. 53–54.
20. D'Arcy W. Thompson (1952) *On Growth and Form,* vol. 2, 2nd ed. (Cambridge: Cambridge University Press), p. 1019. A. G. Cairns-Smith (1982) *Genetic Takeover and the Mineral Origins of Life* (Cambridge: Cambridge University Press), p. 78. R. Lewontin (1978) "Adaptation," *Scientific American* 219(3):212–231; see p. 231.
21. P. Alberch (1980) "Ontogenesis and Morphological Diversification," *American Zoologist* 20:653–657.

22. S. Ohno (1973) "Ancient Linkage Groups and Frozen Accidents," *Nature* 244:259–262.
23. J. D. Harris and N. R. Lemoine (1996) "Strategies for Targeted Gene Therapy," *Trends in Genetics* 12:400–405; see p. 401.
24. R. Dawkins (1986) *The Blind Watchmaker* (London: Longman Scientific), pp. 85–86, 311.
25. Ibid., pp. 317–318.
26. E. Hadorn (1961) *Developmental Genetics and Lethal Factors* (London: Methuen); see p. 196.
27. E. Mayr (1970) *Populations, Species, and Evolution* (Cambridge: Harvard University Press), pp. 162–164.
28. C. Q. Doe et al. (1988) "Expression and Function of the Segmentation Gene *Fushi Tarazu* During *Drosophila* Neurogenesis," *Science* 239:170–175.
29. R. Lewin (1984) "Why Is Development So Illogical?" *Science* 224:1327–1329; see p. 1327.
30. Ibid.
31. Ibid.
32. L. Wolpert (1992) "Gastrulation and the Evolution of Development," *Development,* Supplement, pp. 7–13.
33. *McGraw-Hill Encyclopedia of Science and Technology* (1992), vol. 17, p. 163.
34. Ibid., vol. 6, p. 243.
35. K. P. Able (1980) "Mechanism and Orientation, Navigation and Homing," in *Animal Migration, Orientation, and Navigation,* ed. S. A. Gauthreaux (New York: Academic Press); see chap. 5, quote from p. 327. T. H. Waterman (1989) *Animal Navigation* (New York: Scientific American Library); see pp. 59, 60–61.
36. H. P. Erickson (1993) "Tenascin-C, Tenascin-R, and Tenascin-X: A Family of Talented Proteins in Search of Functions," *Current Biology* 5:869–876; see p. 869.
37. M. L. Goldberg, R. A. Colvin, and A. F. Mellin (1989) "The Drosophila *Zeste* Locus is Nonessential," *Genetics* 123:145–155.
38. J. Brookfield (1992) "Can Genes Be Truly Redundant?" *Current Biology* 2:553–554; see p. 553. See also J. H. Thomas (1993) "Thinking About Genetic Redundancy," *Trends in Genetics,* 9:395–399.
39. P. W. Sternberg (1993) "Intercellular Signaling and Signal Transduction in *C. elegans,*" *Annual Review of Genetics* 27:497–521; see p. 512.
40. S. W. Wilson (1993) "Clues from Clueless," *Current Biology* 3:536–539.
41. C. Kenyon (1995) "A Perfect Vulva Every Time: Gradients and Signaling Cascades in *C. elegans,*" *Cell* 82:171–174; see p. 173.
42. J. van Brunt (1986) "Protein Architecture: Designing from the Ground Up," *Bio/technology* 4:277–283; see p. 277.
43. M. F. Perutz (1985) "The Birth of Protein Engineering," *New Scientist* 106, no. 1460: 12–15; see p. 14.
44. M. Mutter (1985) "The Construction of New Proteins and Enzymes—A Prospect for the Future," Angewandte Chemie (International Edition in English) 24:639–653.
45. S. Dalal, S. Balasubranian, and L. Regan (1997) "Protein Alchemy: Changing β-Sheet into α Helix," *Nature Structural Biology* 4:458–552.
46. C. B. Anfinsen (1964) "On the Possibility of Predicting Tertiary Structure from Primary Sequence," in *New Perspectives in Biology,* ed. M. Sela (New York: Elsevier), pp. 42–50.
47. G. S. Orsini (1977) "The Ancient Roots of a Modern Idea," in *Organic Form: The Life of an Idea,* ed. G. S. Rousseau (London: Routledge & Kegan Paul); see pp. 8–12.
48. M. I. Simon (1992) "Summary: The Cell Surface Regulates Information Flow, Material Transport, and Cell Identity," *Cold Spring Harbor Symposium on Quantitative Biology* 57:673–688. E. R. Kandel (1983) "Neurobiology and Molecular Biology: The Second

Encounter," *Cold Spring Harbor Symposium on Quantitative Biology* 48:891–908; see p. 904.

49. L. S. B. Goldstein (1993) "With Apologies to Scheherazade: Tails of 1001 Kinesin Motors," *Annual Review of Genetics* 27:319–351.
50. H. Charbonneau and N. K. Tonks (1992) "1002 Protein Phosphatases," *Annual Review of Cell Biology* 8:463–469.
51. Simon, op. cit., p. 678.
52. C. W. J. Smith, J. G. Patton, and B. Nadal-Ginard (1989) "Alternative Splicing in the Control of Gene Expression," *Annual Review of Genetics* 23:527–577; see p. 564.
53. Ibid., p. 676.
54. Ibid., p. 687.
55. S. L. McIntire et al. (1992) "Genes Necessary for Directed Axonal Elongation or Fasciculation in *C. elegans*," *Neuron* 8:307–322; see p. 307.
56. H. Nawa, T. Le. Yamamori, and P. H. Patterson (1990) "Generation of Neuronal Diversity: Analogies and Homologies with Hematopoiesis," *Cold Spring Harbor Symposium of Quantitative Biology* 55:247–253; see p. 247.
57. "Glia: The Brain's Other Cells," in "Research News," *Science* 266 (1994):970–972.
58. R. Rucker (1983) *Infinity and the Mind* (New York: Bantam); see p. 2.

Chapter 15: The Eye of the Lobster

1. J. R. P. Angel (1979) "Lobster Eyes as X-ray Telescopes," *Astrophysical Journal* 233:364–373. See also B. K. Hartline (1980) "Lobster-Eye X-ray Telescope Envisioned," *Science* 207:47.
2. M. F. Land (1976) "Superposition Images Are Formed by Reflection in the Eyes of Some Oceanic Decapod Crustacea," *Nature* 263:764–765.
3. M. F. Land (1978) "Animal Eyes with Mirror Optics," *Scientific American* 239(6):88–99.
4. Ibid.
5. W. J. Dakin (1928) "The Eyes of *Pecten, Spondylus, Amussium,* and Allied Lamellibranchs, with a Short Discussion on Their Evolution," *Proceedings of the Royal Society of London, Series B* 103:355–365; see pp. 359–360.
6. Ibid., p. 361.
7. Ibid., p. 364.
8. M. F. Land (1966) "Activity in the Optic Nerve of *Pecten maximus* in Response to Changes in Light Intensity, and to Pattern and Movement in the Optical Environment," *Journal of Experimental Biology* 45:83–99.
9. E. M. del Pino (1989) "Marsupial Frogs," *Scientific American* 260(5):76–84.
10. J. S. Levinton (1992) "The Big Bang of Animal Evolution," *Scientific American* 267(11):52–59; see p. 59.
11. H. R. Dunker (1978) "Development of the Avian Respiratory and Circulation Systems," in *Respiratory Function in Birds, Adult and Embryonic,* ed. J. Piiper, (New York: Springer-Verlag), pp. 260–273.
12. C. Darwin (1962) *The Origin of Species,* 6th ed. (New York: Collier Books), p. 182.
13. K. Schmidt-Nielsen (1975) *Animal Physiology* (New York: Cambridge University Press), p. 61.
14. P. C. W. Davies, (1992) *The Mind of God* (London: Penguin), p. 149.
15. G. Gibbs (1994) "The Demon Grasshoppers," *New Zealand Geographic,* January–March, pp. 90–117.
16. L. Cuénot (1944) "L'anti Hasard," *Revue Scientifique,* pp. 339–346.
17. Ibid.; see p. 345, trans. M. P. Schutzenberger.

18. See L. Watson (1973) *Supernature* (London: Hodder & Stoughton). Also R. Sheldrake (1981) *A New Science of Life: The Hypothesis of Formative Causation* (London: Blond & Briggs). And see also the short note in *New Scientist,* July 26, 1997, p. 39, reporting recent experiments carried out by Sheldrake on a "mysterious sixth sense."
19. Darwin, op. cit., p. 484.

Conclusion: The Long Chain of Coincidence

1. B. Appleyard (1992) *Times Saturday Review,* April 25, p. 12.
2. C. Sagan (1985) *Cosmos* (New York: Ballantine), p. 30.
3. S. J. Dick (1982) *Plurality of Worlds* (Cambridge: Cambridge University Press), p. 61.
4. C. Sagan (1985) *Cosmos* (New York: Ballantine). "One Voice in the Cosmic Fugue" is the title of chap. 2.
5. J. S. Shklovskii and C. Sagan (1977) *Intelligent Life in the Universe* (London: Pan Books), p. 22.
6. Sagan, op. cit., p. 242.
7. G. Feinberg and R. Shapiro (1980) *Life Beyond Earth* (New York: Morrow); see p. 247 and pp. 246–250 for discussion of life in ammonia.
8. Ibid., pp. 252–256.
9. Ibid., pp. 382–384.
10. F. Hoyle (1957) *The Black Cloud* (London: Heinemann); see chap. 10.
11. Sagan, op. cit.; see pp. 29–30.
12. Feinberg and Shapiro, op. cit., p. 235.
13. Ibid., pp. 274–275.
14. J. Needham (1929) *The Sceptical Biologist* (London: Chatto & Windus), p. 217; see pp. 210–218 for a review of Henderson's *Fitness.*
15. H. F. Osborne (1894) *From the Greeks to Darwin* (New York: Macmillan), p. 61.
16. N. K. Smith (1935) *Hume's Dialogues Concerning Natural Religion* (Oxford: Oxford University Press), pp. 224–226.
17. P. W. Atkins (1981) *The Creation* (San Francisco: W. H. Freeman), pp. 36, 123, 125.
18. L. J. Henderson (1958) *The Fitness of the Environment* (Boston: Beacon Press), pp. 275–276.

Epilogue

1. J. Lear (1988) *Aristotle: The Desire to Understand* (Cambridge: Cambridge University Press).
2. *Encyclopaedia Britannica* (1994), 15th ed., article entitled "Electricity and Magnetism," in vol. 18, pp. 189–193. See p. 191 for Oersted's discovery while lecturing students.

Appendix: Miscellaneous Additional Evidence of the Fitness of Earthly Life

1. D. E. Green and R. F. Goldberger (1967) *Molecular Insights into the Living Process* (New York: Academic Press), p. 24. Note that because of its closed-ring hemiacetal form, glucose has the lowest frequency free-aldehyde conformation and the slowest rate of Schiff base formation. Consequently, glucose is the most stable sugar in the presence of protein and other amino-bearing groups such as nucleoproteins.
2. A. E. Needham (1965) *The Uniqueness of Biological Materials* (Oxford: Pergamon Press), p. 43.
3. Green and Goldberger, op. cit., pp. 35–36.

4. Needham, op. cit., pp. 43–44.
5. Green and Goldberger, op. cit.; see p. 25.
6. Needham, op. cit., p. 77.
7. Ibid., p. 78.
8. Ibid.
9. Ibid.
10. B. Hendry (1981) *Membrane Physiology and Cell Excitation* (London: Croom Helm); see p. 18.
11. Ibid., pp. 19–21.
12. F. H. Westheimer (1987) "Why Nature Chose Phosphates," *Science* 235:1173–1178; see p. 1173. See also comments in Green and Goldberger, op. cit., p. 27: "The P-O-P bonds of ATP represent a convenient packet of chemical energy in the sense that it is sufficient to drive all the chemical reactions for which it is required while not being so large that its power is wasted. Another important prerequisite that a molecule must meet as a satisfactory energy store is stability. There are many compounds with 'energy rich' bonds which are highly unstable in water at body temperature. Such compounds would not be very useful. The P-O-P bonds of ATP are remarkably stable under physiological conditions; thus ATP poses no problems of storage."
13. Westheimer, op. cit., p. 1176.
14. Ibid., p. 1178.
15. Green and Goldberger, op. cit., pp. 25–26.
16. M. Toporek (1968) *Basic Chemistry of Life* (Englewood Cliffs, N.J.: Prentice-Hall); see table 1, p. 137.
17. L. J. Henderson (1958) *The Fitness of the Environment* (Boston: Beacon Press), pp. 159–160.
18. W. S. Hoffman (1970) *The Biochemistry of Clinical Medicine* (Chicago: Year Book Medical Publishers); see chap. 12, esp. pp. 548–549.
19. J. J. R. Fraústo da Silva and R. J. P. Williams (1991) *The Biological Chemistry of the Elements* (Oxford: Oxford University Press), p. 138.
20. B. D. Rose (1977) *Clinical Physiology of Acid-Base and Electrolyte Disorders* (New York: McGraw-Hill), p. 176.
21. Henderson, op. cit., p. 158.
22. M. Suzuki and N. Yagi (1994) "DNA Recognition Code of Transcription Factors in the Helix Turn Helix, Probe Helix, Hormone Receptor and Zinc Finger Families," *Proc. Natl. Acad. Sci. USA* 91:12357–12361.
23. M. Ptashne (1986) *A Genetic Switch* (Palo Alto, Calif.: Blackwell Scientific Publications).
24. Ibid.; see Appendix 1, pp. 109–115, esp. p. 112: "Consider a repressor that binds to its operator with a dissociation constant of 10^{-10} M (the dissociation constant is a measure of the strength of binding between two molecules, the lower the figure the greater the intensity of binding) and to a random DNA sequence with a dissociation constant of 10^{-4} M. Now double the size of the repressor and the operator, so that twice as many . . . contacts are made. The larger version, to a first approximation, would bind to operator and to random DNA with a dissociation constant of 10^{-20} M (ten orders of magnitude stronger) and 10^{-8} M respectively. (Twice as many contacts implies twice the energy [-DG] and recall that the dissociation constant is related to DG exponentially: Dissociation constant = $e^{-DG/RT}$. For every 2.8 kcal change in DG, the dissociation constant changes 100 fold.)"

Index

Page numbers in *italics* refer to illustrations.